River mechanics

PIERRE Y. JULIEN

CAMBRIDGE
UNIVERSITY PRESS

PUBLISHED BY THE PRESS SYNDICATE OF THE UNIVERSITY OF CAMBRIDGE
The Pitt Building, Trumpington Street, Cambridge, United Kingdom

CAMBRIDGE UNIVERSITY PRESS
The Edinburgh Building, Cambridge CB2 2RU, UK
40 West 20th Street, New York, NY 10011-4211, USA
477 Williamstown Road, Port Melbourne, VIC 3207, Australia
Ruiz de Alarcón 13, 28014 Madrid, Spain
Dock House, The Waterfront, Cape Town 8001, South Africa

http://www.cambridge.org

First published 2002

Printed in the United Kingdom at the University Press, Cambridge

Typeface Times Roman 10/13 pt.   *System* LATEX 2$_\varepsilon$ [TB]

A catalog record for this book is available from the British Library.

Library of Congress Cataloging in Publication Data

Julien, Pierre Y.
   River mechanics / Pierre Y. Julien.
      p.   cm.
   Includes bibliographical references and index.
   ISBN 0-521-56284-8 (pc.)
   1. River engineering.   I. Title.
   TC405 .J85   2002
   627′.12–dc21   2001035678

ISBN 0 521 56284 8 hardback
ISBN 0 521 52970 0 paperback

River mechanics

This textbook offers a thorough mechanical analysis of rivers from upland areas to oceans. It scrutinizes state-of-the-art methods, underlining both theory and engineering applications.

Each chapter includes a presentation of fundamental principles, followed with an engineering analysis and instructive problems, examples, and case studies illustrating engineering design. The emphasis is on river equilibrium, river dynamics, bank stabilization, and river engineering. Channel stability and river dynamics are examined in terms of river morphology, lateral migration, aggradation, and degradation. The text provides a detailed treatment of riverbank stabilization and engineering methods. Separate chapters cover physical and mathematical models of rivers. This textbook also contains essential reading for understanding the mechanics behind the formation and propagation of devastating floods, and offers knowledge crucial to the design of appropriate countermeasures to reduce flood impact, prevent bank erosion, improve navigation, increase water supply, and maintain suitable aquatic habitat.

More than 100 exercises (including computer problems) and nearly 20 case studies enhance graduate-student learning, while researchers and practitioners seeking broad technical expertise will find it a valuable reference.

Pierre Y. Julien is Professor of Civil Engineering at Colorado State University.

Dedicated to
Helga and Patrick

Contents

Preface

Rivers have fascinated humanity for centuries. Most prosperous cities around the world have been founded along rivers. Today, river engineers are still designing structures to draw benefits from the fluvial system for developing societies. It is clear that river engineering is not based solely on a simple understanding of local hydrodynamic forces, but also on an encompassing knowledge of the watershed that supplies water and sediment to dynamic river systems. Expertise in river mechanics combines knowledge of watershed climatology, geomorphology, and hydrology, with a deep understanding of hydrodynamic forces governing the motion of water and sediment in complex river systems. State-of-the-art teaching of river mechanics clearly requires study material that emphasizes both theoretical concepts and practical engineering technology. Ideally, scientists should develop new concepts that may be applicable to engineering design, and practitioners should understand why certain structures work and why others fail.

This textbook has been prepared for engineers and scientists seeking broad-based technical expertise in river mechanics. It has been specifically designed for graduate students, for scholars actively pursuing scientific research, and for practitioners keeping up with developments in river mechanics. The prerequisites simply include basic knowledge of undergraduate fluid mechanics and partial differential equations. The textbook *Erosion and Sedimentation*, by the same author and Cambridge University Press, serves as prerequisite material for the graduate course on river mechanics taught at Colorado State University.

Rather than a voluminous encyclopedia, this textbook scrutinizes a selected number of methods to meet pedagogical objectives underlining both theory and engineering applications. This text has been designed to be covered within a regular 45-lecture-hour graduate-level course.

The chapters of this book contain, besides theory and lecture material, various exercises, general problems, data sets, computer problems, examples, and case studies. They illustrate specific aspects of the profession from theoretical derivations through exercises, to practical solutions to real problems through the analysis of case studies. Most problems can be solved with a few algebraic equations; others require the use of computers. The computer problems offer students the opportunity to develop skills for solving physical problems with computers. No specific computer code or language is required. Instead of using existing software packages, I stimulate student creativity and originality in developing the students' own computer programs. Throughout, a solid diamond (♦) denotes equations and problems of particular significance. Problems with a double diamond (♦♦) are considered most important.

I am grateful to D. B. Simons, E. V. Richardson, H. W. Shen, H. Rouse, M. Frenette, J. L. Verrette, Y. Ouellet, C. F. Nordin, S. Schumm, S. R. Abt, and J. Ruff, who influenced my teaching over the years. I am also particularly thankful to P. G. Combs and D. C. Baird for sharing their practical expertise in river engineering. This book also benefits from numerous suggestions formulated by a generation of graduate students at Colorado State University. They helped me tailor this textbook to meet their needs under the constraints of quality, concision, and affordability. Jenifer Davis diligently typed successive draft versions of the manuscript, and Jean Parent prepared the figures. Finally, it has been a renewed pleasure to collaborate with Florence Padgett, Ellen Carlin, Zach Dorsey, and the Cambridge University Press production staff.

Notation

Symbols

a	acceleration
a	reference elevation
a	pier width
a_{cent}	centrifugal acceleration
a_{cor}	Coriolis acceleration
a_i	incremental cross-section area
a_{i+1}	coefficient of h_{i+1}
a_t	partial watershed area
a_θ	particle-stability coefficient
$\tilde{\mathbf{a}}$	wave amplitude
a, b	coefficients of the resistance equation
a, b	transform coefficients for duration curves
A	surface area
A_a	amplitude factor
A_{sb}	surface area of a settling basin
A_t	watershed drainage area
$\tilde{\mathbf{A}}, \tilde{\mathbf{B}}$	wave coefficients
b_r	river-bend coefficient
c	wave celerity
c_G	group velocity
C	Chézy coefficient
C_a	reference concentration
C_s	Courant number
C_{fl}	Courant–Friedrich–Levy coefficient
C_k	grid dispersion number
C_{0i}	upstream sediment concentration
C_r	runoff coefficient
C_u	velocity Courant number

$C_v, C_w, C_{ppm}, C_{mg/l}$	sediment concentration
d_{10}, d_{50}	particle size distribution, % finer by weight
d_m	effective riprap size
d_s	particle size
d_*	dimensionless particle diameter
D	pipe diameter
D_d	degree-days
D_p	drop height of a grade-control structure
e	void ratio
E	specific energy
E_{tons}	expected soil loss in tons
$\tilde{E}$	total energy of a wave
$E()$	exceedance probability
ΔE	specific energy lost in a hydraulic jump
f	Darcy–Weisbach friction factor
f_1	Lacey silt factor
$f(t)$	infiltration rate
F	force
$\tilde{F}$	fetch length of wind waves
F_B	buoyancy force
F_c	centrifugal force
F_D	drag force
F_g	gravitational force
F_h	hydrodynamic force
F_i	inertial force
F_L	lift force
F_M	momentum force
F_p	pressure force
F_s	shear force in a bend
F_S	submerged weight of a particle
F_w	weight of water
F_W	weight of a particle
$F()$	nonexceedance probability
$F(t)$	cumulative infiltration
$F_a(t)$	actual cumulative infiltration
$F_p(t)$	potential cumulative infiltration
Fr	Froude number
g	gravitational acceleration
G	specific gravity of sediment
Gr	gradation coefficient

G_u	universal gravitation constant
h	flow depth
h_c	critical flow depth
h_d	downstream flow depth
h_n	normal flow depth
h_p	pressure head at the wetting front
h_r	rainfall depth
h_s	cumulative snowmelt
h_t	tailwater depth
h_u	upstream flow depth
h_w	partial elevation drop on a watershed
Δh	local change in flow depth
ΔH	energy loss over a meander wavelength
H	Bernoulli sum
H_w	elevation drop on a watershed
$H_s = 2\tilde{a}$	wave height
i	rainfall intensity
i_b	riverbed infiltration rate
i_e	excess rainfall intensity
i_f	snowmelt rate
j	space index
k	time index
k_0	resistance parameter for laminar overland flow
k_s	surface roughness
k_s'	grain roughness
k_t	total resistance to laminar overland flow
$\tilde{k}$	wave number
K	saturated hydraulic conductivity
K_1, K_2	coefficients of the pier scour equation
K_c	riprap coefficient
K_d	dispersion coefficient
K_{num}	numerical dispersion coefficient
K_p	plunging jet coefficient
K_{sj}	submerged jet coefficient
l_1 to l_4	moment arms
l_c, l_d	moment arms in radial stability of river bends
L_a	abutment length
L_f	depth of the wetting front
L_p	pier length
L_r	river length

L_r	length ratio
L_{sb}	settling-basin length
L_M	runoff-model grid-cell size
L_R	grid size of rainfall precipitation
L_S	correlation length of a storm
L_W	length scale of a watershed
L_Δ	length of arrested saline wedge
m	exponent of the resistance equation
m_s	sediment mass eroded from a single storm
m_E	mass of the Earth
m_M	mass of the Moon
M	mass
M	specific momentum
M_1, M_2	first and second moments of a distribution
M, N	particle-stability coefficients
n	Manning coefficient n
$\tilde{n}$	wave number index
N	number of points per wavelength
N	number of storms
$O()$	order of an approximation
p	pressure
$p()$	probability density function
p_0	porosity
p_{0e}	effective porosity
p_{0i}	initial water content
p_{0r}	residual water content
Δp_c	fraction of material coarser than d_{sc}
Δp_i	sediment size fraction
Δp_0	change in water content at the wetting front
P	wetted perimeter
$\tilde{P}$	total power of a wave
ΔP	power loss in a hydraulic jump
$P()$	probability
P_0	power loss
P_Δ	grid Peclet number
q	unit discharge
q_{bv}	unit sediment discharge by volume
$q_{\text{bv}}^* = q_{\text{bv}}/\omega_0 d_s$	dimensionless unit sediment discharge
q_l	lateral unit discharge
q_m	maximum unit discharge

q_s	unit sediment discharge
q_t	total unit sediment discharge
Q	river discharge
Q_{bv}	bed sediment discharge
Q_e	watershed size correction factor
Q_p	peak discharge
Q_s	sediment discharge
r	radial coordinate
r^*	dimensionless radius of curvature
r, θ, z	cylindrical coordinate system θ downstream, r lateral, and z upward
R	radius of curvature of a river
ΔR_e	excess rainfall
R_h	hydraulic radius
R_m	minimum radius of curvature of a channel
R_E	radius of the Earth
Re	Reynolds number
Re_*	grain shear Reynolds number
$\text{Re} = Vh/\nu$	Reynolds number
$\text{Re}_* = u_* d_s/\nu$	grain shear Reynolds number
$\text{Ro} = \omega/\kappa u_*$	Rouse number
S	slope
S_e	effective saturation
S_0, S_f, S_w	bed, friction, and water-surface slope
S_o, S_f, S_w	radial water-surface slope
S_r^*	dimensionless radial slope
S_r, S_{wr}	radial water-surface slope
S_{DR}	sediment delivery ratio
t	time
Δt	time increment
Δt_s	time increment for sediment
t_a	cumulative time with positive air temperature
t_e	time to equilibrium
t_f	cumulative duration of snowmelt
t_r	rainfall duration
$t_r^* = t_r/\bar{t}_r$	normalized storm duration
t_t	transversal mixing time
t_v	vertical mixing time
T	period of return of extreme events
T	wave period

T_E	trap efficiency
T_s	windstorm duration
u, v	velocity along a vertical profile
$\bar{u}$	average flow velocity
u_*	shear velocity
U_w	wind velocity
v_s	velocity against the stone
v_x, v_y, v_z	velocity components
V	cross-section averaged velocity
V_c	critical velocity
V_Δ	densimetric velocity
V_θ	downstream velocity in cylindrical coordinates
W	channel width
W_m	meander width
W_o	overland plane width
x, y, z	coordinates usually x downstream, y lateral, and z upward
x_r, y_r, z_r	length ratios of hydraulic models
Δx	grid spacing
X	runoff length
X_c	reach length
X_e	equilibrium runoff length
Y	sediment yield
z_b	bed elevation
z_w	water-surface elevation
z^*	dimensionless depth
Δz	scour depth

Greek Symbols

α	coefficient of the stage–discharge relationship
α_b	deflection angle of barges
α_e	energy correction factor
β	exponent of the stage–discharge relationship
β	bed particle motion angle
β_m	momentum correction factor
γ	specific weight of water
γ_m	specific weight of a water–sediment mixture
γ_{md}	dry specific weight of a water–sediment mixture
γ_s	specific weight of sediment
$\Gamma(x)$	gamma function

δ	angle between streamline and particle direction
ξ	ratio of exceedance probabilities
$\tilde{\xi}$	displacement in the x direction
η	sideslope stability number
$\tilde{\eta}$	displacement in the y-direction
λ	streamline deviation angle
λ	wavelength
λ_f	snowmelt intensity
$\lambda_r = t_r/t_e$	hydrograph equilibrium number
Λ	meander wavelength
μ	dynamic viscosity of water
ν	kinematic viscosity of water
ϕ	angle of repose of bed material
ϕ	latitude
ϕ	dimensionless soil mass eroded from a single storm
Φ	velocity potential
ρ	mass density of water
ρ_m	mass density of a water–sediment mixture
ρ_{md}	dry mass density of a water–sediment mixture
ρ_s	mass density of sediment
ρ_{sea}	mass density of seawater
$\Pi = \ln[-\ln E(x)]$	double logarithm of exceedance probability
ω	settling velocity
ω_E	angular velocity of the Earth
Ω	sinuosity
Ω_R	coefficient of secondary flows in bends
θ	downstream orientation of channel flow
θ	angular coordinate
θ_j	jet angle measured from the horizontal
θ_m	maximum orientation of channel flow
θ_p	flow orientation angle against a pier
θ_r	raindrop angle
θ_0, Θ_0	downstream bed angle
Θ_1	sideslope angle
$\Theta = (t - t_r)/t_e$	dimensionless time
σ	stress components
σ_d	standard deviation of dispersed material
σ_g	gradation coefficient
$\tilde{\sigma}$	angular frequency of surface waves
τ	shear stress

τ_0, τ_b	bed shear stress
τ_{0x}, τ_{0y}	downstream and lateral bed shear stress
τ_{bn}	bed shear stress at a normal depth
τ_c	critical shear stress
τ_r	radial shear stress
τ_r^*	dimensionless radial shear stress
τ_s	side shear stress
τ_{sc}	critical shear stress on a sideslope
τ_w	wind shear stress
τ_{zx}	shear stress applied in the x direction in a plane perpendicular to z
τ_*	Shields parameter
τ_{*c}	critical value of the Shields parameter
$\psi = q/i_e L$	dimensionless discharge
ψ, θ	weighting coefficients
Ψ	reduced variable
ζ_n^k	Fourier coefficient

Superscripts and Diacriticals

$\hat{a}$	coefficient of the logarithm resistance equation
$\tilde{n}$	wave properties
$\hat{C}$	parameters of the universal soil-loss equation
$\bar{e}$	average value
h^k	time index k

Subscripts

a_r, a_θ	cylindrical coordinate components
a_x, a_z	Cartesian components
τ_c	critical shear stress
h_{j+1}	space index at $j+1$
L_m, Q_m	model value
L_p, Q_p	prototype value
L_r, Q_r	similitude scaling ratio
ρ_m, γ_m	properties of a water–sediment mixture
ρ_{md}, γ_{md}	properties of a dry water–sediment mixture
ρ_s, γ_s	sediment properties

1

Introduction to river mechanics

It has long been understood that water flows downhill. This maybe the only statement to be remembered until a river dries out and crops wilt. Droughts unfortunately threaten humanity with the constraint that, without water, life cannot be sustained. On the other hand, the devastating consequences of excess water through floods stem from the fact that humanity, crops, and cattle are not well adapted to submerged life. Although nomadic tribes coped with the continuously changing pulses of fluvial systems, sedentary conditions forced humanity to protect against floods and droughts. In arid lands, perennial streams with regulated flow and a year-round supply of water are so much more valuable to humanity and wildlife than are natural sequences of short floods that succeed long droughts in dry ephemeral streams. River engineers are facing the daunting challenge of optimizing the urban and environmental resources pertaining to rivers while minimizing the damages caused by floods and droughts.

Perhaps the origin of river engineering started with Yu (4000 B.C.) who was selected to be emperor of China on the premise of long-lasting dikes for the protection of fertile Chinese plains against floods. For centuries, Chinese emperors were classified into "good dynasties" or "bad dynasties" depending on whether or not they succeeded in their struggle to harness large rivers. At approximately the same time, in Mesopotamia, an extensive irrigation system was developed between the Tigris and the Euphrates Rivers. Flood-control levees were constructed to protect fertile lands from destructive inundations. In these early periods of civilization, humanity's cultural development was dominated by fear of thunder, lightning, rain, floods, storms, cyclones, and earthquakes. The lack of understanding of cause-and-effect relationships to explain natural phenomena such as floods has characterized earlier civilizations. Nonetheless, humanity was compelled to develop hydraulic engineering and tame rivers in order to prevent famine and to survive. Today, several hydraulic structures from past civilizations serve as landmarks of excellence.

Natural philosophy emerged during Greek antiquity. Thales of Miletus (circa 600 B.C.) explored natural laws through philosophical meditation in replacing

1

mythological traditions with logical thoughts and reflections based on observations of nature. Basic principles underlying natural processes were deduced
by rational approaches, including reflection and speculation. Hypotheses and
assumptions were formulated by natural philosophers. For instance, Thales
believed that "water was the origin of all things" and "the earth rested on
water." Plato (428–348 B.C.) speculated on matters of physics and metaphysics
alike, without interest in possible discrepancies between theory and reality.
Democritus of Thrace (465 B.C.) believed everything to be inherently mechanical in nature and admitted nothing fortuitous or providential.

In opposition to Plato's speculative ideas, the philosophy of Aristotle (384–
322 B.C.) contemplated nature through facts, and his writings on logic, physics,
biology, metaphysics, and ethics promote continuous advances and evolution
of knowledge in each field. He recognized two types of action: a motivation,
to which the speed of movement is directly proportional, and a resistance, to
which motion is inversely proportional. He also believed that the motivation was
proportional to the density of the body and that the resistance was proportional to
the density of the medium through which it moved. These statements essentially
describe the concepts of linear momentum and resistance to motion. Archimedes
(287–212 B.C.) was the greatest mathematician of antiquity, with chief interest
in geometry, centers of gravity, hydrostatics, a theory of floating bodies, and
anticipated foundations for differential and integral calculus.

Another important milestone, achieved by Hippocrates of Cos (460–380?
B.C.), is worth mention. He proved the existence of evaporation by weighing a
vessel filled with water over a long period of time. The most splendid achieve-
ments are twofold: (1) The concept of studying nature through experiments
was born; and (2) the concept that quantitative information could be gathered
from measurements was developed. It took nearly 2,000 years after Hippocrates
before experiments and measurements supplanted speculations and hypotheses.
The real treasures of natural philosophy found between 600 and 300 B.C. were
highly reputed, but they had few practical effects, and society could hardly
benefit from this new knowledge. Humanity had to wait until the Renaissance,
circa 1500 A.D., to appreciate the long-lasting values of the Greek civilization.

As much as Greeks were interested in pure rational knowledge, Romans
were the true pragmatic engineers of antiquity. Marcus Vitruvius Pollio (first
century B.C.) and Sextus Julius Frontinus (40–103 A.D.) were concerned with
the construction of the aqueducts that supplied water to Rome. There is still
debate as to whether water conduits had been calculated, but colossal aqueduct
and water-distribution systems were designed on the basis of only experience
and rough estimates. It is intriguing that the simple concept of conservation of
mass was not understood at the time of the design. Frontinus could measure

flow depth, and he keenly observed that a steeper slope results in higher flow velocity. However, he incorrectly considered that the discharge corresponds to the cross section given by the measured canal width and flow depth. He observed that velocity increases the discharge, but he could not recognize the quantitative proportionality between velocity and discharge. Correct understanding of the relationship between discharge Q, cross-sectional area A, and velocity V in terms of $Q = AV$ is due to Hero of Alexandria (first century A.D.).

Almost 1,500 years elapsed until the discharge relationship was correctly rediscovered by Leonardo da Vinci (1452–1519) and Benedetto Castelli (1577–1644?). The Renaissance period marks the rebirth of civilization after the Middle Ages. The development of printing contributed to rapid dissemination of knowledge. Leonardo da Vinci understood the principles of experimental science and advocated the necessity of observation: "I will treat of such a subject. But first of all I shall make a few experiments and then demonstrate why bodies are forced to act in this manner. This is the method that one has to pursue in the investigation of phenomena of nature. It is true that nature begins by reasoning and ends by experience; but, nevertheless, we must take the opposite route: as I have said, we must begin with experiment and try through it to discover the reason."

Flow kinematics became better understood under Benedetto Castelli, who wrote "Sections of the same river discharge equal quantities of water in equal times, even if the sections themselves are unequal. Given two sections of a river, the ratio of the quantity of water which passes the first section to that which passes the second is in proportion to the ratio of the first to the second sections and to that of the first and second velocities. Given two unequal sections by which pass equal quantities of water, the sections are reciprocally proportional to the velocities."

The seventeenth century brought remarkable advances in mechanics and mathematics. Dynamic concepts of inertia and momentum became clear under René Descartes (1596–1650), who wrote "I assume that the movement which is once impressed upon a given body is permanently retained, if it is not removed by some other course; that is, whatever has commenced to move in a vacuum will continue to move indefinitely at the same velocity."

Pressure concepts in fluids at rest were described by Blaise Pascal (1623–1662), who postulated that pressure was transmitted equally in all directions. Christian Huygens (1629–1695) defined the principle of centrifugal force and is sometimes credited with the principle of conservation of energy. Isaac Newton (1642–1727) clearly formulated three laws of motion as a concise synthesis of concepts explicitly formulated by Descartes, Wallis, Huygens, and Wren. His contribution is a concise definition of mass, momentum, inertia, and force.

He also studied fluid resistance between a fluid and a solid to conclude that resistance is proportional to the relative velocity of adjacent zones. Energy concepts and present-day calculus evolved around the contribution of Gottfried Wilhelm von Leibniz (1646–1716). He introduced the concept of a live force that is proportional to the second power of velocity, now known as kinetic energy, and raised a lively debate between kinetic energy and momentum proportional to the first power of velocity.

Hydrodynamics can be attributed to outstanding mathematical developments in the eighteenth century. Daniel Bernoulli (1700–1782) dealt with fluid statics and dynamics. It is nevertheless Leonard Euler (1707–1783) who rigorously derived the Bernoulli equation and the differential forms of the equations of continuity and acceleration in frictionless fluids.

Resistance to flow remained obscure until the nineteenth century when experiments on flow in small pipes resulted from the studies of Gotthilf Heinrich Ludwig Hagen (1797–1884), Jean-Louis Poiseuille (1799–1869), Julius Weisbach (1806–1871), Henry Phillibert Gaspard Darcy (1803–1858), Wilhelm Rudolf Kutter (1818–1888), Emile Oscar Ganguillet (1818–1894), and Robert Manning (1816–1897). The Navier–Stokes equations for the analysis of viscous fluid motion became possible from the contributions of Jean-Claude Barré de Saint-Venant (1797–1886), Louis Marie Henri Navier (1785–1836), Baron Augustin Louis de Cauchy (1789–1857), Simeon Denis Poisson (1781–1840) and George Gabriel Stokes (1819–1903). Turbulence challenged generations of scientists including Joseph Boussinesq (1842–1929), Osborne Reynolds (1842–1912), Ludwig Prandtl (1875–1953), and Theodor von Kármán (1881–1963), who contributed to unveil part of its inherent complexity.

In comparison with those of hydraulics, the advances in sediment transport, which is essential to understanding river mechanics, have been extremely slow. Two contributions before the twentieth century are noteworthy: (1) the contributions of Albert Brahm on the relationship between the bedflow velocity and the 1/6 power of the immersed weight of bed material and (2) the concept of tractive force by Paul Francois Dominique du Boys (1847–1924) and his relationship to bed-sediment transport.

Today, in-class discussions can emerge from a simple question such as, Why do rivers form? It is interesting to note the required physical processes that lead to river formation. The concept of a gravitational-force component should first come to mind. The need for erodible material or alluvium emanates from the discussion. The concept of an alluvial river usually is gradually becoming clear. However, all of this does not explain why rivers form. Do we understand the mechanics of formation of alluvial rivers? The effects of flow convergence and divergence allude to the concepts of continuity of water. Aggradation and

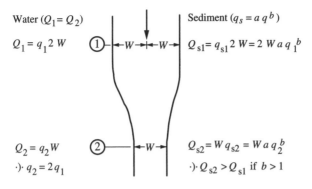

Figure 1.1. Water and sediment balance for converging flow.

degradation results from conservation of sediment. Does converging flow tend to cause aggradation or degradation? We can formulate an intuitive understanding by using a simple sediment rating curve of the type $q_s = aq^b$, where, q_s is the unit sediment discharge and q is the unit discharge; see Fig. 1.1.

The results of converging flow are to cause degradation when $b > 1$ and aggradation when $b < 1$. Is there any reason from our understanding of erosion and sedimentation that supports that $b > 1$? If so, I guess we have answered our question. In a simplified form, rivers form because sediment concentration increases with unit discharge. Flow convergence thus causes scour, and this clearly illustrates that river mechanics stems from an understanding of hydrodynamics and sediment transport.

On Earth, the study of the water and sediment discharge to the oceans from the rivers around the world shows that the annual suspended sediment discharge is 13.5×10^9 metric tons per year. Some important rivers are listed in Table 1.1. Approximately half of the sediment discharge to the oceans originates from rivers in Southeast Asia. In comparison, the total freshwater flow to the oceans from all rivers of the world combines to 1.2×10^6 m³/s. The average sediment concentration of flows to oceans is $\sim$360 mg/l.

The proposed physical analysis of river mechanics is based on the concept of water and sediment transport down the rivers under the action of gravity from the upland areas to the oceans. The surface area of the land that drains into a particular river delineates the watershed, also termed the drainage basin or catchment. Chapter 2 outlines the physical properties of water and sediment and the governing equations of motion. Chapter 3 reviews the sources and the yields of water and sediment at the watershed scale. Chapter 4 treats the steady-flow conditions in canals and rivers. Chapter 5 delves into the mechanics of unsteady flows in rivers. Chapter 6 describes the downstream hydraulic

Table 1.1. *Water and sediment loads of selected rivers (after Jansen et al., 1979)*

| River | Station | Catchment area (10^6 km^2) | Discharge | | | | Sediment as ppm of discharge[a] (mg l^{-1}) |
| | | | Water | | Sediment | | |
			m^3 s^{-1}	mm yr^{-1}	10^6 tons yr^{-1}	μm yr^{-1}	
Amazon	Mouth	7.0	100,000	450	900	90	290
Mississippi	Mouth	3.9	18,000	150	300	55	530
Congo	Mouth	3.7	44,000	370	70	15	50
La Plata/Parana	Mouth	3.0	19,000	200	90	20	150
Ob	Delta	3.0	12,000	130	16	4	40
Nile	Mouth	2.9	3,000	30	80	15	630
Yenisei	Mouth	2.6	17,000	210	11	3	20
Lena	Mouth	2.4	16,000	210	12	4	25
Amur	Mouth	2.1	11,000	160	52	15	150
Yangtse Kiang	Mouth	1.8	22,000	390	500	200	1,400
Wolga	Mouth	1.5	8,400	180	25	10	100
Missouri	Mouth	1.4	2,000	50	200	100	3,200
Zambesi	Mouth	1.3	16,000	390	100	50	200
St. Lawrence	Mouth	1.3	14,000	340	10	6	20
Niger	Mouth	1.1	5,700	160	40	25	220
Murray-Darling	Mouth	1.1	400	10	30	20	2,500
Ganges	Delta	1.0	14,000	440	1,500	1,000	3,600

River	Location						
Indus	Mouth	0.96	6,400	210	400	300	2,000
Orinoco	Mouth	0.95	25,000	830	90	65	110
Orange River	Mouth	0.83	2,900	110	150	130	1,600
Danube	Mouth	0.82	6,400	250	67	60	330
Mekong	Mouth	0.80	15,000	590	80	70	170
Hwang Ho	Mouth	0.77	4,000	160	1,900	1,750	15,000
Brahmaputra	Bahadurabad	0.64	19,000	940	730	800	1,200
Dnieper	Mouth	0.46	1,600	110	1.2	2	25
Irrawaddi	Mouth	0.41	13,000	1,000	300	500	750
Rhine	Delta	0.36	2,200	190	0.72	1	10
Magdelena (Colombia)	Calamar	0.28	7,000	790	220	550	1,000
Vistula (Poland)	Mouth	0.19	1,000	160	1.5	5	50
Kura (USSR)	Mouth	0.18	580	100	37	150	2,000
Chao Phya (Thailand)	Mouth	0.16	960	190	11	50	350
Oder (Germany/Poland)	Mouth	0.11	530	150	0.13	1	10
Rhone (France)	Mouth	0.096	1,700	560	10	75	200
Po (Italy)	Mouth	0.070	1,500	670	15	150	300
Ishikari (Japan)	Mouth	0.016	230	450	6	270	850
Tiber (Italy)	Mouth	0.013	420	1,000	1.8	100	140
Tone (Japan)	Mouth	0.012	480	1,250	3	180	200
Waipapa (New Zealand)	Kanakanala	0.0016	46	900	11	5,000	7,500

[a] ppm: parts in 10^6

geometry and equilibrium in alluvial rivers. Chapter 7 discusses the concepts of river dynamics and response to perturbations from equilibrium conditions. Chapter 8 particularly deals with river stability and presents methods to stabilize river banks. Chapter 9 presents several river engineering techniques from flood control to bridge crossings and waterways. Chapter 10 focuses on physical modeling techniques, with a particular analysis on the underlying theoretical concepts. Chapter 11 introduces the reader to numerical methods used to solve river engineering problems. Finally, Chapter 12 summarizes theory and applications of river engineering problems associated with waves and tides, usually observed in river estuaries.

2

Physical properties and equations

As a natural science, the variability of river processes must be examined through the measurement of physical parameters. This chapter first describes dimensions and units (Section 2.1), physical properties of water (Section 2.2), and sediment (Section 2.3). The equations governing the motion of water and sediment from upland areas to oceans include kinematics of flow (Section 2.4), the equation of continuity (Section 2.5), the equation of motion (Section 2.6), and the concept of hydraulic and energy grade lines (Section 2.7). In this chapter and in the rest of the book, a solid diamond (♦) denotes equations and problems of particular significance. Problems with a double diamond (♦♦) are considered most important.

2.1   Dimensions and units

Physical properties are usually expressed in terms of the following fundamental dimensions: mass (M), length (L), time (T), and temperature (T°). Throughout the text, the unit of mass is preferred to the corresponding unit of force. The fundamental dimensions are measurable parameters that can be quantified in fundamental units.

In the SI system of units, the fundamental units of mass, length, time, and temperature are the kilogram (kg), the meter (m), the second (s), and degrees Kelvin ($^{\circ}$K). Alternatively, the Celsius scale ($^{\circ}$C) is commonly preferred because it refers to the freezing point of water at 0°C and the boiling point at 100°C.

A Newton (N) is defined as the force required for accelerating 1 kg at 1 m/s^2. Knowing that the acceleration that is due to gravity at the Earth's surface, g, is 9.81 m/s^2, we obtain the weight of a kilogram from Newton's second law: $F = \text{mass} \times g = 1 \text{ kg} \times 9.81 \text{ m/s}^2 = 9.81 \text{ N}$. The unit of work (or energy) is the joule (J), which equals the product of 1 N × 1 m. The unit of power is a watt (W), which is 1 J/s. Prefixes are used in the SI system to indicate multiples or fractions of units by powers of 10:

9

$$\mu \text{ (micro)} = 10^{-6}, \quad k \text{ (kilo)} = 10^3,$$
$$m \text{ (milli)} = 10^{-3}, \quad M \text{ (mega)} = 10^6,$$
$$c \text{ (centi)} = 10^{-2}, \quad G \text{ (giga)} = 10^9.$$

For example, sand particles are coarser than 62.5 micrometers, or μm; gravels are coarser than 2 millimeters, abbreviated 2 mm, and one megawatt (MW) equals one million watts (1,000,000 or 10^6 W).

In the English system of units, the time unit is a second, the fundamental units of length and mass are, respectively, the foot (ft), equal to 30.48 cm, and the slug, equal to 14.59 kg. The force required for accelerating a mass of one slug at 1 ft/s^2 is a pound force (lb). Throughout this text, a pound refers to a force, not a mass. The temperature, in degrees Celsius, $T_C°$, is converted to the temperature in degrees Fahrenheit, $T_F°$, by $T_F° = 32.2\,°F + 1.8\,T_C°$.

Variables are classified as geometric, kinematic, dynamic, and dimensionless variables, as shown in Table 2.1. Geometric variables involve length dimensions only and describe the geometry of a system through length, area, and volume. Kinematic variables describe the motion of fluid and solid particles, and these variables can be depicted by only two fundamental dimensions, namely L and T. Dynamic variables involve mass terms in the fundamental dimensions. Force, pressure, shear stress, work, energy, power, mass density, specific weight, and dynamic viscosity are common examples of dynamic variables. Several conversion factors are listed in Table 2.2.

2.2   Properties of water

The physical properties of a nearly incompressible fluid such as water are sketched in Fig. 2.1.

Mass density of water, ρ. The mass of water per unit volume is referred to as the mass density ρ. The maximum mass density of water at 4 °C is 1,000 kg/m^3 and varies slightly with temperature, as shown in Table 2.3. In comparison, the mass density of sea water is 1,025 kg/m^3 and, at sea level, the mass density of air is 1.29 kg/m^3 at 0 °C. The conversion factor is 1 slug/ft^3 = 515.4 kg/m^3.

Specific weight of water, γ. The gravitational force per unit volume of fluid, or simply the fluid weight per unit volume, defines the specific weight γ. At 10 °C, water has a specific weight, $\gamma = 9,810$ N/m^3 or 62.4 lb/ft^3 (1 lb/ft^3 = 157.09 N/m^3). Specific weight varies slightly with temperature, as given in Table 2.3. Mathematically, the specific weight γ equals the product of the mass

Table 2.1. *Geometric, kinematic, dynamic, and dimensionless variables*

Variable	Symbol	Fundamental dimensions	SI Units
Geometric (L)			
Length	L, x, h, d_s	L	m
Area	A	L^2	m^2
Volume	$\forall$	L^3	m^3
Kinematic (L, T)			
Velocity	v_x, u, u^*	LT^{-1}	m/s
Acceleration	a, a_x, g	LT^{-2}	m/s^2
Kinematic viscosity	ν	L^2T^{-1}	m^2/s
Unit discharge	q	L^2T^{-1}	m^2/s
Discharge	Q	L^3T^{-1}	m^3/s
Dynamic (M, L, T)			
Mass	m	M	1 kg
Force	$F = ma, mg$	MLT^{-2}	1 kg m/s^2 = 1 N
Pressure	$p = F/A$	$ML^{-1}T^{-2}$	1 N/m^2 = 1 Pa
Shear stress	$\tau_{xy}, \tau_0, \tau_c$	$ML^{-1}T^{-2}$	1 N/m^2 = 1 Pa
Work or energy	$E = F \cdot d$	ML^2T^{-2}	1 N m = 1 J
Mass density	ρ, ρ_s	ML^{-3}	kg/m^3
Specific weight	$\gamma, \gamma_s = \rho_s g$	$ML^{-2}T^{-2}$	N/m^3
Dynamic viscosity	$\mu = \rho\nu$	$ML^{-1}T^{-1}$	1 kg/m s = 1 N s/m^2 = 1 Pa s
Dimensionless			
Slope	S_0, S_f	—	—
Specific gravity	$G = \gamma_s/\gamma$	—	—
Reynolds number	$Re = uh/\nu$	—	—
Grain-shear			
Reynolds number	$Re_* = u_* d_s/\nu$	—	—
Froude number	$Fr = u/\sqrt{gh}$	—	—
Shields parameter	$\tau_* = \tau/(\gamma_s - \gamma)d_s$	—	—
Concentration	C_v, C_w	—	—

Note: Pa stands for pascal.

density ρ times the gravitational acceleration $g = 32.2$ ft/s^2 = 9.81 m/s^2:

$$\gamma = \rho g. \tag{2.1}$$

Dynamic viscosity μ. As a fluid is brought into deformation, the velocity of the fluid at any boundary equals the velocity of the boundary. The ensuing rate of fluid deformation causes a shear stress τ_{zx} that is proportional to the dynamic viscosity μ and the rate of deformation of the fluid, dv_x/dz:

$$\tau_{zx} = \mu \frac{dv_x}{dz}. \tag{2.2}$$

Table 2.2. *Conversion of units*

Unit	kg, m, s	N, Pa, W
1 acre	4,046.87 m^2	
1 acre-foot (acre-ft)	1,233.5 m^3	
1 atmosphere (atm)	101,325 kg/m s^2	101.3 kPa
1 bar	100,000 kg/m s^2	100 kPa
1 barrel (U.S., dry) (bbl)	0.1156 m^3	
1 British thermal unit (Btu) $= 778$ lb ft	1,055 kg m^2/s^2	1,055 N m
1 cubic foot per second (ft^3/s)	0.0283 m^3/s	
1 day	86,400 s	
1 degree Celsius ($^\circ$C) $= (T_{F^\circ} - 32^\circ)\ 5/9$	1 degree Kelvin (K)	
1 degree Fahrenheit ($^\circ$F) $= 32 + 1.8\ T_{C^\circ}$	0.555556 degree Kelvin	
1 drop	61.6 mm^3	
1 dyne (dyn)	0.00001 kg m/s^2	1×10^{-5} N
1 dyne per square centimeter (dyn/cm^2)	0.1 kg/m s^2	0.1 Pa
1 fathom (fath)	1.8288 m	
1 foot (ft)	0.3048 m	
1 gallon (U.S., liquid) (gal)	0.0037854 m^3	
1 horsepower (hp) $= 550$ lb ft/s	745.70 kg m^2/s^3	745.7 W
1 inch (in.)	0.0254 m	
1 inch of mercury (in. Hg)	3,386.39 kg/m s^2	3,386.39 Pa
1 inch of water	248.84 kg/m s^2	248.84 Pa
1 joule (J)	1 kg m^2/s^2	1 N m $= 1$ J
1 kip	4,448.22 kg m/s^2	4,448.22 N
1 knot	0.5144 m/s	
1 liter (l)	0.001 m^3	
1 micrometer (μm)	1×10^{-6} m	
1 mile (nautical)	1,852 m	
1 mile (statute)	1609.34 m	
1 million gallons per day (mgd) $= 1.55$ ft^3/s	0.04382 m^3/s	
1 Newton (N)	1 kg m/s^2	
1 ounce (avoirdupois) (oz)	0.02835 kg	
1 fluid ounce (U.S.)	2.957×10^{-5} m^3	
1 pascal (Pa)	1 kg/m s^2	1 N/m^2
1 pint (U.S., liquid) (pt)	0.0004732 m^3	
1 poise (P)	0.1 kg/m s	0.1 Pa s
1 pound-foot (lb-ft)	1.356 kg m^2/s^2	1.356 N m
1 pound per square foot (lb/ft^2 or psf)	47.88 kg/m s^2	47.88 Pa
1 pound per square inch (lb/in.^{2} or psi)	6,894.76 kg/m s^2	6,894.76 Pa
1 pound-force (lb)	4.448 kg m/s^2	4.448 N
1 pound-force per cubic foot (lb/ft^3)	157.09 kg/m^2 s^2	157.09 N/m^3
1 quart (U.S., liquid) (qt)	0.00094635 m^3	
1 slug	14.59 kg	
1 slug per cubic foot (slug/ft^3)	515.4 kg/m^3	
1 stoke (S) $= 1$ cm^2/s	0.0001 m^2/s	
1 ton (U.K., long)	1,016.05 kg	
1 ton (metric) (t)	1,000 kg	
1 ton (short) $= 2,000$ lb	8,900 kg m/s^2	8.9 kN
1 watt (W)	1 kg m^2/s^2	
1 yard (yd)	0.9144 m	
1 year (yr)	31,536,000 s	

Note: Those units for which no abbreviations are given are spelled out.

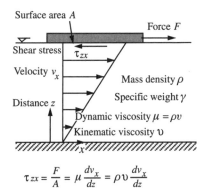

$$\tau_{zx} = \frac{F}{A} = \mu \frac{dv_x}{dz} = \rho \upsilon \frac{dv_x}{dz}$$

Figure 2.1. Newtonian fluid properties.

The fundamental dimension of the dynamic viscosity μ is M/LT, which is a dynamic variable. As indicated in Table 2.3, the dynamic viscosity of water decreases with temperature. Fluids without yield stress for which the dynamic viscosity remains constant regardless of the rate of deformation are called Newtonian fluids. The dynamic viscosity of clear water at 20 °C is 1 centipoise: 1 cP = 0.01 P = 0.001 N s/m² = 0.001 Pa s. The conversion factor is 1 lb s/ft² = 47.88 N s/m² = 47.88 Pa s.

Kinematic viscosity v. When the dynamic viscosity of a fluid μ is divided by the mass density ρ of the same fluid, the mass terms cancel out. This results in kinematic viscosity v with dimensions L^2/T, which is also shown in Table 2.3, decreasing with temperature. The viscosity of clear water at 20 °C is 1 centistokes = 1 cS = 0.01 cm²/s = 1 × 10⁻⁶ m²/s. The conversion factor is 1 ft²/s = 0.0929 m²/s. The change in kinematic viscosity of water, v, with temperature $T°$ in degrees Celsius can be roughly estimated from

$$v = \frac{\mu}{\rho} = [1.14 - 0.031\,(T° - 15) + 0.00068\,(T° - 15)^2] \times 10^{-6} \text{ m}^2/\text{s}. \quad (2.3)$$

It is important to remember that both the density and the viscosity of water decrease with temperature. The maximum water density is found at 4 °C, and water either colder or warmer than 4 °C will be found near the surface. The density of ice increases as the temperature decreases. This causes the ice cover to crack during cold nights and expand to apply large forces on the banks of lakes, reservoirs, and wide rivers during warm days.

2.3 Properties of sediment

The physical properties of sediment are classified into single particles (Subsection 2.3.1), sediment mixture (Subsection 2.3.2), and sediment suspension (Subsection 2.3.3).

Table 2.3. *Approximate physical properties of clear water and ice at atmospheric pressure*

Temperature °C	Density ρ kg/m^3	Specific weight γ N/m^3	Dynamic viscosity μ N s/m^2	Kinematic viscosity ν m^2/s
−30°C	921	9,035	Ice	Ice
−20°C	919	9,015	Ice	Ice
−10°C	918	9,005	Ice	Ice
0°C	999.9	9,809	1.79×10^{-3}	1.79×10^{-6}
4°C	1,000	9,810	1.56×10^{-3}	1.56×10^{-6}
5°C	999.9	9,809	1.51×10^{-3}	1.51×10^{-6}
10°C	999.7	9,807	1.31×10^{-3}	1.31×10^{-6}
15°C	999	9,800	1.14×10^{-3}	1.14×10^{-6}
20°C	998	9,790	1.00×10^{-3}	1.00×10^{-6}
25°C	997	9,781	8.91×10^{-4}	8.94×10^{-7}
30°C	996	9,771	7.97×10^{-4}	8.00×10^{-7}
35°C	994	9,751	7.20×10^{-4}	7.25×10^{-7}
40°C	992	9,732	6.53×10^{-4}	6.58×10^{-7}
50°C	988	9,693	5.47×10^{-4}	5.53×10^{-7}
60°C	983	9,643	4.66×10^{-4}	4.74×10^{-7}
70°C	978	9,594	4.04×10^{-4}	4.13×10^{-7}
80°C	972	9,535	3.54×10^{-4}	3.64×10^{-7}
90°C	965	9,467	3.15×10^{-4}	3.26×10^{-7}
100°C	958	9,398	2.82×10^{-4}	2.94×10^{-7}
°F	slug/ft^3	lb/ft^3	lb s/ft^2	ft^2/s
0°F	1.78	57.40	Ice	Ice
10°F	1.78	57.34	Ice	Ice
20°F	1.78	57.31	Ice	Ice
30°F	1.77	57.25	Ice	Ice
32°F	1.931	62.40	3.75×10^{-5}	1.93×10^{-5}
40°F	1.938	62.43	3.23×10^{-5}	1.66×10^{-5}
50°F	1.938	62.40	2.73×10^{-5}	1.41×10^{-5}
60°F	1.936	62.37	2.36×10^{-5}	1.22×10^{-5}
70°F	1.935	62.30	2.05×10^{-5}	1.06×10^{-5}
80°F	1.93	62.22	1.80×10^{-5}	0.930×10^{-5}
100°F	1.93	62.00	1.42×10^{-5}	0.739×10^{-5}
120°F	1.92	61.72	1.17×10^{-5}	0.609×10^{-5}
140°F	1.91	61.38	0.981×10^{-5}	0.514×10^{-5}
160°F	1.90	61.00	0.838×10^{-5}	0.442×10^{-5}
180°F	1.88	60.58	0.726×10^{-5}	0.385×10^{-5}
200°F	1.87	60.12	0.637×10^{-5}	0.341×10^{-5}
212°F	1.86	59.83	0.593×10^{-5}	0.319×10^{-5}

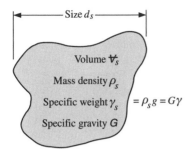

Figure 2.2. Physical properties of a single particle.

2.3.1 Single particle

The physical properties of a single solid particle of volume $\forall_s$ are sketched in Fig. 2.2. The mass density of a solid particle, ρ_s, describes the solid mass per unit volume. The mass density of quartz particles, 2,650 kg/m^3 (1 slug/ft^3 = 515.4 kg/m^3), does not vary significantly with temperature and is assumed constant in most calculations. It must be kept in mind, however, that heavy minerals such as iron, copper, etc., have much larger values of mass density.

Specific weight of solid particles, γ_s. The particle specific weight γ_s corresponds to the solid weight per unit volume of solid. Typical values of γ_s are 26.0 kN/m^3 or 165.4 lb/ft^3. The conversion factor is 1 lb/ft^3 = 157.09 N/m^3. The specific weight of a solid, γ_s, also equals the product of the mass density of a solid particle, ρ_s, times the gravitational acceleration g; thus

$$\gamma_s = \rho_s \, g. \qquad (2.4)$$

Specific gravity G. The ratio of the specific weight of a solid particle to the specific weight of fluid at a standard reference temperature defines the specific gravity G. With common reference to water at 4 °C, the specific gravity of quartz particles is

$$G = \frac{\gamma_s}{\gamma} = \frac{\rho_s}{\rho} = 2.65. \qquad (2.5)$$

The specific gravity is a dimensionless ratio of specific weights, and thus its value remains independent of the system of units.

Submerged specific weight of a particle, $\tilde{\gamma}_s$. Owing to the Archimedes principle, the specific weight of a solid particle, γ_s, submerged in a fluid of specific weight γ equals the difference between the two specific weights; thus

$$\tilde{\gamma}_s = \gamma_s - \gamma = (G - 1)\gamma. \qquad (2.6)$$

Sediment size d_s. The most important physical property of a sediment particle is its size. Table 2.4 shows the grade scale commonly used in

Table 2.4. *Sediment grade scale and approximate properties*

Class name	Particle diameter d_s (mm)	Angle of repose ϕ (deg)	Critical shear stress τ_c (N/m^2)	Critical shear velocity u_{*c} (m/s)	Settling velocity ω_0 (mm/s)
Boulder					
Very large	>2,048	42	1790	1.33	5,430
Large	>1,024	42	895	0.94	3,839
Medium	>512	42	447	0.67	2,715
Small	>256	42	223	0.47	1,919
Cobble					
Large	>128	42	111	0.33	1,357
Small	>64	41	53	0.23	959
Gravel					
Very coarse	>32	40	26	0.16	678
Coarse	>16	38	12	0.11	479
Medium	>8	36	5.7	0.074	338
Fine	>4	35	2.71	0.052	237
Very fine	>2	33	1.26	0.036	164
Sand					
Very coarse	>1.000	32	0.47	0.0216	109
Coarse	>0.500	31	0.27	0.0164	66.4
Medium	>0.250	30	0.194	0.0139	31.3
Fine	>0.125	30	0.145	0.0120	10.1
Very fine	>0.062	30	0.110	0.0105	2.66
Silt					
Coarse	>0.031	30	0.083	0.0091	0.67[a]
Medium	>0.016	30	0.065	0.0080	0.167[a]
Fine	>0.008		Cohesive		0.042[a]
Very fine	>0.004		material		0.010[a]
Clay					
Coarse	>0.0020				2.6×10^{-3}[a]
Medium	>0.0010				6.5×10^{-4}[a]
Fine	>0.0005				1.63×10^{-4}[a]
Very fine	>0.00024				4.1×10^{-5}[a]

[a] Possible flocculation

sedimentation. Note that the size scales are arranged in geometric series with a ratio of two units (1 in. = 25.4 mm).

The size of particles can be determined in a number of ways: The nominal diameter refers to the diameter of a sphere with the same volume as that of the particle, usually measured by the displaced volume of a submerged particle, the sieve diameter is the minimum length of the square sieve opening through which a particle will fall, and the fall diameter is the diameter of an equivalent sphere of specific gravity $G = 2.65$ having the same terminal settling velocity in water at 24 °C.

A wet-sieve method keeps the sieve screen and sand completely submerged. The sediment is washed onto the wet sieve and agitated somewhat vigorously in several directions until all particles smaller than the sieve openings have a chance to fall through the sieve. Material passing through the sieve with its wash water is then poured onto the next-smaller-size sieve. Particles retained on each sieve and those passing through the 0.062-mm sieve are transferred to containers that are suitable for drying the material and for obtaining the net weight of each fraction.

The dry-sieve method is less laborious than the wet-sieve method because a mechanical shaker can be used with a nest of sieves for simultaneous separation of all sizes of interest. It requires only that the dry sand be poured over the coarsest sieve and that the nest of sieves be shaken for 10 min on a shaker that has both lateral and vertical movements.

2.3.2 Sediment mixture

The properties of a sediment mixture are sketched in Fig. 2.3. The total volume $\forall_t$ is the total of the volume of solids $\forall_s$ and the volume of voids $\forall_v$.

Particle-size distribution. An example of particle-size distribution in Fig. 2.4 shows the percentage by weight of material finer than a given sediment size. The sediment size d_{50} for which 50% by weight of the material is finer is called the median grain size. Likewise d_{90} and d_{10} are values of grain size for which 90% and 10% of the material are finer, respectively.

Gradation coefficients σ_g and Gr. The gradation of the sediment mixture is a measure of nonuniformity of sediment mixtures. It can be described by

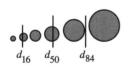

$$\sigma_g = \left(\frac{d_{84}}{d_{16}}\right)^{1/2} \qquad (2.7a)$$

$d_{16} \quad d_{50} \quad d_{84}$

or by the gradation coefficient

$$\mathrm{Gr} = \frac{1}{2}\left(\frac{d_{84}}{d_{50}} + \frac{d_{50}}{d_{16}}\right). \qquad (2.7b)$$

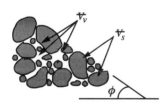

Figure 2.3. Properties of a sediment mixture.

Both gradation coefficients reduce to unity for uniform sediment mixtures, i.e., when $d_{84} = d_{50} = d_{16}$. The gradation coefficient increases with nonuniformity, and high-gradation coefficients describe well-graded mixtures.

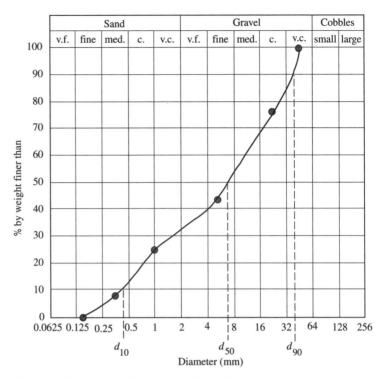

Figure 2.4. Particle-size distribution: v.f., very fine; c., coarse; v.c., very coarse.

Angle of repose ϕ. Typical values of the angle of repose ϕ of granular material are shown in Fig. 2.5. The angle of repose varies with grain size and angularity of the material. Typical values of the angle of repose are also given in Table 2.4 for material coarser than medium silt.

Critical shear stress τ_c *and shear velocity* u_{*_c}. Approximate values of critical shear stress τ_c for noncohesive particles can be obtained from the extended Shields diagram. The values from Julien (1995) are given in Table 2.4 as approximate reference values. The corresponding critical shear velocity u_{*_c} is defined as $u_{*_c} = \sqrt{\tau_c/\rho}$. Note that both τ_c and u_{*_c} do not change significantly for sands and silts. To get crude approximations, a shear-stress value of $\tau = 0.1$ Pa is sufficient to move silts but not sands, and $\tau_c = 1$ Pa is sufficient to move sands but not gravels.

2.3.3   Sediment suspension

The properties of a sediment suspension are sketched in Fig. 2.6, with the volume of void $\forall_v$ equal to the volume of water $\forall_w$.

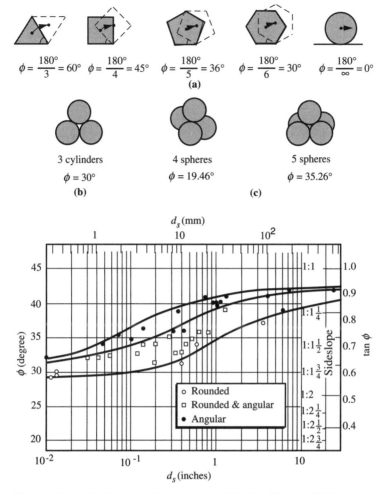

Figure 2.5. Angle of repose of granular material (after Simons, 1957).

Volumetric sediment concentration C_v. The volumetric sediment concentration C_v is defined as the volume of solids $\forall_s$ over the total volume $\forall_t$. When the voids are completely filled with water, $\forall_v = \forall_w$, we obtain

$$C_v = \frac{\forall_s}{\forall_s + \forall_w}. \tag{2.8a}$$

The most common unit for sediment concentration is milligrams per liter, which describes the ratio of the mass of sediment particles to the volume of the water–sediment mixture. Other units include kilograms per cubic meter ($1\ mg/l = 1\ g/m^3$), the volumetric sediment concentration C_v, the concentration

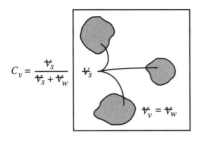

Figure 2.6. Properties of a suspension.

in parts in 10^6 (ppm) C_{ppm}, and the concentration by weight C_w. We can easily demonstrate the following identities:

$$C_w = \frac{\text{sediment weight}}{\text{total weight}}$$

$$= \frac{C_v G}{1 + (G - 1) C_v} \qquad (2.8b)$$

in which $G = \gamma_s / \gamma$ and

$$C_{ppm} = 10^6 C_w. \qquad (2.8c)$$

Note that the percentage by weight C_{ppm} is given by 1,000,000 times the weight of sediment over the weight of the water–sediment mixture. The corresponding concentration in milligrams per liter is then given by

$$C_{mg/l} = \frac{1 \text{ mg/l } G \, C_{ppm}}{G + (1 - G) \, 10^{-6} \, C_{ppm}} = \rho \, G C_v = 10^6 \text{ mg/l } G C_v. \qquad (2.8d)$$

The conversion factors from C_{ppm} to $C_{mg/l}$ are given in Table 2.5. Note that there is less than a 10% difference between C_{ppm} and $C_{mg/l}$ at concentrations of $C_{ppm} < 145{,}000$.

Table 2.5. *Equivalent concentrations for* C_v, C_w, C_{ppm}, $C_{mg/l}$, P_0, *and* e

C_v	C_w	C_{ppm}	$C_{mg/l}$	P_0	e
Suspension					
0.0001	0.00026	265	265		
0.001	0.00264	2,645	2650		
0.0025	0.00659	6,598	6625		
0.005	0.01314	13,141	13250		
0.0075	0.01963	19,632	19875		
0.01	0.02607	26,069	26500		
0.025	0.06363	63,625	66250		
Hyperconcentration					
0.05	0.12240	122,401	132,500	0.95	19
0.075	0.17686	176,863	198,750	0.925	12.3
0.1	0.22747	227,467	265,000	0.9	9
0.25	0.46903	469,027	662,500	0.75	3
0.5	0.72603	726,027	1,325,000	0.50	1
0.75	0.88827	888,268	1,987,500	0.25	0.33

Note: Calculations based on $G = 2.65$.

Settling velocity ω_0. The settling velocity ω_0 of sediment particles in clear water at 10 °C is calculated from

$$\omega_0 = \frac{8v}{d_s}\left\{\left[1 + \frac{(G-1)g}{72v^2}d_s^3\right]^{0.5} - 1\right\},$$ (2.9)

where d_s is the particle diameter, v is the kinematic viscosity, G is the specific gravity, and g is the gravitational acceleration. The values of the settling velocity in clear water are given in Table 2.4.

Specific weight of a mixture, γ_m. The specific weight of a submerged mixture is the total weight of solid and water in the voids per unit total volume. The specific weight of a mixture, γ_m, is a function of the volumetric concentration C_v as

$$\gamma_m = \frac{\gamma_s \forall_s + \gamma \forall_v}{\forall_s + \forall_v} = \gamma_s(C_v) + \gamma(1 - C_v).$$ (2.10)

The specific mass ρ_m of a submerged mixture is the total mass of solid and water in the voids per unit total volume. The specific mass of a mixture is given by $\rho_m = \gamma_m/g$.

Porosity p_0. The porosity p_0 is a measure of the volume of void $\forall_v$ per total volume $\forall_t = \forall_v + \forall_s$. The volume of solid particles $\forall_s = (1 - p_0)\forall_t$ is thus

$$p_0 = \frac{\forall_v}{\forall_t} = \frac{e}{1+e},$$ (2.11)

where the void ratio e is the ratio of the volume of void $\forall_v$ to the volume of solid $\forall_s$. The values of porosity and void ratios at various hyperconcentrations are listed in Table 2.5.

Dry specific weight of a mixture, γ_{md}. The dry specific weight of a mixture is the weight of solid per unit total volume, including the volume of solids and voids. The dry specific weight of a mixture, γ_{md}, is a function of porosity p_0 as

$$\gamma_{md} = \gamma_s(1 - p_0) = \gamma G(C_v).$$ (2.12)

The dry specific weight of sand deposits is approximately ~ 14.75 kN/m^3 or 93 lb/ft^3. The dry specific mass of a mixture is the mass of solid per unit total volume. The dry specific mass of a mixture is $\rho_{md} = \gamma_{md}/g$.

2.4 River flow kinematics

Flow kinematics describes fluid motion in terms of velocity and acceleration. In rivers, two orthogonal coordinate systems are common: (1) global right-hand Cartesian (x,y,z) systems, with x in the main downstream direction, y in the lateral direction to the left bank, and z upward; and (2) local cylindrical (r,θ,z) systems, in which r is the river radius of curvature in a horizontal plane, as shown in Fig. 2.7.

The rate of change in the position of a fluid element is a measure of its velocity. Velocity is defined as the ratio between the displacement ds and the corresponding increment of time dt. Velocity is a vector quantity $\mathbf{v}$ that varies in both space (x,y,z) and time t. Its magnitude v at a given time equals the square root of the sum of squares of its orthogonal components:

$$v = \sqrt{v_x^2 + v_y^2 + v_z^2},$$

where $v_x = dx/dt$, $v_y = dy/dt$, and $v_z = dz/dt$.

A line tangent to the velocity vector at every point at a given instant is known as a streamline. The path line of a fluid element is the locus of the element through time, e.g., the path followed by a single buoy on a river. A streak line is defined as the line connecting all fluid elements that have passed successively at a given point in space, e.g., instantaneous position of all buoys released over time from a single point on a river.

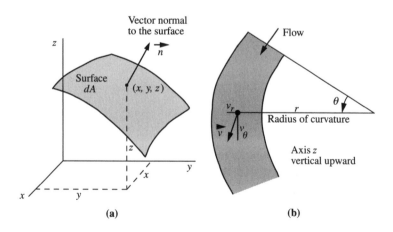

Figure 2.7. Cartesian and cylindrical coordinates.

The differential velocity components over an infinitesimal distance ds (dx,dy,dz) and time increment dt at a point (x,y,z) are

$$dv_x = \frac{\partial v_x}{\partial t}dt + \frac{\partial v_x}{\partial x}dx + \frac{\partial v_x}{\partial y}dy + \frac{\partial v_x}{\partial z}dz, \tag{2.13a}$$

$$dv_y = \frac{\partial v_y}{\partial t}dt + \frac{\partial v_y}{\partial x}dx + \frac{\partial v_y}{\partial y}dy + \frac{\partial v_y}{\partial z}dz, \tag{2.13b}$$

$$dv_z = \frac{\partial v_z}{\partial t}dt + \frac{\partial v_z}{\partial x}dx + \frac{\partial v_z}{\partial y}dy + \frac{\partial v_z}{\partial z}dz. \tag{2.13c}$$

$$\underbrace{\qquad}_{\text{local}} \quad \underbrace{\qquad\qquad\qquad}_{\text{convective}}$$

The Cartesian acceleration components are obtained directly after the velocity equations are divided by the time increment dt,

$$a_x = \frac{dv_x}{dt} = \frac{\partial v_x}{\partial t} + v_x\frac{\partial v_x}{\partial x} + v_y\frac{\partial v_x}{\partial y} + v_z\frac{\partial v_x}{\partial z}, \tag{2.14a}$$

$$a_y = \frac{dv_y}{dt} = \frac{\partial v_y}{\partial t} + v_x\frac{\partial v_y}{\partial x} + v_y\frac{\partial v_y}{\partial y} + v_z\frac{\partial v_y}{\partial z}, \tag{2.14b}$$

$$a_z = \frac{dv_z}{dt} = \frac{\partial v_z}{\partial t} + v_x\frac{\partial v_z}{\partial x} + v_y\frac{\partial v_z}{\partial y} + v_z\frac{\partial v_z}{\partial z}, \tag{2.14c}$$

$$\underbrace{\qquad}_{\text{local}} \quad \underbrace{\qquad\qquad\qquad}_{\text{convective}}$$

in cylindrical coordinates, with $v_r = dr/dt$, $v_\theta = rd\theta/dt$, $v_z = dz/dt$, and the properties of curvilinear vectors give additional convective terms in centrifugal acceleration v^2/r in Eq. (2.15a) and the Coriolis acceleration $(v_r v_\theta)/r$ in Eq. (2.15b):

$$a_r = \frac{dv_r}{dt} = \frac{\partial v_r}{\partial t} + v_r\frac{\partial v_r}{\partial r} + \frac{v_\theta}{r}\frac{\partial v_r}{\partial \theta} - \frac{v_\theta^2}{r} + v_z\frac{\partial v_r}{\partial z}, \tag{2.15a}$$

$$a_\theta = \frac{dv_\theta}{dt} = \frac{\partial v_\theta}{\partial t} + v_r\frac{\partial v_\theta}{\partial r} + \frac{v_\theta}{r}\frac{\partial v_\theta}{\partial \theta} + \frac{v_r v_\theta}{r} + v_z\frac{\partial v_\theta}{\partial z}, \tag{2.15b}$$

$$a_z = \frac{dv_z}{dt} = \frac{\partial v_z}{\partial t} + v_r\frac{\partial v_z}{\partial r} + \frac{v_\theta}{r}\frac{\partial v_z}{\partial \theta} + v_z\frac{\partial v_z}{\partial z}. \tag{2.15c}$$

$$\underbrace{\qquad}_{\text{local}} \quad \underbrace{\qquad\qquad\qquad}_{\text{convective}}$$

It is shown in Eqs. (2.14) and (2.15) that the total acceleration can be separated into local- and convective-acceleration terms. Flows in which local-acceleration terms vanish at any point are called steady. Flows are uniform when all convective-acceleration terms vanish. Steady-uniform flows describe the particular case of motion without any acceleration component.

2.5   Conservation of mass

The equation of continuity, or law of conservation of mass, states that mass cannot be created nor destroyed. The continuity equation can be written in either differential form, discussed in this chapter, or integral form, to be discussed in subsequent chapters (e.g., Section 5.1).

In differential form, consider the infinitesimal control volume in Fig. 2.8 filled with a fluid and a homogeneous concentration of sediment.

The difference between the mass fluxes entering and leaving the differential control volume equal the rate of increase of internal mass. For instance, in the x direction, the net mass flux leaving the control volume is $[(\partial \rho_m v_x)/(\partial x)]\,dx$ times the area $dy\,dz$. The change in internal mass is $(\partial \rho_m / \partial t)\,dx\,dy\,dz$. Repeating the procedure in the y and the z directions yields the following differential relationships:

Cartesian coordinates (x, y, z):

$$\frac{\partial \rho_m}{\partial t} + \frac{\partial}{\partial x}(\rho_m v_x) + \frac{\partial}{\partial y}(\rho_m v_y) + \frac{\partial}{\partial z}(\rho_m v_z) = 0. \qquad (2.16a) \blacklozenge$$

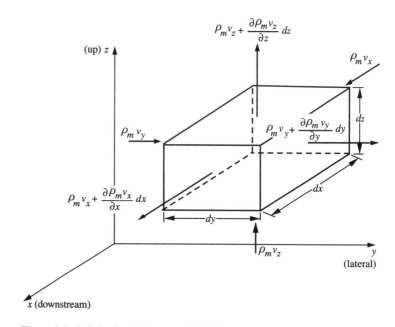

Figure 2.8. Infinitesimal element of a fluid.

Cylindrical coordinates (r, θ, z):

$$\frac{\partial \rho_m}{\partial t} + \frac{1}{r}\frac{\partial}{\partial r}(\rho_m r v_r) + \frac{1}{r}\frac{\partial}{\partial \theta}(\rho_m v_\theta) + \frac{\partial}{\partial z}(\rho_m v_z) = 0. \qquad (2.16b) \blacklozenge$$

For the particular case in which sediment diffusion is not significant, the conservation of solid mass is also defined after ρ_m is replaced with C_v obtained after Eq. (2.10) is substituted into Eq. (2.16). For sediment-transport problems in which turbulent diffusion and dispersion are significant, sediment-continuity equation (10.5) in Julien (1995), including turbulent-mixing coefficients, should be used. For homogeneous incompressible suspensions without settling, the mass density is independent of space and time ($\rho_s, \rho, \rho_m = $ const); consequently $\partial \rho_m / \partial t = 0$ and the divergence of the velocity vector in Cartesian coordinates must be zero, i.e.,

$$\frac{\partial v_x}{\partial x} + \frac{\partial v_y}{\partial y} + \frac{\partial v_z}{\partial z} = 0. \qquad (2.17)$$

When dealing with open-channel flows at low sediment concentrations, we can neglect compressibility effects, and we find that Eq. (2.17) is applicable.

2.6 Equations of motion

The analysis of fluid motion results from the application of forces on a fixed control volume. Given that the force F equals the product of mass m and acceleration a, the approach for fluids of mass density $\rho = (m/\forall)$ stems from $a = (F/m) = (F/\rho\forall)$. The forces acting on a Cartesian element of fluid and sediment (dx, dy, dz) are classified as either internal forces or external forces. The internal accelerations, or body forces per unit mass, acting at the center of mass of the element are denoted by g_x, g_y, and g_z. The external forces per unit area applied on each face of the element are subdivided into normal- and tangential-stress components. The normal stresses σ_x, σ_y, and σ_z are designated as positive for tension. Six shear stresses, τ_{xy}, τ_{yx}, τ_{xz}, τ_{zx}, τ_{yz}, and τ_{zy}, with two orthogonal components are applied on each face, as shown in Fig. 2.9. The first subscript indicates the direction normal to the face, and the second subscript designates the direction in which the stress is applied. The identities $\tau_{xy} = \tau_{yx}$, $\tau_{xz} = \tau_{zx}$, and $\tau_{yz} = \tau_{zy}$ result from the sum of moments of shear stresses around the centroid.

The cubic element in Fig. 2.9 is considered in equilibrium when the sum of the forces per unit mass in each direction, x, y, and z, equals the corresponding

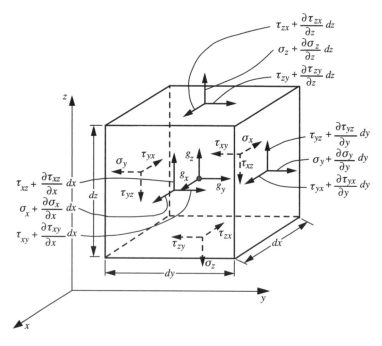

Figure 2.9. Surface stresses on a fluid element.

Cartesian acceleration components a_x, a_y, and a_z:

$$a_x = g_x + \frac{1}{\rho_m}\frac{\partial \sigma_x}{\partial x} + \frac{1}{\rho_m}\frac{\partial \tau_{yx}}{\partial y} + \frac{1}{\rho_m}\frac{\partial \tau_{zx}}{\partial z},$$ (2.18a)

$$a_y = g_y + \frac{1}{\rho_m}\frac{\partial \sigma_y}{\partial y} + \frac{1}{\rho_m}\frac{\partial \tau_{xy}}{\partial x} + \frac{1}{\rho_m}\frac{\partial \tau_{zy}}{\partial z},$$ (2.18b)

$$a_z = g_z + \frac{1}{\rho_m}\frac{\partial \sigma_z}{\partial z} + \frac{1}{\rho_m}\frac{\partial \tau_{xz}}{\partial x} + \frac{1}{\rho_m}\frac{\partial \tau_{yz}}{\partial y}.$$ (2.18c)

These equations of motion are general without any restriction as to compressibility, viscous shear, turbulence, or other effects. The normal stresses can be rewritten as a function of the pressure p and the additional normal stresses, τ_{xx}, τ_{yy}, and τ_{zz}, accompanying deformation:

$$\sigma_x = -p + \tau_{xx},$$ (2.19a)

$$\sigma_y = -p + \tau_{yy},$$ (2.19b)

$$\sigma_z = -p + \tau_{zz}.$$ (2.19c)

Table 2.6. *Equations of motion*

Cartesian coordinates

x component

$$a_x = \frac{\partial v_x}{\partial t} + v_x \frac{\partial v_x}{\partial x} + v_y \frac{\partial y_x}{\partial y} + v_z \frac{\partial v_x}{\partial z} = g_x - \frac{1}{\rho_m} \frac{\partial p}{\partial x} + \frac{1}{\rho_m} \left(\frac{\partial \tau_{xx}}{\partial x} + \frac{\partial \tau_{yx}}{\partial y} + \frac{\partial \tau_{zx}}{\partial z} \right) \quad \text{(2.20a)} \blacklozenge$$

y component

$$a_y = \frac{\partial v_y}{\partial t} + v_x \frac{\partial v_y}{\partial x} + v_y \frac{\partial v_y}{\partial y} + v_z \frac{\partial v_y}{\partial z} = g_y - \frac{1}{\rho_m} \frac{\partial p}{\partial y} + \frac{1}{\rho_m} \left(\frac{\partial \tau_{xy}}{\partial x} + \frac{\partial \tau_{yy}}{\partial y} + \frac{\partial \tau_{zy}}{\partial z} \right) \quad \text{(2.20b)} \blacklozenge$$

z component

$$a_z = \frac{\partial v_z}{\partial t} + v_x \frac{\partial v_z}{\partial x} + v_y \frac{\partial v_z}{\partial y} + v_z \frac{\partial v_z}{\partial z} = g_z - \frac{1}{\rho_m} \frac{\partial p}{\partial z} + \frac{1}{\rho_m} \left(\frac{\partial \tau_{xz}}{\partial x} + \frac{\partial \tau_{yz}}{\partial y} + \frac{\partial \tau_{zz}}{\partial z} \right) \quad \text{(2.20c)} \blacklozenge$$

Cylindrical coordinates

r component

$$\frac{\partial v_r}{\partial t} + v_r \frac{\partial v_r}{\partial r} + \frac{v_\theta}{r} \frac{\partial v_r}{\partial \theta} - \frac{v_\theta^2}{r} + v_z \frac{\partial v_r}{\partial z} = g_r - \frac{1}{\rho_m} \frac{\partial p}{\partial r}$$
$$+ \frac{1}{\rho_m} \left[\frac{1}{r} \frac{\partial}{\partial r}(r\tau_{rr}) + \frac{1}{r} \frac{\partial \tau_{\theta r}}{\partial \theta} - \frac{\tau_{\theta\theta}}{r} + \frac{\partial \tau_{zr}}{\partial z} \right] \quad \text{(2.21a)} \blacklozenge$$

θ component

$$\frac{\partial v_\theta}{\partial t} + v_r \frac{\partial v_\theta}{\partial r} + \frac{v_\theta}{r} \frac{\partial v_\theta}{\partial \theta} + \frac{v_r v_\theta}{r} + v_z \frac{\partial v_\theta}{\partial z} = g_\theta - \frac{1}{\rho_m r} \frac{\partial p}{\partial \theta}$$
$$+ \frac{1}{\rho_m} \left[\frac{1}{r^2} \frac{\partial}{\partial r}(r^2 \tau_{r\theta}) + \frac{1}{r} \frac{\partial \tau_{\theta\theta}}{\partial \theta} + \frac{\partial \tau_{z\theta}}{\partial z} \right] \quad \text{(2.21b)} \blacklozenge$$

z component

$$\frac{\partial v_z}{\partial t} + v_r \frac{\partial v_z}{\partial r} + \frac{v_\theta}{r} \frac{\partial v_z}{\partial \theta} + v_z \frac{\partial v_z}{\partial z} = g_z - \frac{1}{\rho_m} \frac{\partial p}{\partial z} + \frac{1}{\rho_m} \left[\frac{1}{r} \frac{\partial (\tau_{rz})}{\partial r} + \frac{1}{r} \frac{\partial \tau_{\theta z}}{\partial \theta} + \frac{\partial \tau_{zz}}{\partial z} \right] \quad \text{(2.21c)} \blacklozenge$$

After the acceleration components a_x, a_y, and a_z from Eqs. (2.14) are considered, the equations of motion in Cartesian and cylindrical coordinates can be written as shown in Table 2.6.

2.7 Hydraulic and energy grade lines

Let us consider a small element of fluid in a wide-rectangular channel at a bed slope S_0, as shown in Fig. 2.10. The flow is one dimensional (1D) in the x direction; thus $v = v_x$ and $v_y = v_z = 0$. The shear stress that is due to element stretching is $\tau_{xx} = 0$. The effects of bank shear τ_{yx} in a wide channel can be neglected, $\tau_{yx} = 0$, but the bed shear stress is significant, $\tau_{zx} \neq 0$. At small bed-slope angles, $\sin\theta \simeq \tan\theta$ and $g_x = g\sin\theta \cong gS_0$, Eq. (2.20a) thus reduces to

$$\frac{\partial v_x}{\partial t} + v_x \frac{\partial v_x}{\partial x} \cong gS_0 - \frac{1}{\rho} \frac{\partial p}{\partial x} + \frac{1}{\rho} \frac{\partial \tau_{zx}}{\partial z}. \quad \text{(2.22)}$$

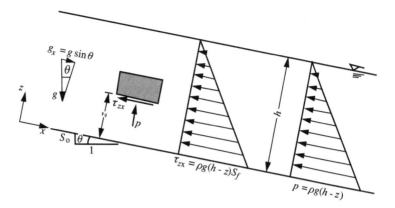

Figure 2.10. Pressure and shear-stress distributions in a fluid column.

We obtain the pressure distribution by integrating Eq. (2.20c), given that $a_z = 0$ and that shear-stress variations are small. The resulting hydrostatic-pressure approximation that is defined from $g_z = g \cos \theta \simeq -g$ from the bed elevation is

$$\int_p^0 dp = \rho \int_z^h -g dz \quad \text{or} \quad p = \rho g(h - z) \tag{2.23}$$

at the bed, $z = 0$, the pressure is $p_0 = \rho g h$ and the relative pressure vanishes at the free surface, $p = 0$ at $z = h$.

The bed shear stress τ_0 is obtained from the definition of the friction slope S_f for steady-uniform flow as $\tau_0 = \rho g h S_f$. Like pressure, the shear-stress vanishes at the free surface and varies linearly over the depth. The shear-stress distribution is thus

$$\tau_{zx} = \rho g(h - z)S_f. \tag{2.24} \blacklozenge$$

Relation (2.22) is greatly reduced after Eqs. (2.23) and (2.24) are substituted into it, owing to $(\partial z/\partial x) = 0$ and $(\partial S_f/\partial z) = (\partial h/\partial z) = 0$; thus

$$\frac{\partial v_x}{\partial t} + v_x \frac{\partial v_x}{\partial x} \cong g S_0 - g \frac{\partial h}{\partial x} - g S_f. \tag{2.25}$$

We can solve equation of motion (2.25) in dimensionless form for S_f after dividing by g:

$$S_f \cong S_0 - \frac{\partial h}{\partial x} - \frac{v_x \partial v_x}{g \partial x} - \frac{\partial v_x}{g \partial t}. \tag{2.26}$$

This very important formulation is usually attributed to Saint-Venant (1871).

The practical significance is that point velocities v_x can be replaced with the mean flow velocities V, considering that the momentum correction factor is

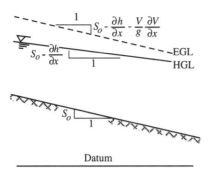

Figure 2.11. Energy grade line (EGL) and hydraulic grade line (HGL).

close to unity:

$$S_f \cong S_0 - \underbrace{\frac{\partial h}{\partial x}}_{} - \underbrace{\frac{V}{g}\frac{\partial V}{\partial x}}_{} - \frac{1}{g}\frac{\partial V}{\partial t}.$$

$\underbrace{\text{bed slope}}$ (2.27) ◆◆

$\underbrace{\text{free-surface slope}}$

$\underbrace{\hspace{4cm}\text{energy slope}\hspace{4cm}}$

Finally, the graphical representation in Fig. 2.11 after the consideration that $S_0 = -(\partial z/\partial x)$. Thus the first term on the right-hand side of relation (2.27) physically describes the bed slope. The first two terms on the right-hand side of relation (2.27) describe the free-surface slope. The first three terms on the right-hand side of relation (2.27) describe the slope of the energy grade line.

Problem 2.1

Determine the mass density, specific weight, dynamic viscosity, and kinematic viscosity of clear water at 20 °C (a) in SI units and (b) in the English system of units.

Answers: (a) $\rho = 998$ kg/m^3, $\gamma = 9790$ N/m^3, $\mu = 1.0 \times 10^{-3}$ N s/m^2, $\nu = 1 \times 10^{-6}$ m^2/s, (b) $\rho = 1.94$ slug/ft^3, $\gamma = 62.3$ lb/ft^3, $\mu = 2.1 \times 10^{-5}$ lb s/ft^2, $\nu = 1.1 \times 10^{-5}$ ft^2/s.

◆Problem 2.2

Determine the sediment size, mass density, specific weight, submerged specific weight, and angle of repose of small quartz cobbles (a) in SI units and (b) in the English system of units.

♦♦Problem 2.3

The volumetric sediment concentration of a sample is $C_v = 0.05$. Determine the corresponding (a) concentration by weight, C_w, (b) concentration in parts in 10^6, C_{ppm}, (c) concentration in milligrams per liter, $C_{mg/l}$, (d) porosity p_0, and (e) void ratio e.

Answers: The answers to Problem 2.3 are in Table 2.5.

♦Problem 2.4

The porosity of a sandy loam is 0.45. Determine the corresponding soil properties: (a) volumetric concentration, (b) void ratio e, (c) specific weight γ_m, (d) specific mass ρ_m, (e) dry specific weight γ_{md}, and (f) dry specific mass ρ_{md}.

♦Problem 2.5

Calculate the gradation coefficients σ_g and Gr from the particle-size distribution shown in Fig. 2.4.

Answer:

$$\sigma_g = \sqrt{\frac{32}{0.5}} = 8, \, Gr \approx \frac{1}{2}\left[\frac{32}{8} + \frac{8}{0.5}\right] = 10,$$

i.e., a well-graded mixture.

♦♦Problem 2.6

Apply the continuity equation to 1D runoff at a unit discharge q and flow depth h for rainfall intensity i_e on an impervious plane. Demonstrate that $(\partial h/\partial t) + (\partial q/\partial x) = i_e$, where x denotes the downstream direction and t denotes time.

Problem 2.7

Demonstrate Eq. (2.8b) from the definition that C_w is the sediment weight/total weight.

3

River basins

This chapter covers the characteristic of river basins (also called watersheds) that affect surface runoff and sediment yield. The main watershed characteristics are illustrated in Section 3.1, followed by rainfall precipitation (Section 3.2), interception and infiltration (Section 3.3), excess rainfall (Section 3.4), and surface runoff (Section 3.5). Soil eroded from upland areas is usually the source of most sediments that are transported by rivers to reservoirs and estuaries. Methods are presented to calculate upland erosion (Section 3.6) and to estimate sediment yield from watersheds (Section 3.7).

3.1 River-basin characteristics

The hydrologic cycle describes processes that contribute to the source and the yield of water and sediment from upland areas to the fluvial system. Figure 3.1 depicts a portion of a watershed during a precipitation event. Shown in this figure are the processes of condensation, precipitation, interception, evaporation, transpiration, infiltration, subsurface flow, exfiltration, deep percolation, groundwater flow, surface flow, surface-detention storage, channel precipitation, evaporation, and streamflow. All of these processes play a role in hydrology; however, precipitation, infiltration, overland flow, and streamflow are most important in surface runoff, upland erosion, and river mechanics.

River basins or watersheds define areas of the Earth's surface where rainwater drains into a particular stream. The terms basin and catchment are synonymously used in the literature. Watershed characteristics can often be described in geographical terms, including physiography and topography, geology and pedology, and climatology and forestry. Watershed boundaries are delineated by drainage divides, usually located at high points, that separate different drainage areas.

Watershed physiography is described primarily by topographic maps. The elevation relates to the type of precipitation in terms of rain and snow. The surface slope indicates the rate at which potential energy (PE) is transformed

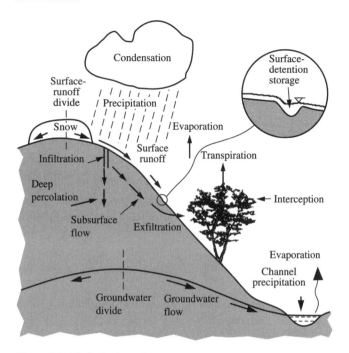

Figure 3.1. Hydrologic cycle.

into the kinetic energy (KE) of surface waters. The surface slope is also a dominant parameter in the calculation of soil erosion and sediment transport. The topography of small watersheds can nowadays be depicted with a digital elevation model with data available at 30-m resolution. Geographic information systems (GISs) provide standard procedures for slope calculations from the mean elevation at each pixel. Limitations lie in the size of the data files and the computing power of computers performing hydrologic calculations.

Digital elevation models provide the following physiographical characteristics of a watershed: (1) drainage area, (2) extreme elevations, (3) hypsometric curves of the percentage of the drainage area at or below a given elevation (see Fig. 3.2), (4) average value and distribution of the terrain slope, (5) percentage of drainage area in lakes and reservoirs, and (6) drainage length from the physically remotest point on the watershed to the outlet.

The mainstream length of a river varies with drainage area, and the relationship shown in Fig. 3.3 indicates that mainstream length increases approximately with the square root of the drainage area.

Geologic information indicates the overall erodibility of large watersheds. For instance, the Yellow River in China carries a large sediment load from the Loess plateau, and the St. Lawrence River drains the granitic Laurentian shield

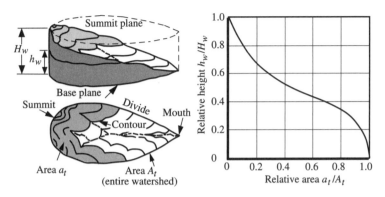

Figure 3.2. Hypsometric curve of a watershed.

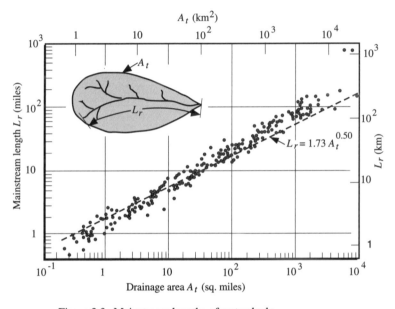

Figure 3.3. Mainstream lengths of watersheds.

at an average sediment concentration of 20 mg/l. In large rivers, the location of faults can be useful in detecting changes in bed elevation through tectonic activity and also in tracking possible lateral shifting of rivers through ancient times. Paleohydrology provides broad guidance to fluvial geomorphologists as to what might have been ancient fluvial conditions. Without quantitative measurements of ancient tectonic activities, such information, however, often remains quite speculative.

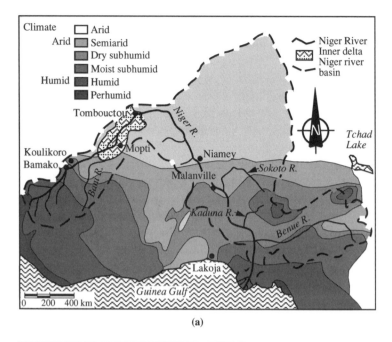

(a)

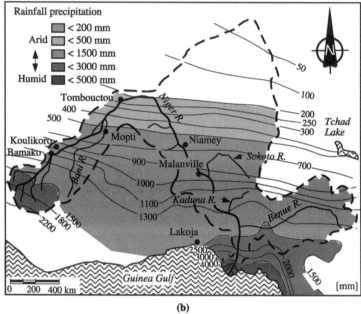

(b)

Figure 3.4. Niger River: (a) climate, (b) rainfall precipitation, and (c) potential evapotranspiration.

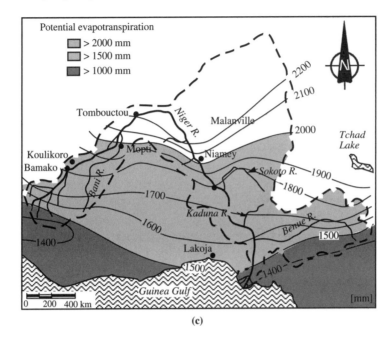

(c)

Figure 3.4. (*cont.*)

Soil types are usually classified according to agronomic standards, such as the Soil Conservation Service (SCS) classification. The digitized information at some fine resolutions enables the estimation of infiltration characteristics and surface-roughness parameters for the calculation of surface runoff.

Climatic conditions are also known to change over sufficiently large watersheds, and many change over time, e.g., desertification. For instance, the climate of the Niger river watershed ranges from arid to humid [Fig. 3.4(a)], and the vegetation changes from that of a desert to that of a rain forest. Qualitative changes in vegetative cover can often be corroborated with mean annual rainfall precipitation and mean annual potential evapotranspiration, as shown in Figs. 3.4(b) and 3.4(c).

3.2   Rainfall precipitation

In the United States, the mean annual rainfall precipitation increases southward from 50 mm in the north to over 4,000 mm near the Mississippi River delta. Such spatial variability in climate and vegetation is typical of very large rivers (over 1×10^6 km^2 of drainage area). This contrasts with the relative homogeneity in climate and vegetation of most small watersheds. In terms of large events,

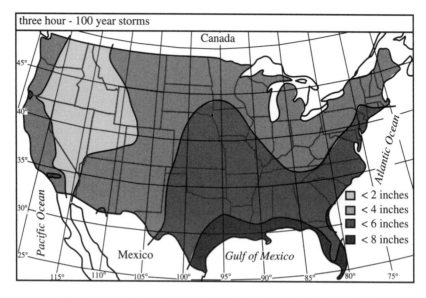

Figure 3.5. 3-h storm with 100-yr period of return for the United States.

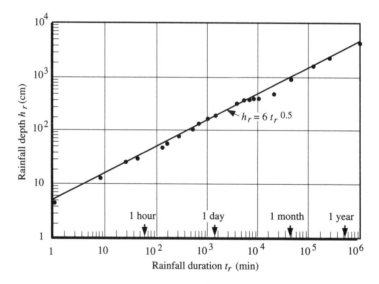

Figure 3.6. Maximum precipitation as a function of rainfall duration.

Fig. 3.5 shows the distribution of the rainfall precipitation with a duration of 3 h and a period of return of 100 yr in the continental United States. Note the spatial variability that ranges from less than 2 in. in Nevada to more than 7 in. in Louisiana. Figure 3.6 shows the world's largest precipitation events in terms of rainfall depth as a function of rainfall duration.

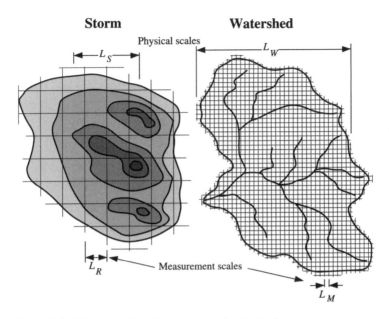

Figure 3.7. Characteristics of rainstorms and watersheds.

Consider a convective rainstorm passing over a watershed, as in Fig. 3.7. There are four length parameters associated with rainstorms and watersheds in the context of raster-based hydrologic models: (1) the correlation length of the rainstorm cell, L_S; (2) the grid size of the rainfall precipitation data, L_R; (3) the characteristic length of the watershed, L_W; and (4) the runoff-model grid-cell size, L_M. The radar and the model resolution parameters, L_R and L_M, are defined by the modeler, whereas the basin- and the storm-size parameters, L_W and L_S, are determined by the field site and the natural storm size in the study area. In addition, it is important to note that the correlation length for watershed characteristics is of the order of tens to hundreds of meters, whereas the spatial correlation length for rainfall, L_S, is hundreds to thousands of meters.

An example of a rainfall event measured by a network of 46 rain gauges with an average spacing of 10.6 km is that in the Denver area, as shown in Fig. 3.8. This small mesoscale rainstorm has a diameter of 20–30 km, and the spatial cross correlation between 2-min rainfall data in a frame of reference moving with the rainstorm is shown in the same figure. The correlation coefficient becomes close to zero at a rain-gauge separation distance of ∼16 km. This means that, for this particular storm, 2-min rainfall measurements with an average spacing of 15 km would be essentially uncorrelated. May and Julien (1998) demonstrated the importance of using a system of coordinates that moves with

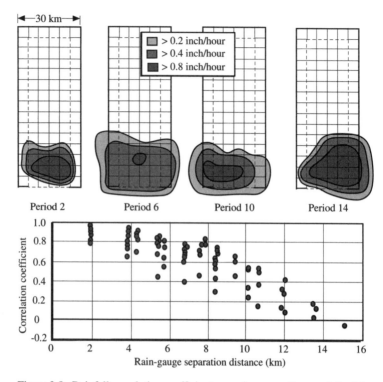

Figure 3.8. Rainfall correlation coefficients vs. rain-gauge distance (after May and Julien, 1998).

the center of mass of the storm cell in order to have physically representative correlation coefficients. The reason for this is that some rainstorms can move at average velocities exceeding 100 km/h.

Conceptually, there are three conditions required for appropriate distributed hydrologic modeling. The first is $L_M \ll L_W$, so that the basin is subdivided into sufficiently small grid elements to describe the spatial variability of basin characteristics. The second condition is $L_R \ll L_S$, which ensures preservation of the spatial gradients of rainfall, particularly for smaller, convective rainstorms. The third requirement is that $L_R \leq L_M$, such that rainfall is placed in the correct watershed grid cell. The third requirement is most difficult to satisfy with either rain gauges or radar. State-of-the-art rainfall precipitation estimates combine radar measurements for spatial variability and rain gauges for ground-truth calibration of radar measurements.

Recent polarimetric weather radar techniques offer an opportunity to record spatially distributed rainfall events with unprecedented resolution. The

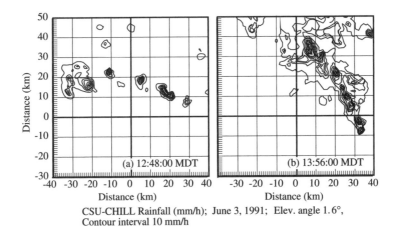

CSU-CHILL Rainfall (mm/h); June 3, 1991; Elev. angle 1.6°,
Contour interval 10 mm/h

Figure 3.9. Colorado rainfall rates measured by the CSU-CHILL radar (after
Ogden and Julien, 1994).

weather radar data shown in Fig. 3.9 was recorded at the Colorado State
University (CSU-CHILL) radar facility, near Greeley, Colorado. The CSU-
CHILL radar is a dual linearly polarized coherent radar that operates at 2.75
GHz, which corresponds to a wavelength of 10.7 cm in the S band. The beam
width of the 8-m-diameter antenna is approximately 1.0°. Attenuation does
not affect S-band weather radars, even in the heaviest rainfall, making them
ideal for weather observation. The radar transmits an average power of 1 kW,
with peak power during transmit pulses of 1 MW, and can measure Doppler
velocities. Ground clutter, high reflectivity, ice and graupel phase, representative precipitation values at different elevations (or CAPPIs), and ground
calibration remain active research areas to improve the accuracy of radar
measurements.

Figure 3.9 shows multiparameter rainfall rates within the 80 km × 80 km
data domain at 12:48 MDT and 13:56 MDT. The radar was located at the origin
of the plots. The rainfall fields were converted from radar conical coordinates
to rectangular coordinates by interpolation to a 1 km × 1 km grid size. The
scale of individual storm cells was approximately 4–10 km, with an average
storm size near 8 km. The rainfall data from 27 radar scans were processed,
and the correlation length L_S was determined to be 2.3 km by both covariance
models. Note that the precipitation patterns of convective rainstorms change
very rapidly in both time and in space. Discretized rainfall precipitation can
serve as input to raster-based hydrologic models such as the CASC2D (Julien
et al., 1995).

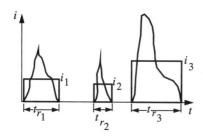

Figure 3.10. Definition sketch of hyetographs.

Point rainfall precipitation can be described as a random time series of discrete storm events, each having finite duration and constant intensity. The principal variables of rainfall precipitation sketched on Fig. 3.10 are (1) the storm duration t_r, (2) the storm intensity i, and (3) the time arrival between successive storms. The time of arrival of successive storms has often been described as a Poisson process. The emphasis is on the storm duration t_r, average intensity i, and the rainstorm depth h_r obtained from $h_r = it_r$.

Considering that the average storm duration is

$$\bar{t}_r = \frac{1}{N} \sum_1^N t_r,$$

where N is the number of storms, the normalized storm duration $t_r^* = t_r/\bar{t}_r$ is distributed exponentially. The particular characteristic of exponential distribution is that the probability density function $p(\ )$ and the exceedance probability function $E(\ )$ are identical:

$$E(t_r^*) = p(t_r^*) = e^{-t_r^*}. \tag{3.1}$$

The corresponding cumulative distribution function $F(t_r)$ or nonexceedance probability, is obtained from

$$F(t_r) = F(t_r^*) = 1 - E(t_r^*). \tag{3.2}$$

An example of the distribution of rainfall precipitation data is provided in Fig. 3.11.

Similarly, the storm intensity i can be normalized after being divided by the average intensity

$$\bar{i} = \frac{1}{N} \sum_1^N i$$

over N storms. The normalized intensity $i^* = i/\bar{i}$ is exponentially distributed with a probability density function $p(\ )$ equal to the exceedance probability $E(\ )$ as

$$E(i^*) = p(i^*) = e^{-i^*}, \tag{3.3}$$

$$F(i) = F(i^*) = 1 - E(i^*). \tag{3.4}$$

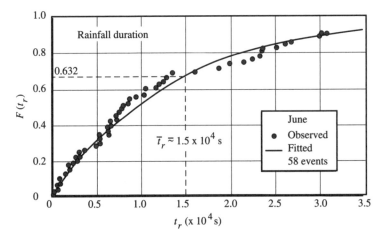

Figure 3.11. Cumulative distribution function of rainfall duration.

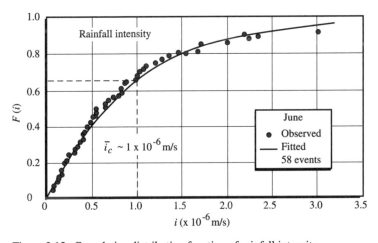

Figure 3.12. Cumulative distribution function of rainfall intensity.

For instance, the observed and the fitted exponential distribution functions for rainstorm intensity, $F(i^*)$, are shown in Fig. 3.12.

For the example shown in Figs. 3.11 and 3.12 with 6 yrs of rainfall data, monthly periods were found to be sufficiently long to assume constant values of average rainfall duration and intensity for each month. Detailed statistical analyses can show whether rainstorm duration and intensity are nearly independent for the period considered. Example 3.1 illustrates how to determine the observed and the fitted exponential distribution functions.

Example 3.1   Application to distribution of rainfall duration. The rainfall durations of $N = 10$ rainstorms measured during one month are 50, 5, 25, 210, 320, 150, 45, 40, 25, and 40 min. Determine the distribution of these events and compare with the exponential distribution. Note that the procedure should normally be applied to large samples.

In Table E.3.1.1, the events are listed in column 1 and ranked in decreasing order in column 2. Their increasing rank in column 3 is divided by $N + 1 = 11$ in column 4 to determine the measured nonexceedance probability $F_m()$. The measured exceedance probability $E_m() = 1 - F()$ is given in column 5.

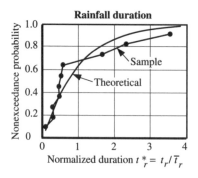

Figure E.3.1.1. Distribution of rainfall duration.

The rainfall duration t_r is normalized as $t_r^* = t_r/\bar{t}_r$ after being divided by the average rainfall duration $\bar{t}_r = 91$ min. The calculated nonexceedance probability $E_c() = 1 - e^{-t_r^*}$. The comparison between measured and calculated nonexceedance probability curves is shown in Fig. E.3.1.1. Note that this sample is very small in this case.

Case Study 3.1   The Big Thompson River flood, United States. The two reports of Grozier et al. (1976) and McCain et al. (1979) document the Big Thompson River flood of July 31–August 1, 1976. As much as 12 in. (305 mm)

Table E.3.1.1. *Nonexceedance probability curves for rainfall duration*

Duration t_r (min)	Ranked duration	Rank	Measured nonexceedance probability $F_m()$	Measured exceedance probability $E_m()$	$t_r^* = \dfrac{t_r}{\bar{t}_r}$	Calculated nonexceedance probability $F_c()$
50	5	1	0.09	0.91	0.05	0.05
5	25	2	0.18	0.82	0.27	0.24
25	25	3	0.27	0.73	0.27	0.24
210	40	4	0.36	0.64	0.44	0.35
320	40	5	0.45	0.55	0.44	0.35
150	45	6	0.54	0.46	0.49	0.39
45	50	7	0.64	0.36	0.55	0.42
40	156	8	0.73	0.27	1.65	0.81
25	210	9	0.82	0.18	2.31	0.90
40	320	10	0.91	0.09	3.51	0.97
Average $\bar{t}_r = 91$ min, $N = 10$						

of rain fell on the Big Thompson River basin during the evening of July 31, 1976, causing a devastating flood on the Big Thompson River and its tributaries between Estes Park and Loveland, Colorado. Larimer County officials reported 139 lives lost and property damage of $16.5 million.

During the evening hours of July 31, a series of violent thunderstorms (Fig. CS.3.1.1) released large volumes of rain along a path several miles wide from Estes Park to the Wyoming border, as sketched in Fig. CS.3.1.2. The Big Thompson River basin west of Drake was severely hit by the storm, and devastating flooding occurred along the Big Thompson River between Estes Park

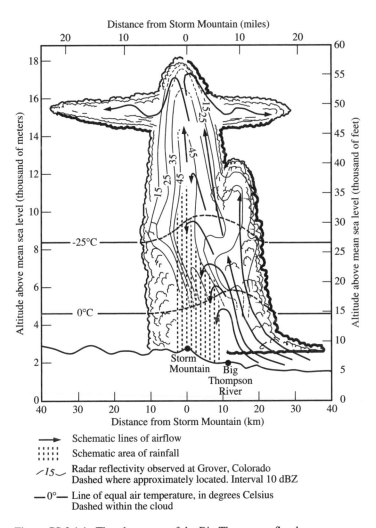

Figure CS.3.1.1. Thunderstorms of the Big Thompson flood.

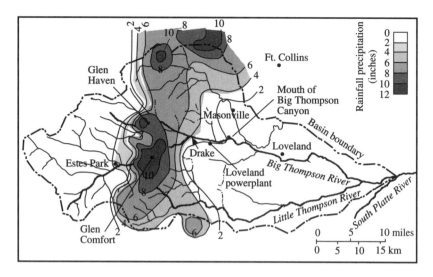

Figure CS.3.1.2. Rainfall precipitation of the Big Thompson flood.

and Loveland and along the North Fork of the Big Thompson River from Glen Haven to its mouth at Drake.

The isohyetal map of the total precipitation from July 31 to August 2, 1976, is shown in Fig. CS.3.1.2. Eastern Colorado was under conditions favorable for heavy rain on July 31, 1976, for a number of reasons. The surface map of that morning showed a slowly moving cold front in the state. Such fronts display lines of convergence that lift air to form thunderstorms. Also favorable was the easterly wind just north of the front, moving air upslope and aiding the frontal lifting. The low-level air was very moist, well above the seasonal normals, and the moisture aloft was also unusually high. Thunderstorms move with the speed and the direction of the winds aloft, and the 500-mbar (millibar) level is usually adequate for judging such movement. The 500-mbar wind was only ~5 knots and was not expected to change much during the day. This was the case with the thunderstorms near Estes Park. They moved very slowly while putting out large amounts of water over a period of several hours, as shown in Fig. CS.3.1.3.

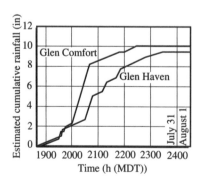

Figure CS.3.1.3. Cumulative rainfall precipitation.

Rainfall began at approximately 18:30 MDT on July 31, 1976, and ended at approximately 23:30 MDT that evening. Additional rainfall was observed on August 1 and 2. Precipitation totals were as much as 10 in. (254 mm) between Estes Park and Drake and more than 12 in. (305 mm) in the Glen Haven area. Very little rainfall contributed to the flood east of Drake and west of Estes Park.

Flood runoff in the Big Thompson basin derived from an area of approximately 60 square miles (155 km^2) centered on the Big Thompson River from Lake Estes to Drake. The topography of the area is characterized by steep north- and south-facing slopes with rugged rock faces along the ridges and a thin soil mantle at lower elevations that supports a moderate stand of coniferous trees. Because of the steep slopes and small storage capacity of the soils, the storm runoff quickly reached nearby surface channels.

The flood lasted only a few hours. The reported peak stages on the Big Thompson River occurred as follows: 20:00 at Glen Comfort, 21:00 at Drake, 21:30 at the Loveland power plant, and approximately 23:00 at the mouth of the canyon ~8 miles (13 km) west of Loveland. The relative timing of the peak stages was such that the peak on the Big Thompson River just downstream from Drake occurred before the peak from the North Fork arrived at Drake. The flood peak moved through the 7.3-mile (11.7-km) length of channel between Drake and the canyon mouth in ~2 h with no apparent reduction in discharge.

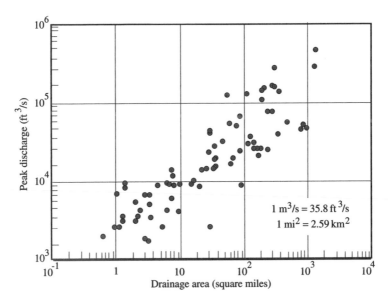

Figure CS.3.1.4. Big Thompson peak discharge vs. drainage area.

East of the canyon mouth, the Big Thompson River valley widens rapidly and the flood discharge was quickly reduced by valley storage and overflow to numerous reservoirs. The peak discharge at the mouth of the Big Thompson River near LaSalle was ~2,500 ft³/s (70.8 m³/s), occurring at noon on August 1, as compared with 31,200 ft³/s (883 m³/s) ~35 miles (56 km) upstream at the mouth of the canyon. The peak discharges at various locations are shown as a function of drainage area in Fig. CS.3.1.4.

3.3   Interception and infiltration

The excess rainfall volume available for surface runoff is the volume of precipitation in excess of the rainfall losses. Losses include interception, evapotranspiration, surface detention, storage, and infiltration.

Some rainfall is intercepted by vegetation before it reaches the ground. The amount of interception varies with the type, density, and stage of growth of the vegetation, intensity of the rainfall, and wind speed. A dense forest canopy may intercept as much as 25% of the annual rainfall in climates with frequent, light rainfalls, low wind speeds, and evergreen vegetation. On a single storm basis, the interception storage is generally a small percentage of the total rainfall event, except in dense forests.

Evapotranspiration is the combination of evaporation and transpiration. Evaporation refers to the phase change of water from liquid to vapor from wet surfaces. Water evaporation from plant surfaces is termed transpiration. All evaporation from a leaf surface is not transpiration as intercepted water is also evaporated. On an annual basis, evapotranspiration generally involves a large fraction of the total precipitation, e.g., refer to Figs. 3.4 for a comparison between arid and humid climates. In spite of the high losses on an annual basis, evapotranspiration is usually neglected in cases in which severe rainstorms are considered.

Detention storage is the volume of water required for filling land depressions before surface runoff begins. Detention storage is the depth of water required for initiating surface runoff. Actual measurements of surface storage and detention are practically nonexistent. Detention storage is estimated at 0.2–0.6 in. (0.5–1.5 cm) in pervious areas such as open fields, woodlands, and lawn grass. Values of 0.05–0.3 in. (0.13–7.6 cm) relate to paved surfaces and roofs in urban areas.

Infiltration is the process of water's permeating into soil pores. In general, the infiltration rate is dependent on soil physical properties, vegetative cover, antecedent soil-moisture conditions, rainfall intensity, and the slope of the infiltrating surface. Bare soils tend to have lower infiltration rates than soils protected by

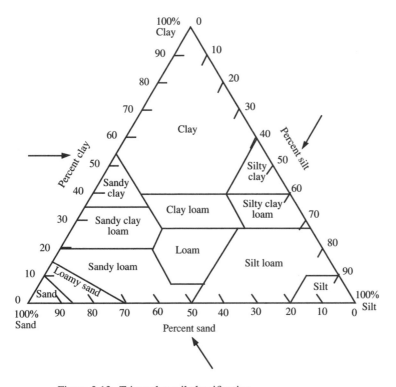

Figure 3.13. Triangular soil classification.

a vegetative cover. The impact of falling raindrops breaks down soil aggregates, and small particles are carried into the soil pores, thus sealing the surface and reducing the infiltration rate. The antecedent moisture condition also alters the infiltration rate, and wet soils have lower infiltration rates than dry soils. On steep slopes, the water tends to run off rapidly and there is less infiltration. The soil classification in Fig. 3.13 depends on the percentage of sand, silt, and clay in the soil. For instance, a soil with 60% sand, 30% silt, and 10% clay is a sandy loam.

Green and Ampt (1911) developed an approximate infiltration model based on Darcy's law. They assumed vertical flow in a column initially at a uniform water content p_{0i} and saturation at the surface p_{0e}. The saturated wetting front moves gradually into the soil zone unaffected by infiltration, as sketched in Fig. 3.14.

The infiltration rate $f(t)$ varies with time as the piston-type wetting front advances into the soil:

$$f(t) = \frac{K(h + L_f + h_p)}{L_f},$$
(3.5)

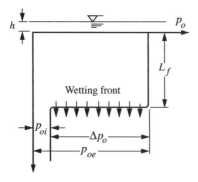

Figure 3.14. Piston-type infiltration approximation.

where K is the hydraulic conductivity of the wetted soil part of the soil profile, L_f is the depth of the wetting front, h_p is the pressure head for wetting at the wetting front, and h is the depth of ponding of water on the soil surface.

In general, the flow depth h is small compared with the length of the wetting front L_f and may be neglected in Eq. (3.5). The change in water content across the wetting front, Δp_0, depends on the initial water content p_{0i}, the thoroughly drained (or residual) water content, p_{0r}, the effective saturation S_e, and the total porosity p_0. The relationship is $\Delta p_0 = p_{0e} - p_{0i} = p_{0e} - S_e p_{0e} = (1 - S_e)p_{0e}$, where the effective porosity is given by $p_{0e} = p_0 - p_{0r}$ and $S_e = p_{0i}/p_{0e}$. By noting that $F(t) = L_f \Delta p_0$ or $L_f = F(t)/\Delta p_0$ and taking h as zero, we find that the infiltration rate $f(t)$ varies with time as

$$f(t) = K\left[\frac{h_p \Delta p_0}{F(t)} + 1\right], \qquad (3.6) \ \blacklozenge\blacklozenge$$

where $F(t)$ is the cumulative infiltration depth at time t. The cumulative infiltration is found by integration of Eq. (3.6) as

$$F(t) = Kt + h_p \Delta p_0 \ln\left(1 + \frac{F(t)}{h_p \Delta p_0}\right) \qquad (3.7)$$

Equations (3.6) and (3.7) apply for the case in which the ponded depth is negligible. If this is not the case, h_p should be replaced with $h_p + h$.

Rawls et al. (1983) present values for p_0, p_{0e}, h_p, and K as functions of soil type in Table 3.1. In practice, because Eq. (3.7) cannot be solved explicitly for $F(t)$, an iterative procedure is required. A trial value for $F(t)$ is substituted into the right-hand side of the equation, which is then compared with the left-hand side. This process is repeated until agreement between the two values is obtained. A good first estimate is Kt for $F(t)$. Possibly an easier way to calculate $f(t)$ from Eq. (3.6) is to solve Eq. (3.7) for t at various values of $F(t)$. The cumulative infiltration $F(t)$ can then be used in Eq. (3.6) to determine $f(t)$ at corresponding time.

Table 3.1. *Green–Ampt infiltration parameters (after Rawls et al. 1983)[a]*

Soil texture	p_0	p_{0e}	h_p (cm)	K (cm/h)
Sand	0.437	0.417	4.95	11.78
Loamy sand	0.437	0.401	6.13	2.99
Sandy loam	0.453	0.412	11.01	1.09
Silt loam	0.501	0.486	16.68	0.65
Loam	0.463	0.434	8.89	0.34
Sandy clay loam	0.398	0.330	21.85	0.15
Clay loam	0.464	0.309	20.88	0.10
Silty clay loam	0.471	0.432	27.30	0.10
Sandy clay	0.430	0.321	23.90	0.06
Silty clay	0.479	0.423	29.22	0.05
Clay	0.475	0.385	31.63	0.03

[a] Rawls et al. (1983) contains more information on these parameters including their standard deviations and values for various soil horizons.

Example 3.2 illustrates how to calculate infiltration by use of the Green–Ampt method. Typical potential infiltration curves for initially dry soils in terms of $f(t)$, $F(t)$, and $f(t)/K$ for cumulative rainfall infiltration up to 5 cm are shown in Figs. 3.15(a), 3.15(b), and 3.15(c), respectively.

Example 3.2   Calculation of the infiltration potential. Calculate the infiltration curve for a silt loam at 30% effective saturation.

Solution: Given $S_e = 0.3$, Table 3.1 provides the following characteristics for a silt loam as $p_{0e} = 0.486$, $h_p = 16.7$ cm, and $K = 0.65$ cm/h. We then calculate $\Delta p_0 = (1 - S_e)\, p_{0e} = (1 - 0.3)\,0.486 = 0.34$. For instance, we compute the cumulative infiltration $F(t) = 1$ cm by rearranging Eq. (3.7):

$$t = \frac{F(t) - h_p \Delta p_0 \ln\left[1 + \dfrac{F(t)}{h_p \Delta p_0}\right]}{K}$$

$$= \frac{1 - 16.7(0.342)\ln\left[1 + \dfrac{1}{16.7(0.34)}\right]}{0.65} = 0.12 \text{ h}.$$

From Eq. (3.6), the infiltration rate at $t = 0.12$ h is

$$f(0.12 \text{ h}) = K\left[\frac{h_p \Delta p_0}{F(t)} + 1\right] = 0.65\left[\frac{16.7(0.34)}{1} + 1\right] = 4.36 \text{ cm/h}.$$

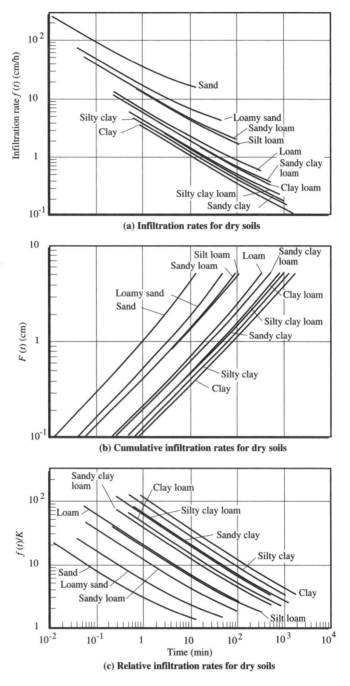

(a) **Infiltration rates for dry soils**

(b) **Cumulative infiltration rates for dry soils**

(c) **Relative infiltration rates for dry soils**

Figure 3.15. Infiltration characteristics for dry soils.

Table E.3.2.1. *Green–Ampt infiltration calculations*

$F(t)^a$ (cm)	t^b (min)	$f(t)^c$ (cm/h)
0.0	0.0	∞
0.25	0.5	15.4
0.5	1.9	8.0
0.75	4.2	5.6
1.0	7.3	4.4
1.5	15.6	3.4
2.0	26.4	2.5
3.0	54.6	1.9
4.0	90.0	1.6
5.0	130.0	1.4

[a] Assumed.
[b] From Eq. (3.7) with $S_e = 0.3$, $p_{0e} = 0.486$, $h_p = 16.7$ cm, $K = 0.65$ cm/h, and $\Delta p_0 = 0.34$
[c] From Eq. (3.6).

Table E.3.2.1 illustrates the values of infiltration calculated from the assumed values of F. A plot of f versus t from columns 3 and 2 yields the infiltration curve.

3.4 Excess rainfall

Excess rainfall represents the supply of water to the surface-runoff process. Excess rainfall represents the amount of rainfall in excess of interception, evapotranspiration, and infiltration. When the rainfall rate exceeds the infiltration rate, detention storage begins to fill. Runoff will begin where the detention storage is filled. When the rainfall rate drops below the infiltration rate, water in surface storage is gradually depleted until surface runoff ceases. The water in detention storage then infiltrates. Example 3.3 provides the detailed calculation of excess rainfall by use of the Green–Ampt infiltration equation.

Example 3.3 Calculation of excess rainfall. Use the Green–Ampt equation to calculate the excess rainfall for the rainstorm in the first two columns of Table E.3.3.1. The maximum detention storage is 0.75 cm, and the soil is a silt loam soil with 30% effective saturation.

Table E.3.3.1. *Excess rainfall calculations*

t^a (min)	Rainfall[a] $i\Delta t$ (cm)	$f_p(t)^b$ (cm/h)	$\Delta F_p(t)^c$ (cm)	Infiltration[d] $\Delta F_a(t)$ (cm)	$F_p(t)^e$ (cm)	Detention storage[f] S (cm)	Excess rainfall[g] ΔR_e (cm)
0	0.00				0.00	0.00	0.00
15	0.25	Very large	Very large	0.25	0.25	0.00	0.00
30	0.30	15.40	3.85	0.30	0.55	0.00	0.00
45	0.43	7.36	1.84	0.43	0.98	0.00	0.00
60	0.66	4.42	1.10	0.66	1.64	0.00	0.00
75	1.55	2.90	0.73	0.73	2.37	0.75	0.07
90	4.85	2.21	0.55	0.55	2.92	0.75	4.30
105	0.91	1.91	0.48	0.48	3.40	0.75	0.43
120	0.51	1.74	0.43	0.43	3.83	0.75	0.08
135	0.36	1.61	0.40	0.40	4.23	0.71	0.00
150	0.28	1.52	0.38	0.38	4.61	0.61	0.00
165	0.23	1.45	0.36	0.36	4.97	0.48	0.00
180	0.20	1.39	0.35	0.35	5.32	0.33	0.00
Totals	10.53		=		5.32	0.33	+4.88

[a] The hyetograph is in the first two columns; rain starts at $t = 15$ min.
[b] Potential infiltration rate from F_p at the previous time increment.
[c] Potential infiltration volume $\Delta F_p = f_p \Delta t$.
[d] The actual infiltration volume is the smaller of column 4 and column 2 + column 7.
[e] $F_p(t) = F_p(t-1) + \Delta F_a$.
[f] Surface storage.
[g] Excess rainfall.

Solution: The parameters for the Green–Ampt equation from Table 3.1 are $p_{0e} = 0.486$, $h_p = 16.68$ cm, and $K = 0.65$ cm/h. The change in water content as a result of the passing of the wetting front is calculated as $\Delta p_0 = (1 - S_e)p_{0e} = (1 - 0.3)0.486 = 0.34$.

The infiltration rate $f(t)$ and the cumulative infiltration $F(t)$ are related by Eq. (3.7) as

$$f(t) = K\left[\frac{h_p \Delta p_0}{F(t)} + 1\right] = 0.65\frac{\text{cm}}{\text{h}}\left[\frac{16.68(0.34)}{F(t)} + 1\right].$$

The rainfall hyetograph is defined in the first two columns of Table E.3.3.1. Note that the rainfall starts at $t = 15$ min. Column 3 represents the potential infiltration rate $f_p(t)$ calculated from the preceding equation with F_p from the previous time increment. Column 4 gives the potential infiltration volume $\Delta F_p(t)$ calculated as $f_p(t)\Delta t$. The actual infiltration volume $\Delta F_a(t)$ for the time increment is the smallest of columns 4 vs. column 2 plus column 7 for the increment.

Early in the storm, i.e., during the first 60 min, the potential infiltration rate exceeds the rainfall rate and the actual infiltration volume is limited to the rainfall volume. The cumulative infiltration $F_p(t)$ at a particular time is equal to $F_p(t)$ from the previous time increment plus $\Delta F_a(t)$ for the current time increment. In the time increment from 60 to 75 min, the potential infiltration rate falls below the rainfall rate and some detention storage and rainfall excess are generated. Because at this time the surface storage is empty, the rainfall first satisfies infiltration with 0.73 cm, then 0.75 cm to detention storage, and 0.07 cm in surface storage or rainfall excess ready for surface runoff. This process is continued until the end of the storm. It can be seen that, for the time interval from 120 to 135 min, $\Delta F(t)$ is 0.40 cm whereas the rainfall is only 0.36 cm. Thus 0.04 cm of water is taken from the detention storage. The actual infiltration $\Delta F_a(t)$ cannot exceed the rainfall depth for the time increment $\Delta f_p \Delta t$ plus the depth of water in surface storage. Mass balance must be preserved at all times. The cumulative sum of excess rainfall, column 8 of Table E.3.3.1, plus the incremental value of cumulative infiltration $\Delta F_p(t)$ and detention storage S must equal the cumulative rainfall precipitation in column 2.

3.5 Surface runoff

Runoff refers to the surface flow occurring during and immediately after precipitation events. Base flow refers to seepage and groundwater flow between precipitation events. Surface runoff is added to the base flow to determine the total flow. The base flow from small watersheds can often be neglected in the computation of surface runoff from large rainstorms. Excess rainfall generates surface runoff as overland flow and channel flow. To capture the essential features of the rainfall–runoff relationship, consider a rectangular excess hyetograph of constant intensity i and duration t_r, as sketched in Fig. 3.10. This section aims at defining the corresponding runoff hydrograph, which requires knowledge of resistance to overland flow (Subsection 3.5.1) and stage–discharge relationships (Subsection 3.5.2). Surface runoff is then calculated for overland flow (Subsection 3.5.3) and snowmelt runoff (Subsection 3.5.4).

3.5.1 Resistance to overland flow

Resistance to flow defines the relationship between flow depth h and depth-averaged flow-velocity $\bar{u}$. Resistance to flow can be written in terms of the

Table 3.2. *Resistance to overland flow*

Surface	Laminar flow k_0	Turbulent Flow		
		Manning n	Chézy C ($ft^{1/2}$/s)	Darcy–Weisbach f
Concrete or asphalt	24–108	0.01–0.013	78–38	0.03–0.4
Bare sand	30–120	0.01–0.016	65–33	0.04–0.5
Graveled surface	90–400	0.012–0.03	38–18	—
Bare clay–loam soil (eroded)	100–500	0.012–0.033	36–16	—
Sparse vegetation	1,000–4,000	0.053–0.13	11–5	0.1–1000
Short grass prairie	3,000–10,000	0.10–0.20	6.5–3.6	0.5–13,000
Bluegrass sod	7,000–100,000	0.17–0.48	4.2–1.8	1–10,000

Darcy–Weisbach friction factor f, Manning coefficient n, or Chézy coefficient C. The corresponding definitions of C, n, and f are, respectively,

$$\bar{u} = C \, h^{1/2} S^{1/2}, \tag{3.8a}$$

$$\bar{u} = \frac{1}{n} h^{2/3} S^{1/2} \text{ (in S.I. units),} \tag{3.8b}$$

$$\bar{u} = \sqrt{\frac{8g}{f}} h^{1/2} S^{1/2}. \tag{3.8c} \blacklozenge$$

Note that only the Darcy–Weisbach coefficient f is dimensionless; the Manning coefficient n has the dimensions $T/L^{1/3}$ and the numerator is replaced with 1.49 in the English system of units. The Chézy coefficient C has the dimensions $L^{1/2}/T$.

Table 3.2 lists typical values of the Darcy–Weisbach friction factor f, the Chézy coefficient C, given the identity $C = \sqrt{8g/f}$, the Manning coefficient n, and the laminar resistance coefficient k_0 for various overland-flow conditions. Woolhiser (1975) presented a method to evaluate resistance coefficients for overland flow.

For smooth impervious surfaces, Fig. 3.16 shows the Darcy–Weisbach friction factor f as a function of the Reynolds number $Re = [(\bar{u}h)/v] = (q/v)$ given the flow depth h, the mean flow velocity $\bar{u}$ or unit discharge q and kinematic viscosity v. It is important to note that two flow regimes are observed in this figure: (1) laminar flow when $Re < 1{,}000$, characterized by a constant value of k_0; and (2) turbulent flow when $Re > 1{,}000$, approximated by the Blasius equation $f \cong 0.223/Re^{0.25}$. It is also interesting that rainfall intensity increases the value of k_0 in the laminar-flow regime. Raindrop impact thus only increases

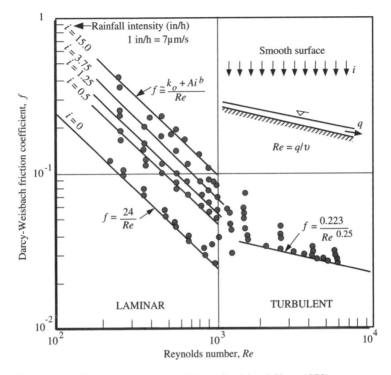

Figure 3.16. Resistance to overland flow (after Li and Shen, 1973).

resistance to flow in the laminar-flow regime. Raindrop effects are negligible in turbulent flows, i.e., when Re > 1,000.

Laminar flows with raindrop impact can be described by the Darcy–Weisbach equation in which the friction factor f relates to (1) the Reynolds number Re, (2) the friction coefficient k_0, and (3) the empirical coefficients A and b for raindrop impact. The friction coefficient k_t includes both effects that are due to surface roughness and rainfall intensity. The following relationship is generally used:

$$f = \frac{k_t}{\mathrm{Re}} = \frac{k_0 + Ai^b}{\mathrm{Re}}. \tag{3.9}$$

The values of k_0 have been tabulated by Woolhiser for various surface characteristics, and the value $k_0 = 24$ is representative of a smooth surface. For rainfall intensity given in feet per second, $A \cong 4.32 \times 10^5$ s/ft ($A \cong 1.42 \times 10^6$ s/m for i in meters per second) and $b \cong 1$.

The Darcy–Weisbach friction factor f is shown in Fig. 3.17 for the case of vegetated upland surfaces. It is found that resistance to flow on vegetated

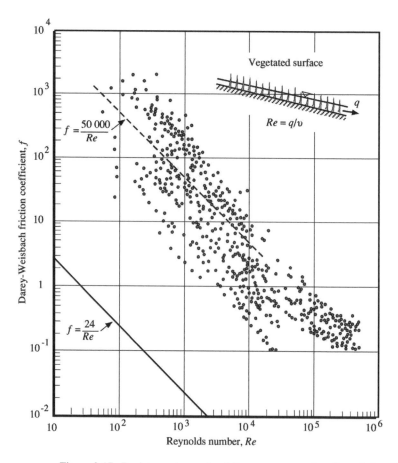

Figure 3.17. Resistance to overland flow on vegetated surfaces (after Chen, 1976).

surfaces is much higher than that for smooth impervious surfaces. The laminar regime extends up to Re $< 10^5$, and turbulent flows are found when Re $> 10^5$. The values of k_0 in Table 3.2 indicate that k_0 can be several orders of magnitude larger than for a smooth surface. The effects of raindrop impact become negligible compared with the effects of vegetation.

3.5.2   Stage–discharge relationship

In general, resistance to flow can be written as a stage–discharge relationship,

$$q = \alpha h^\beta, \tag{3.10} \blacklozenge$$

Table 3.3. *Resistance relationships $q = \alpha h^\beta$ for overland flow*

Flow Type	Resistance Coefficient	α	β	t_e
Laminar	k_t = constant	$\dfrac{8gS}{k_t\nu}$	3	$\left(\dfrac{k_t\nu L}{8gSi^2}\right)^{1/3}$
Turbulent				
Darcy–Weisbach	f = constant	$\sqrt{\dfrac{8gS}{f}}$	1.5	$\left(\dfrac{fL^2}{8gSi}\right)^{1/3}$
Chézy	C = constant	$CS^{1/2}$	1.5	$\left(\dfrac{L^2}{C^2Si}\right)^{1/3}$
Manning (S.I. units)	n = constant	$S^{1/2}/n$	1.67	$\left(\dfrac{nL}{S^{1/2}i^{0.667}}\right)^{0.6}$

where the unit discharge q is a power function of flow depth h. The resistance coefficient α and exponent β for overland flow on rectangular planes are given in Table 3.3 for laminar and turbulent flows as functions of k_t, f, C, and n in Table 3.2. In Table 3.3, g is the gravitational acceleration, S is the surface slope, ν is the kinematic viscosity, and i is the rainfall intensity.

Similarly, flow depth h, flow velocity $\bar{u}$, bed shear stress τ_0, and Froude number Fr can be determined as functions of discharge from

$$h = \left(\frac{q}{\alpha}\right)^{\frac{1}{\beta}}, \tag{3.11a}$$

$$\bar{u} = \alpha\left(\frac{q}{\alpha}\right)^{\frac{\beta-1}{\beta}}, \tag{3.11b}$$

$$\tau_0 = \gamma h S = \gamma S\left(\frac{q}{\alpha}\right)^{1/\beta}, \tag{3.11c}$$

$$\text{Fr} = \frac{\bar{u}}{\sqrt{gh}} = \alpha^{3/2\beta}q^{\frac{2\beta-3}{2\beta}}g^{-1/2} \tag{3.11d}$$

It is important to understand that constant values of n do not correspond to constant values of k_t, C, or f. In fact, if k_t is constant as a function of discharge q or Reynolds number Re, this implies that the Manning coefficient n will change with discharge. For instance, in the laminar-flow regime with constant k_t, the Manning coefficient n depends on the Reynolds number; after combining Eqs. (3.8b), (3.8c), and (3.9), we obtain

$$n = \left(\frac{k_t}{8g}\right)^{5/9}\frac{\nu^{1/9}}{S^{1/18}}\text{Re}^{-4/9}. \tag{3.12a}$$

We find that, for overland flows in upland areas, the Manning coefficient n is inversely proportional to the Reynolds number. At very low flow rates on flat slopes in upland areas, the Manning coefficient n can become extremely high, even greater than 1.

Similarly, we can combine Eq. (3.8b) with Eq. (3.8c) to find the following relationship between the Manning coefficient n, the Darcy–Weisbach factor f, and the Reynolds number Re:

$$f = \frac{8g\,S^{0.1}n^{1.8}}{\nu^{0.2}\,\mathrm{Re}^{0.2}}. \tag{3.12b}$$

We find that at a constant slope S and fluid viscosity ν, a line of constant Manning coefficient n is described by $f \sim \mathrm{Re}^{-0.2}$, which is approximately equivalent to the Blasius relationship in Fig. 3.16 for turbulent flows at Re > 1,000.

3.5.3 Overland-flow hydrographs

Analytical expressions for overland-flow hydrographs are derived for a plane-rectangular surface of length L and width W at a constant slope S_0 under a constant excess-rainfall intensity i_e. It is assumed that the overland-flow plane is initially dry ($h = 0$ and $q = 0$) before the beginning of precipitation at time $t = 0$. The flow depth increases linearly with time $h = i_e t$ during the rising limb until the flow depth conveys the equilibrium discharge ($q_m = i_e L$). As sketched in Fig. 3.18, at a given time t, the upstream portion of the plane $X < X_e$ reached complete equilibrium. In the complete-equilibrium domain, $X < X_e$, the flow is steady and nonuniform and the unit discharge q at any point x increases as $q = i_e x$ but does not change with time. In the partial-equilibrium domain, $X > X_e$, the flow is unsteady and uniform and the unit discharge at any point is constant in space at $q = i_e X_e$ but changes with time given the corresponding flow depth $h = i_e t$.

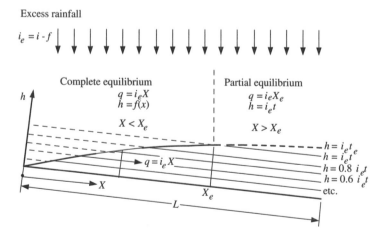

Figure 3.18. Sketch of overland-flow depth.

The continuity relationship (see Problem 2.6) describing the conservation of fluid mass is applied to wide-rectangular flow:

$$\frac{\partial h}{\partial t} + \frac{\partial q}{\partial x} = i_e. \qquad (3.13) \blacklozenge$$

Note that this continuity relationship describes a steady flow in the complete-equilibrium domain ($X < X_e$) when $\partial h/\partial t = 0$ and a nonuniform flow for partial equilibrium results from $\partial q/\partial x = i_e$. As rainfall duration increases, the length X_e increases and the complete-equilibrium domain becomes larger with time. From combining $h = it$, $q = iX_e$, and $q = \alpha h^\beta$, we obtain the position X_e as a function of time as

$$X_e = \frac{\alpha}{i_e}(i_e t)^\beta . \qquad (3.14)$$

The time to equilibrium t_e is obtained when X_e reaches the downstream end of the plane; the entire plane thus becomes under steady-flow conditions. At that time, $t = t_e$ and the equilibrium discharge $q_m = i_e L = \alpha(i_e t_e)^\beta$ is solved for t_e to give

$$t_e = i_e^{[(1/\beta)-1)]}\left(\frac{L}{\alpha}\right)^{1/\beta}$$

$$= \frac{1}{i_e}\left(\frac{i_e L}{\alpha}\right)^{1/\beta}. \qquad (3.15) \blacklozenge\blacklozenge$$

We can determine the time to equilibrium of upland areas from the excess-rainfall intensity i_e, the length of the plane L, and α and β from Table 3.3. Values of t_e for different resistance relationships are presented in Table 3.3.

Complete-equilibrium hydrographs are those for which the rainfall duration t_r exceeds the time to equilibrium t_e; hence the hydrograph dimensionless time $\lambda_r = t_r/t_e$ is greater than unity. The surface-runoff hydrograph can be subdivided into three parts: the rising limb, equilibrium, and falling limb. The

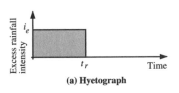

(a) Hyetograph

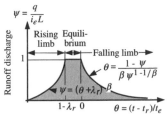

(b) Complete hydrograph ($\lambda_r = \frac{t_r}{t_e} > 1$)

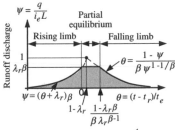

(c) Partial hydrograph ($\lambda_r = \frac{t_r}{t_e} < 1$)

Figure 3.19. Overland-flow hydrographs.

rising limb of runoff hydrographs is characterized by $q = \alpha(i_e t)^\beta$ and the complete equilibrium by $q = i_e L$; the falling limb is rather complex. In general terms, surface runoff over a rectangular plane of length L can be written in dimensionless form as $\psi = [q/(i_e L)]$ as a function of the dimensionless time $\Theta = [(t - t_r)/t_e]$, where t_e is the time to equilibrium. Figure 3.19 illustrates the shape of surface-runoff hydrographs on rectangular planes for any flow-resistance relationship. The rising limb of the complete-equilibrium hydrograph is given by $\psi = (\Theta + \lambda_r)^\beta$. The equilibrium discharge simply equals $\psi = 1$, and the falling limb is given by

$$\Theta = \frac{1 - \psi}{\beta \psi^{\frac{\beta-1}{\beta}}}, \qquad (3.16)$$

as shown in Fig. 3.19(b). For partial-equilibrium hydrographs, the salient features of the hydrographs are shown in Fig. 3.19(c).

Example 3.4   Calculation of surface-runoff hydrographs. Calculate the surface-runoff hydrograph for a 1-h rainstorm of constant excess-rainfall intensity of 1 in./h on a rectangular plane, 100 ft wide, 400 ft long, on a 20% slope. Also calculate the maximum flow depth, velocity, and shear stress at the corresponding Froude number. The plane is sparsely vegetated.

In S.I. units, $i_e = 7 \times 10^{-6}$ m/s, $L = 122$ m, and $t_r = 3,600$ s. Considering the kinematic viscosity $\nu = 1 \times 10^{-6}$ m^2/s, the maximum Reynolds number is

$$\mathrm{Re}_{max} = i_e L/\nu = \frac{7 \times 10^{-6} \times 122\,\mathrm{m\,s}}{\mathrm{s}\,1 \times 10^{-6}\,\mathrm{m}^2} = 860.$$

Thus the flow is laminar. For a sparsely vegetated field, a value of $k_0 \cong 2,000$ is selected from Table 3.2. From Table 3.3, $\beta = 3$ for laminar flow and

$$\alpha = 8gS/k\nu = \frac{8 \times 9.81\,\mathrm{m} \times 0.2\,\mathrm{s}}{\mathrm{s}^2\,2,000 \times 1 \times 10^{-6}\,\mathrm{m}^2} = \frac{7848}{\mathrm{m\,s}}.$$

The time to equilibrium is calculated from

$$t_e = \left(\frac{k_t \nu L}{8g S i_e^2}\right)^{1/3} = \left[\frac{2,000 \times 1 \times 10^{-6} \times 122\,\mathrm{m}^3}{8 \times 9.81\,\mathrm{m} \times 0.2 \times \mathrm{s}}\frac{\mathrm{s}^2}{(7 \times 10^{-6})^2\,\mathrm{m}^2}\right]^{1/3} = 682\,\mathrm{s}.$$

The complete-equilibrium hydrograph $t_r > t_e$ is calculated in three parts. First, the rising limb is calculated from $\psi = (\Theta + \lambda_r)^\beta$ or

$$q = i_e L(t/t_e)^3 = \frac{7 \times 10^{-6} \times 122\,\mathrm{m}^2}{(682)^3\,\mathrm{s}^4}t^3 = 2.69 \times 10^{-12}\,t^3.$$

Second, after 682 s, $q = 8.54 \times 10^{-4}\,\text{m}^2/\text{s}$ and remains constant until $t = 3,600\,\text{s}$. Third, the unit discharge decreases with time after $t > 3,600\,\text{s}$ according to $\Theta = (1 - \Psi)/\beta\Psi^{(\beta-1)/\beta}$ or

$$t = t_r + t_e \frac{(i_e L - q)}{3(i_e L)^{1/3}q^{2/3}} = 3,600\,\text{s} + \frac{682\,\text{s}\,(8.54 \times 10^{-4} - q)}{3(8.54 \times 10^{-4})^{1/3}q^{2/3}}.$$

The falling limb of the hydrograph is thus obtained from substituting values of discharge, $0 < q < 8.54 \times 10^{-4}$, into this equation to calculate the time t at which the discharge will occur.

During complete equilibrium, the maximum flow depth, velocity, shear stress, and Froude number are obtained from Eqs. (3.11):

$$h = \left(\frac{q}{\alpha}\right)^{1/\beta} = \left(\frac{8.54 \times 10^{-4}}{7,848}\right)^{1/3} = 4.8\,\text{mm},$$

$$\bar{u} = \alpha\left(\frac{q}{\alpha}\right)^{\frac{\beta-1}{\beta}} = 7,848\left(\frac{8.54 \times 10^{-4}}{7,848}\right)^{2/3} = 0.18\,\text{m/s},$$

$$\tau_0 = \gamma h S = \frac{9,810\,\text{N}}{\text{m}^3} \times 4.8 \times 10^{-3}\,\text{m} \times 0.2 = 9.4\,\text{Pa},$$

$$\text{Fr} = \frac{\bar{u}}{\sqrt{gh}} = \frac{0.18}{\sqrt{9.81 \times 4.8 \times 10^{-3}}} = 0.83.$$

3.5.4 Snowmelt runoff

Snowmelt runoff is a complex topic that we can analyze by using a complete radiation budget. Approximations like the degree-day method are often used for simplicity. As an example, hourly snowmelt-runoff discharge data from a small experimental plot in Canada were available for comparisons with climatological data (Julien and Frenette, 1986). Cumulative snowmelt h_s, in meters, of equivalent water content was successfully correlated to three factors (see Fig. 3.20): (1) the cumulative number of degree days D_d in degrees Celsius times days, (2) cumulative time t_a in hours when the air temperature is above 0 °C, and (3) cumulative time of snowmelt, t_f, in hours measured from the experimental plot

$$h_p = 7.29 \times 10^{-3} D_d^{1.2}, \tag{3.16a}$$

$$h_s = 4.44 \times 10^{-9} t_a^{3.11}, \tag{3.16b}$$

$$h_s = 6.15 \times 10^{-8} t_f^{2.7}. \tag{3.16c}$$

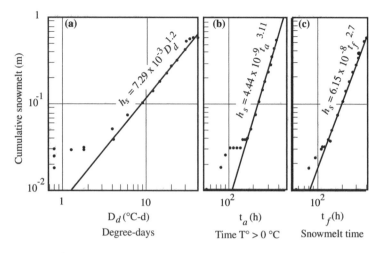

Figure 3.20. Cumulative snowmelt (after Julien, 1982).

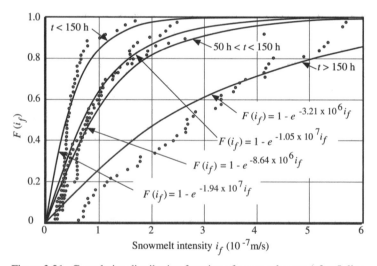

Figure 3.21. Cumulative distribution function of snowmelt rates (after Julien, 1982).

Equation (3.16c) should be given physical preference, but meteorological records may be available to use only Eqs. (3.16a) and (3.16b). The first derivative of Eq. (3.16c) shows that the mean runoff discharge increases with the snowmelt time. The cumulative distribution functions $F(q_f)$ of the hourly runoff intensity q_f measured from this experimental plot are shown in Fig. 3.21. The field measurements fit an exponential probability density function reasonably well, which can be written as

$$F(i_f) = \int_0^{i_f} p(i_f)\mathrm{d}i_f = \int_0^{i_f} \lambda_f e^{-\lambda_f i_f}\mathrm{d}i_f = 1 - e^{-\lambda_f i_f}. \quad (3.17)$$

The snowmelt intensity $\bar{i}_f = 1/\lambda_f$ increases as the melting period progresses. The unit discharge from snowmelt runoff can be calculated from $q_f = i_f L$. The average snowmelt rate $\bar{i}_f$ can be estimated by $\mathrm{d}h_s/\mathrm{d}t_f$ from Eq. (3.16c).

3.6   Upland-erosion losses

Water is the most widespread agent of erosion. Upland erosion by water can be classified into sheet erosion and rill erosion. Sheet erosion is the detachment of land surface material by raindrop impact and thawing of frozen grounds and its subsequent removal by sheet flow.

The surface-erosion process begins when raindrops hit the ground and detach soil particles by splash. The KE released by raindrop impact on the ground is sufficiently large to break the bonds between soil particles. The characteristics of raindrop splash depend on raindrop size and sheet-flow depth; a crown-shaped crater forms a few milliseconds after impact. The impact shear stress can be as large as 100 times the base shear stress from shallow sheet flow. For the example shown in Fig. 3.22, Hartley and Julien (1992) measured shear stress in excess of 10 Pa. This far exceeds the critical shear stress of 2.5 N/m² for cohesive soils. In general, the effect of raindrop impact can be neglected when the sheet-flow depth is larger than three times the raindrop size.

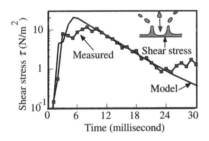

Figure 3.22. Shear stress under raindrop impact (after Hartley and Julien, 1992).

The transport capacity of shallow overland flow, usually called sheet flow, increases with field slope and unit-flow discharge. As sheet flow is concentrated and the unit discharge increases, the increased sediment-transport capacity scours microchannels called rills. Rill erosion is the removal of soil by concentrated sheet flow. Rills are small enough to be removed by normal tillage.

Soil particles detached by raindrop impact are transported downstream by runoff, and the unit sediment discharge on bare soils is a power function of surface slope S, unit discharge q, and rainfall intensity i. Julien and Simons (1985) provided a quantitative evaluation of the exponents of several sediment-transport equations for sheet erosion. For sandy soils, it was found that the following equation can be used:

$$q_s \cong 25{,}500 \, S^{1.66} \, q^{2.035}, \tag{3.18} \blacklozenge$$

where q_s is the unit sediment discharge from sheet and rill erosion in metric ton/m s, S is the slope, and q is the unit discharge in square meters per second. It is appropriate to note that the exponent of discharge is greater than unity and therefore rills are likely to form owing to the discussion followed in the first chapter of this book. The foregoing analysis of upland erosion considers the analysis of a single storm (Subsection 3.6.1), followed by the expected value of soil erosion from a single storm (Subsection 3.6.2) and the universal soil-loss equation (Subsection 3.6.3).

3.6.1 Soil loss from a single event

Upland erosion by overland flow during a single event of constant excess-rainfall intensity i_e and duration t_r is sketched in Fig. 3.23. Because the sediment discharge is a power function of the runoff discharge, the characteristics of the sediment discharge follow the shape of the runoff hydrograph. We obtain, in dimensionless form, the total soil mass eroded per unit width, ϕ, by integrating the sediment discharge, $q_s / \rho v$, over the dimensionless runoff period $t^* = t / \bar{t}_r$:

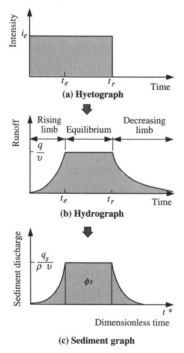

(a) Hyetograph

(b) Hydrograph

(c) Sediment graph

Figure 3.23. Sketches of (a) hyetograph, (b) hydrograph, and (c) sediment graph.

$$\phi_s = \int_0^\infty \frac{q_s}{\rho v} d(t^*) = \left(\frac{1}{\rho v \bar{t}_r} \right) \int_0^\infty q_s \, dt. \tag{3.19}$$

The eroded soil mass from a single storm m_s in metric tons is given by

$$m_s = W_0 \int_0^\infty q_s \, dt = W_0 \rho v \bar{t}_r \phi_s, \tag{3.20}$$

where W_0 is the plane width in meters, ρ is the mass density of water in kilograms per cubic meter, v is the kinematic viscosity of water in square meters per second, t_r is the average rainfall duration in seconds, and ϕ_s is a dimensionless term, determined in the next paragraph, that depends on the type of hydrograph, i.e., partial or complete hydrograph.

For a given constant excess-rainfall intensity i_e over a plane surface of slope S, Julien (1982) integrated Eq. (3.19) and obtained exact analytical solutions as functions of $\lambda_r = t_r/t_e$ for a complete hydrograph ϕ_c when $\lambda_r > 1$ and for a partial-equilibrium hydrograph ϕ_p when $\lambda_r < 1$, as follows:

$$\phi_c = \frac{25{,}500}{\rho v \bar{t}_r} S^{1.66} L^2 i_e^2 t_r \left(1 - \frac{9}{14\lambda_r}\right), \qquad (3.21)$$

$$\phi_p = \frac{25{,}500}{\rho v \bar{t}_r} S^{1.66} L^2 i_e^2 t_r \left(\frac{\lambda_r^3}{2} - \frac{2\lambda_r^6}{7}\right) \qquad (3.22)$$

Simplifications arise from the fact that the erosion rates in Eq. (3.22) become very small for short storms ($\lambda_r < 1$). Also, the erosion rates for complete-equilibrium hydrographs [Eq. (3.21)] become independent of λ_r and increase linearly with rainfall duration t_r when λ is large.

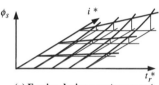

(a) Erosion during one storm event

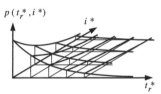

(b) Joint probability density function of the duration intensity parameters of several storms

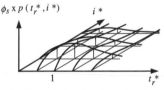

(c) Dimensionless function of erosion

Figure 3.24. Sketch of expected values of soil erosion (after Julien, 1982).

3.6.2 Expected soil loss

The expected value of soil erosion from one storm of unknown duration and intensity is determined with the exponential probability density functions of rainfall duration and intensity. From the exponential distributions of rainfall duration and intensity described in Eqs. (3.1) and (3.3), the expected value of soil erosion by one storm, $\bar{\phi}_s$, is given as sketched in Fig. 3.24:

$$\bar{\phi}_s = \int_0^\infty \int_0^\infty \phi \, e^{-t_r^*} e^{-i^*} \mathrm{d}(t_r^*) \, \mathrm{d}(i^*), \qquad (3.23)$$

where $\bar{t}_r$ is the average storm duration, $\bar{i}$ is the average storm intensity, $t_r^* = t_r/\bar{t}_r$, and $i^* = i/\bar{i}$. This integral must be divided into two parts, as indicated by Eqs. (3.21) and (3.22), which are valid when $t_r > t_e$ and $t_r \le t_e$, respectively.

After defining $t_e^* = t_e/\bar{t}_r$, we obtain

$$\bar{\phi}_s = \int_0^\infty \int_0^{t_e^*} \phi_p e^{-t_r^*} e^{-i^*} \mathrm{d}(t_r^*) \mathrm{d}(i^*) + \int_0^\infty \int_{t_e^*}^\infty \phi_c e^{-t_r^*} e^{-i^*} \mathrm{d}(t_r^*) \mathrm{d}(i^*).$$

(3.24)

The functions ϕ_p and ϕ_c are given by Eqs. (3.21) and (3.22). The exact analytical integration of Eq. (3.23) was obtained by Julien (1982) and Julien and Frenette (1985). The following relation is an approximate solution:

$$\bar{\phi}_s \cong \frac{3.4 \times 10^5}{\rho\, \nu} S^{1.66} L^2 \bar{i}_e^2 \hat{K}\hat{C}\hat{P},$$

(3.25)

where $\hat{K}, \hat{C},$ and $\hat{P}$ are the universal soil-loss equation (USLE) coefficients defined in Subsection 3.6.3.

In regions where infiltration is significant, the expected soil-erosion loss from a single storm $\bar{m}_s$ is multiplied by the runoff coefficient C_r, and the excess-rainfall intensity i_e is approximated by $\bar{i}$ out, or

$$\bar{m}_s = W_0\, \rho\, \nu\, \bar{t}_r\, \bar{\phi}_s \cong W_0\, 3.4 \times 10^5\, S^{1.66} L^2 C_r \bar{i}^2\, \bar{t}_r \hat{K}\hat{C}\hat{P}.$$

(3.26)

The expected amount of soil eroded during a given period of time, e.g., 1 month, is equal to the product of the expected value of soil erosion for one rainfall event $\bar{m}_s$ and the mean number of events $\bar{\nu}$ during that period.

A practical approximation for the expected value of the soil loss in metric tons during a period with an average of $\bar{\nu}$ storms is thus

$$E_{\text{tons}} \cong 3.4 \times 10^5\, \bar{\nu}\, W_0\, S^{1.66} C_r(\bar{i}L)^2\, \bar{t}_r\, \hat{K}\hat{C}\hat{P},$$

(3.27) ◆

where E_{tons} is the expected soil loss in metric tons on a rectangular plane of width W in meters and length L in meters, the average rainfall duration $\bar{t}_r$ is in seconds and the average rainfall intensity $\bar{i}$ is in meters per second, the number of storms is $\bar{\nu}$, the slope S is in meters per meter, the runoff coefficient is C_r, and the USLE parameters are $\hat{K}, \hat{C},$ and $\hat{P}$, which are discussed in the next subsection. The advantage of this formulation is that the unit discharge $q = \bar{i}L$ can be replaced with snowmelt-runoff discharge for the simulation of erosion losses in cold regions.

3.6.3   *Universal soil-loss equation*

The Universal Soil Loss Equation (USLE) was designed to predict the annual average soil-erosion losses from field areas under specified cropping and management systems. The USLE computes the soil loss $\hat{E}$ at a given site as a product of six major factors:

$$\hat{E} = \hat{R}\hat{K}\hat{L}\hat{S}\hat{C}\hat{P},$$ (3.28) ♦♦

where $\hat{E}$ is the soil loss per unit area normally in tons per acre, $\hat{R}$ is the rainfall-erosivity factor, $\hat{K}$ is the soil-erodibility factor, usually in tons per acre, $\hat{L}$ is the field-length factor normalized to a plot length of 72.6 ft, $\hat{S}$ is the field-slope factor normalized to a field slope of 9%, $\hat{C}$ is the cropping–management factor normalized to a tilled area with continuous fallow, and $\hat{P}$ is the conservation-practice factor normalized to straight-row farming up and down the slope.

The rainfall-erodibility factor $\hat{R}$ can be evaluated for each storm (summed over hours) from $\hat{R} = 0.01 \sum i(916 + 331 \log i)$, where the summation is performed over the time increments of the storm, and i is the rainfall intensity in inches per hour. Soil-erosion losses from single storms strongly correlate with the maximum 30-min rainfall intensity. The annual rainfall-erosion index ranges from 0 to 600 in the United States.

The soil-erodibility factor $\hat{K}$ describes the inherent erodibility of the soil expressed in the same units as those of the annual erosion losses, tons per acre. Numerous factors control the erodibility of cohesive soils such as grain-size distribution, texture, permeability, and organic content. Typical values of the factor $\hat{K}$ relate to the general triangular soil classification shown in Fig. 3.13. For each soil type, approximate values of $\hat{K}$ can be found in Table 3.4, given the soil type and the percentage of organic matter.

The slope-length-steepness factor $\hat{L}\hat{S}$ is a topographic factor relating erosion losses from a field of given slope and length when compared with soil losses

Table 3.4. *Soil erodibility factor $\hat{K}$ in tons/acre (after Schwab et al., 1981)*

Textural Class	Organic Matter Content (%)	
	0.5	2
Fine sand	0.16	0.14
Very fine sand	0.42	0.36
Loamy sand	0.12	0.10
Loamy very fine sand	0.44	0.38
Sandy loam	0.27	0.24
Very fine sandy loam	0.47	0.41
Silt loam	0.48	0.42
Clay loam	0.28	0.25
Silty clay loam	0.37	0.32
Silty clay	0.25	0.23

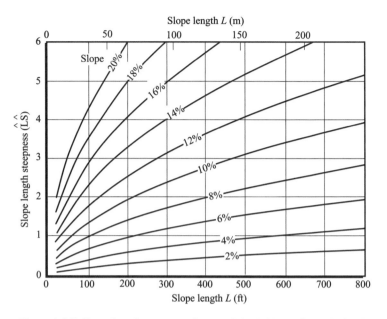

Figure 3.25. Slope-length-steepness factor of the USLE (after Wischmeier and Smith, 1978).

of a standard plot 72.6 ft long inclined at a 9% slope. The values of $\hat{L}\,\hat{S}$ are plotted in Fig. 3.25, with the length and the field slope given.

The cropping–management factor $\hat{C}$ for bare soils is taken as a standard value equal to unity. The factor $\hat{C}$ accounts for soils under different cropping and management combinations such as different vegetation, canopy during growth stage, before and after harvesting, crop residues, mulching, fertilizing, and crop sequence. Typical values of $\hat{C}$ are given in Tables 3.5 for forest (Table 3.5a), pasture, rangeland, and idle land (Table 3.5b), cropland (Table 3.5c), and construction slopes (Table 3.5d). Area-averaged values of $\hat{C}$ can be used when several vegetation types cover a given area.

Table 3.5a. *Cropping–management factor $\hat{C}$ for forest (after Wischmeier and Smith, 1978)*

Percentage of area covered by canopy of trees and undergrowth	Percentage of area covered by duff at least 2 in. deep	Factor $\hat{C}$
100–75	100–90	0.0001–0.001
70–45	85–75	0.002–0.004
40–20	70–40	0.003–0.009

Table 3.5b. *Cropping–management factor Ĉ for pasture, rangeland, and idle land[a] (modified after Wischmeier and Smith, 1978)*

Vegetative canopy (type and height[b])	Type[c]	Cover that is in contact with the soil surface Percentage of ground cover					
		0	20	40	60	80	95+
No appreciable canopy	G	0.45	0.20	0.10	0.042	0.013	0.003
	W	0.45	0.24	0.15	0.091	0.043	0.011
Tall weeds or short brush with average drop fall height of 20 in.	G	0.17–0.36	0.10–0.17	0.06–0.09	0.032–0.038	0.011–0.013	0.003
	W	0.17–0.36	0.12–0.20	0.09–0.13	0.068–0.083	0.038–0.041	0.011
Appreciable brush or bushes, with average drop fall height of 6 1/2 ft	G	0.28–0.40	0.14–0.18	0.08–0.09	0.036–0.040	0.012–0.013	0.003
	W	0.28–0.40	0.17–0.22	0.12–0.14	0.078–0.087	0.040–0.042	0.011
Trees, but no appreciable low brush. Average drop fall height of 13 ft	G	0.36–0.42	0.17–0.19	0.09–0.10	0.039–0.041	0.012–0.013	0.003
	W	0.36–0.42	0.20–0.23	0.13–0.14	0.084–0.089	0.041–0.042	0.011

[a] The listed Ĉ values assume that the vegetation and the mulch are randomly distributed over the entire area.
[b] Canopy height is measured as the average fall height of water drops falling from the canopy to the ground. The canopy effect is inversely proportional to the drop fall height and is negligible if the fall height exceeds 33 ft.
[c] G, Cover at surface is grass, grasslike plants, decaying compacted duff, or litter at least 2 in. deep; W, cover at surface is mostly broadleaf herbaceous plants (such as weeds with little lateral-root network near the surface) or undecayed residues or both.

Table 3.5c. *Cropping–management factor Ĉ for cropland (approximate values from Wischmeier and Smith, 1978)*

Tilled continuous fallow	1.0
Rough fallow	0.30–0.80
Conventional seed bed	0.50–0.90
No tillage	0.05–0.25
Full canopy	0.10–0.20
Residues left on the field	0.10–0.50

Table 3.5d. *Cropping–management factor Ĉ for construction slopes (modified after Wischmeier and Smith, 1978)*

Type of mulch	Mulch rate (tons/acre)	Factor Ĉ
Straw	1.0–2.0	0.06–0.20
Crushed stone 1/4 to 1.5 in.	135	0.05
	240	0.02
	7	0.08
Wood chips	12	0.05
	25	0.02

Table 3.6. *Conservation-practice factor P̂ for contouring, strip-cropping, and terracing (after Wischmeier and Smith, 1978)*

Land slope (percent)	*P* value		Terracing	
	Farming on contour	Contour strip-crop	*a*	*b*
2–7	0.50	0.25	0.50	0.10
8–12	0.60	0.30	0.60	0.12
13–18	0.80	0.40	0.80	0.16
19–24	0.90	0.45	0.90	0.18

[a] For erosion-control planning on farmland.
[b] For prediction of contribution to off-field sediment load.

The conservation-practice factor P equals one for downslope rows and typical values for contouring, strip-cropping, and terracing are given in Table 3.6. Contour practices are most effective on slopes of less than 12%, in which case $\hat{P}$ can be as low as 0.5. Contouring does not reduce erosion losses at slopes exceeding 24%. Strip-cropping and terracing reduce erosion significantly on slopes of less than 12%.

Example 3.5   Application to soil losses on a small watershed. Consider a watershed area covering 600 acres (1 acre = 4,047 m^2) above a proposed floodwater-retarding structure. Compute the annual soil-erosion loss given the conditions shown in Fig. E.3.5.1 in an area where the rainfall erosivity is $\hat{R} = 185$.

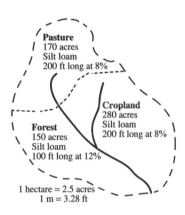

Cropland covers 280 acres of continuous corn with residues left on the field, cultivated up and down an 8% slope that is 200 ft long on a silt loam. The rainfall erosivity $\hat{R} = 185$ and the soil-erodibility factor $\hat{K} \cong 0.46$ for a silt loam in Table 3.4 the topography factor $\hat{L}\hat{S} = 1.4$ from Fig. 3.25 with a length of 200 ft at an 8% slope, the cropping–management factor $\hat{C} \cong 0.4$ for continuous cultivation and $\hat{P} = 1$ for upslope and downslope practices. The annual soil loss is $\hat{E} = 185 \times 0.46 \times 1.4 \times 0.4 = 47$ tons per acre over 280 cropland acres.

Figure E.3.5.1. Example of soil-loss calculation.

Pastureland covers 170 acres, half of which has a canopy cover of short brush (0.5-m fall height), and 80% of the surface is covered by grass and grasslike plants, the soil is a silt loam, and the slopes at 8% are 200 ft long. The parameters are $\hat{R} = 185$, $\hat{K} = 0.46$, $\hat{L}\hat{S} = 1.4$, $\hat{C} = 0.012$ from Table 3.5b, and $\hat{P} = 1$. The annual soil loss is 1.4 tons per acre for 170 acres.

Finally, forestland covers 150 acres, 30% of the area has tree canopy, and 50% of the surface is covered by litter. Silt loam slopes are 100 ft long at a 12% slope. The parameters are $\hat{R} = 185$, $\hat{K} = 0.46$, $\hat{L}\hat{S} = 1.8$ from Fig. 3.25, and $\hat{C} = 0.009$ from Table 3.5a. The annual soil loss $\hat{E} = 185 \times 0.46 \times 1.8 \times 0.009 = 1.37$ tons/acre for these 150 acres. The total soil-erosion loss on this watershed equals 13,600 tons/yr (47 tons/acre $\times$ 280 acres + 1.4 tons/acre $\times$ 170 acres + 1.37 tons/acre $\times$ 150 acres).

3.7   Sediment source and yield

This section on sediment source and sediment yield from watersheds first covers the soil-erosion losses (Subsection 3.7.1) followed by a method to estimate the sediment yield from watersheds (Subsection 3.7.2).

3.7.1   *Soil-erosion losses from large watersheds*

Annual erosion losses can be calculated on watersheds. It is generally found that the topographic factors $\hat{S}$ and the land-use factor $\hat{C}$ determine most of the spatial variability in soil-erosion losses. Comparatively, the rainfall-erodibility and the soil-erosivity parameters remain relatively constant at the basin scale. The spatial variability in the factors $\hat{S}$ and $\hat{C}$ thus requires particular attention in investigations of soil-erosion mapping. When using Geographical Information System (GISs) to delineate the surface slope from raster-based data, the grid-cell size influences the slope calculations. In general, the surface slope decreases as the grid-cell size increases (Molnar and Julien, 1998).

The influence of the grid-cell size on the computation of soil-erosion rates has been studied to determine whether or not average values of the USLE parameters can be used instead of values for each pixel. Julien and Frenette (1987) defined a relative grid-size factor Q_e^* for each matrix from

$$Q_e^* = \frac{N\left[\overline{\hat{R}}\ \overline{\hat{K}}\ \overline{\hat{L}}\ \overline{\hat{S}}\ \overline{\hat{C}}\ \overline{\hat{P}}\right]}{\sum_{i=1}^{N} \hat{R}_i\ \hat{K}_i\ \hat{L}_i\ \hat{S}_i\ \hat{C}_i\ \hat{P}_i}, \tag{3.29}$$

where $\hat{R}_i$ denotes the value of parameter $\hat{R}$ of the USLE on pixel i and $\overline{\hat{R}}$ is the average value of $\hat{R}_i$ over a large area.

The mean value of the relative correction factor $\bar{Q}_e^*$ shown in Fig. 3.26 gradually decreases as the number of pixels, N, increases. This can be attributed to the fact that calculated slopes decrease as grid size increases. The use of average values of USLE parameters tends to underestimate soil-erosion losses when applied to large areas.

When applied to watersheds, the values of the correction factor have been shown to vary primarily with drainage areas, as in Fig. 3.27. As a result, the average value of the watershed-size correction factor $\bar{Q}_e$ can be written as a function of the drainage area A_t in square kilometers as

$$\bar{Q}_e = 0.8\, A_t^{-0.137}, \quad A_t > 0.125 \text{ km}^2. \tag{3.30} \blacklozenge$$

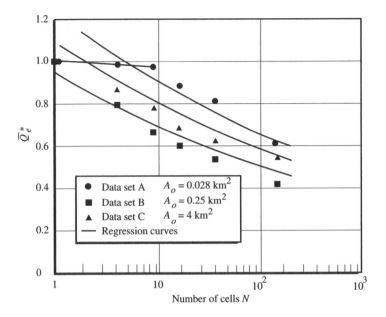

Figure 3.26. Erosion ratio vs. number of cells (after Julien, 1979).

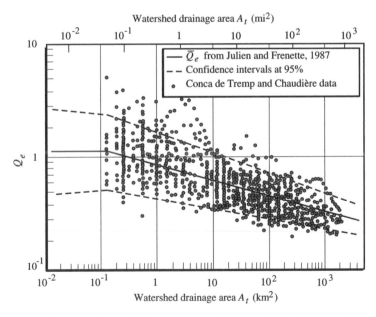

Figure 3.27. Erosion coefficient vs. watershed area (after Julien and del Tanago, 1991).

The correction factor $\bar{Q}_e$ remains constant ($\bar{Q}_e = \bar{Q}_{es}$) when $A_t < 0.125$ km^2, which indicates that the soil-loss equation can be applied to small areas without bias. As the drainage area increases beyond $A_t > 0.125$ km^2, the correction factor $\bar{Q}_e$ decreases gradually, as shown in Fig. 3.27.

Consequently, the annual erosion losses from a large watershed can be estimated from the erosion $\overline{\hat{E}}$ calculated with watershed-averaged values of the USLE coefficients divided by the correction factor $\bar{Q}_e$:

$$\hat{E} \cong \frac{\overline{\hat{E}}}{\bar{Q}_e}. \tag{3.31}$$

This relationship is quite practical because quick estimates of the gross erosion on large watersheds can be obtained from the average characteristics of the watershed. Calculations can be based on either the USLE [Relation (3.28)] or on the surface-runoff parameters in Eq. 3.26. In this case, the watershed slope is calculated from $\bar{S} = \Delta H/1{,}000\sqrt{A_t}$, where ΔH is the elevation difference between the highest and the lowest elevation on the watershed and A_t is the watershed drainage area in square kilometers. The runoff width $W_0 = 10^6 A_t/L$, with A_t in square kilometers and the average runoff length L in meters.

We may also consider the mean precipitation $\bar{h} = \bar{v}\bar{t}_r\bar{i}$ during the given period to reduce Eq. (3.27) to the following relation for annual erosion losses on large watersheds:

$$\hat{E}_{\text{tons}} \simeq \frac{3.4 \times 10^{11}}{\bar{Q}_e} A_t \, \bar{L} \, \bar{S}^{1.66} C_r \, \bar{h} \, \bar{i} \, \overline{\hat{K}} \, \overline{\hat{C}} \, \overline{\hat{P}}, \tag{3.32} \blacklozenge$$

where $\hat{E}$ is the erosion loss in tons, A_t is the drainage area in square kilometers, $\bar{L}$ is the average runoff length in upland areas in meters (~ 100 m), $\bar{S} = \Delta H/1{,}000\sqrt{A_t}$, ΔH is the elevation difference in meters, $\bar{h}$ is the average precipitation in meters during the period considered, C_r is the runoff coefficient, $\bar{i}$ is the average rainfall intensity in meters per second, and $\overline{\hat{K}}$, $\overline{\hat{C}}$, and $\overline{\hat{P}}$ are the watershed-averaged values of the USLE parameters. As shown in Fig. 3.27, the variability around the mean value is approximately a factor of 2. The erosion loss is expected to range between 50% and 200% of the value estimated from relation (3.31).

3.7.2   Sediment yield from large watersheds

The sum of upland and channel erosion in a watershed amounts to the gross erosion. All eroded particles in a watershed, however, do no reach the watershed

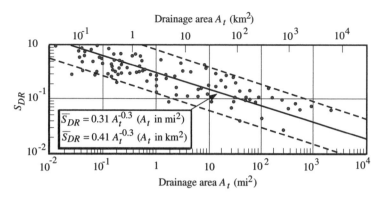

Figure 3.28. Sediment–delivery ratio (modified after Boyce, 1975).

outlet. Particles detached from bare upland areas are trapped in vegetated areas farther downstream. Some material carried in natural streams is deposited in the channels to cause channel aggradation. Some material deposits on the floodplain during major floods and large amounts are permanently trapped in lakes and reservoirs. The total amount of sediment that is delivered to the outlet of the watershed is known as the sediment yield.

The sediment–delivery ratio is defined as the ratio between the sediment yield and the gross erosion on a watershed. The ability of a channel network to convey eroded material to the outlet depends on drainage area, watershed slope, drainage density, and runoff. It is found from Fig. 3.28 that the sediment–delivery ratio decreases primarily with the size of the drainage area. The sediment yield Y from a large watershed can thus be estimated from the following procedure:

$$Y \cong S_{DR} \frac{A_t}{\bar{Q}_e} 3.4 \times 10^{11} \bar{L} \, \bar{S}^{1.66} \, C_r \bar{h}_r \, \bar{i} \, \bar{K} \, \bar{C} \, \bar{P}, \qquad (3.33) \blacklozenge$$

where S_{DR} is the sediment–delivery ratio from Fig. 3.28, A_t is the drainage area in square kilometers, $\bar{Q}_e$ is the watershed-size correction factor from Fig. 3.27, $\bar{S} = \Delta H / 1{,}000\sqrt{A_t}$ with ΔH as the elevation difference in meters between the highest and the lowest elevation on the watershed, $\bar{h}_r$ in meters is the average precipitation during the period considered, $\bar{L}$ is the average runoff length in meters, C_r is the runoff coefficient, $\bar{i}$ is the average rainfall intensity in meters per second, and $\bar{K}, \bar{C}$, and $\bar{P}$ are the watershed average values of the USLE parameters. Finally, the specific degradation of a watershed is obtained from the sediment yield divided by the drainage area. A calculation example is presented in Case Study 3.2.

Case Study 3.2   Soil losses of the Chaudière watershed, Canada. The Chaudière watershed covers $A_t = 5,830$ km^2, as measured from topographic maps (1:250,000). The average cropping–management factor, $\hat{C} \cong 0.35$, is determined from topographic maps (1:50,000) and from forest and agricultural maps (1:250,000). The average soil-erodibility parameters are $\hat{K} \cong 0.17$ and $\hat{P} \cong 1.0$. The characteristic slope is computed from the relationship $\bar{S} = \Delta H/1,000\sqrt{A_t} = 0.0156$, in which $\Delta H = 1190$ m is the elevation difference in meters between the highest and the lowest point on the watershed. The runoff length $\bar{L}$ is the length of sheet and rill flow on upland areas. Based on topographical maps and field observations, this length is estimated at 300 ft, or $\bar{L} = 91.4$ m, and is assumed constant over the watershed.

The data from 22 meteorological stations on the Chaudière watershed and from hourly data summaries (1943–1970) were analyzed to conclude that the mean annual rainfall precipitation is $\sim$770 mm and the runoff coefficient is $\sim$0.7. Assume that the Figs. 3.11 and 3.12 are representative of rainfall condition, $\bar{i} = 1 \times 10^{-6}$ m/s. The average number of rainstorms is $\bar{v}_s = 51$/yr, and the average storm duration is $\bar{t}_r = 1.5 \times 10^4$ s.

We can estimate the expected value of annual erosion loss from this watershed from the average watershed characteristics by using relation (3.31) with $\bar{Q}_e \cong 0.24$ from Fig. 3.27 with $A_t = 5,830$ km^2, $E = (\bar{E}_{\text{tons}}/Q_e) \cong [(3.4 \times 10^{11})/ 0.24] \times 5,830 \times 91.4$ m $\times (0.0156)^{1.66} \times 0.7 \times 0.77$ m $\times 1 \times 10^{-6}$ m/s $\times 0.17 \times 0.35 \times 1 \simeq 2.4 \times 10^6$ tons/yr.

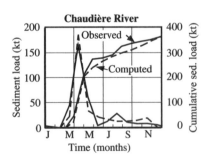

The mean annual sediment yield from this watershed is estimated with the sediment–delivery ratio $S_{\text{DR}} \cong 0.03$ from Fig. 3.28: $Y \cong E \times S_{\text{DR}} = 2.4 \times 10^6 \times 0.03 = 720$ ktons/yr. The mean annual sediment yield measured from 1968–1976 is $\sim$363 ktons/yr. The reader is referred to Julien and Frenette (1986) for a detailed analysis of sediment yield from snowmelt on this watershed. Comparison with the monthly sediment yield in Fig. CS.3.2.1 shows reasonable

Figure CS.3.2.1. Sediment yield of the Chaudière watershed (after Julien and Frenette, 1985).

agreement between measured and calculated sediment yields on a monthly basis.

♦Exercise 3.1

Combine the equations $h = it_e$, $q = iL$, and $q = \alpha h^\beta$ to derive the expressions for the time to equilibrium in Table 3.3.

♦♦Exercise 3.2

Combine Eqs. (3.8b) and (3.8c) to derive the relationship between the Darcy–Weisbach friction factor f and the Reynolds number Re in Eq. (3.12b).

♦Computer Problem 3.1

Find a sample of rainfall duration, intensity, or rainfall depth and compare with the exponential distribution as per the method shown in Example 3.1.

♦♦Problem 3.1

Repeat the infiltration calculations of Example 3.3 for a silty clay. Plot the hyetograph, infiltration rate, detention storage, and excess rainfall as functions of time from Table E.3.3.1. Compare the results for silty loam vs. silty clay.

♦♦Problem 3.2

Consider a 1-h storm at 1-in./h rainfall intensity on a 100-m-long plot at a 5% slope. Determine the maximum flow depth, flow velocity, Froude number, and shear stress if the surface is rough and impervious. (*Hint:* Determine the appropriate resistance equation.) Also calculate and plot the surface-runoff hydrograph.

Problem 3.3

Estimate the time to equilibrium of the largest possible 3-h storm on a rectangular farmland covering 35 acres. Estimate the infiltration on bare soil for silty clay loam.

Problem 3.4

Calculate the average rate of snowmelt in meters per second, from the derivative of Eq. (3.16c) after a snowmelt period of 100 h. Calculate λ_{2f} and compare with the results shown in Fig. 3.21.

♦♦**Problem 3.5**

From the data in Table 1.1, plot on a log–log scale the water discharge, sediment discharge, and concentration as functions of drainage area.

♦**Problem 3.6**

From the expression for sediment yield [Relation (3.33)] and the mean annual flow, $C_r A_t \bar{h}$, determine the expression for mean sediment concentration.

4

Steady flow in rivers

Steady flow refers to flow conditions that do not change with time. Steady flows can be either uniform when the conditions do not change with space or nonuniform when flow conditions change with space. Steady flow in rivers (Section 4.1) includes description of at-a-station hydraulic geometry, followed by a description of steady-uniform flow and resistance to flow. Steady-nonuniform flows (Section 4.2) include an analysis of the momentum equations followed by rapidly varied flow and gradually varied flow. Sediment transport in rivers (Section 4.3) includes a simple description for sediment transport in steady-uniform flow followed with calculations of aggradation and degradation in river reaches.

4.1   Steady river flow

Drainage networks have been studied by geomorphologists and topologists. In general, topologists search mathematical ranking and order among subwatersheds without specific reference to physical entities. The results of several years of experimental studies from the Rainfall Erosion Facility at Colorado State University by Schumm et al. (1987) attempted to relate basin morphology and sediment yield. Although considerable sediment-yield variability is assumed to result from climatic fluctuations and land-use changes, the experiments show that sediment yield is highly variable under steady rainfall conditions. The complex response of channel network evolution seems to be characterized by an exponential decrease in sediment yield as the channel network develops.

Rivers follow the low points along the watershed topographic profiles. With very few exceptions in arid areas, the lowest point of a watershed is located at the river outlet. A watershed map usually shows the hydrographic area and drainage network with lakes, reservoirs, and perhaps also gauging stations, counties, states, and/or countries. The drainage network of a watershed implies a discontinuous increase in drainage area and discharge at river confluences. Except

79

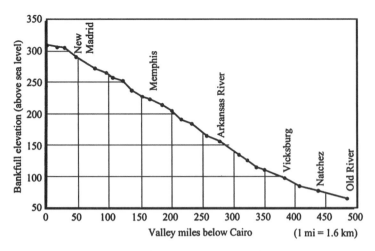

Figure 4.1. Longitudinal profile of the Mississippi River.

for river captures and artificial changes like transmountain water diversions, the drainage network of natural channels does not change significantly with time.

The river length is usually measured in the downstream direction from the uppermost river elevation. Left and right banks are usually referenced to a downstream-looking direction. A longitudinal profile is quite useful to detect bedrock controls, headcuts, nickpoints, and alluvial reaches. The approximate slope of alluvial reaches can be estimated from gradual changes in elevation over long distances. The valley slope corresponds to the floodplain elevation drop over the valley length. The channel slope corresponds to the water-surface elevation drop over the channel length. Channel and valley slopes are defined as positive although the elevation decreases in the positive-downstream direction. The river sinuosity is then defined as the ratio of the channel length to the valley length between two points located on the river. Examples of river reaches and longitudinal profiles are shown in Fig. 4.1 for an alluvial river, and Figs. 4.2 and 4.3 for semialluvial rivers. The longitudinal profile of an alluvial river reach is gradual. Bedrock control in semialluvial river reaches causes discontinuities in longitudinal profiles, bed-material sizes, and flow conditions.

At engineering time scales, it can be considered that the channel characteristics can change significantly during extreme flood events or after active tectonic and volcanic periods. River reaches can be examined one by one to determine their physical characteristics, such as length, sinuosity, width, depth, cross-sectional geometry, surface roughness, and hydraulic-resistance factor. The straightening of river reaches through meander cutoffs can change the

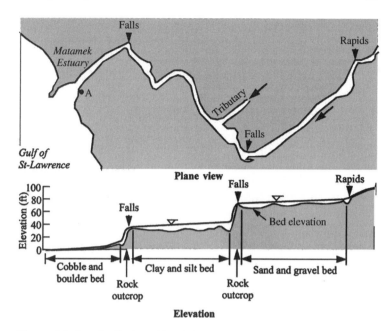

Figure 4.2. Longitudinal profile of the Matamek River (after Frenette and Julien, 1980).

length of rivers and their corresponding sinuosity and slope. River mileage is often measured downstream from a major river confluence or upstream from the mouth of a river. It is important to consider that the exact location of these reference points usually changes over the years. The following discussion focuses on at-a-station hydraulic geometry (Subsection 4.1.1) and on steady-uniform flow in rivers (Subsection 4.1.2).

4.1.1   At-a-station hydraulic geometry

At a given station along the river, a cross-section profile can be drawn in the direction perpendicular to the main flow direction. Cross sections that are not measured perpendicular to the flow direction will appear wider than reality. The cross-section profile shows the flow-depth distribution across the river as well as the elevation of the banks including the floodplain. It is important to reference the elevation to a benchmark that provides an exact horizontal-reference-plane elevation such as the mean sea level. Elevation scales are usually linked to the geodetic reference elevation [above mean sea level (ASL)] or the National Geodetic Vertical Distance (NGVD), which are both absolute vertical elevations. In some cases, the elevation is given with respect to a low-water

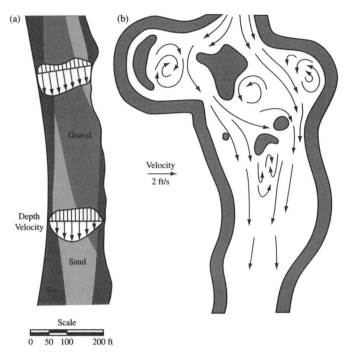

Figure 4.3. Morphology and bed material of the Matamek River (after Frenette and Julien, 1980).

reference plane (LWRP), which is the water-surface elevation in the river that is exceeded 97% of the time. Note that the LWRP is not horizontal, but slopes with the river. An example of a cross-sectional profile is given in Fig. 4.4. In this particular case, the downstream bedrock control retains water in the channel when the discharge reduces to zero.

Indications of the substrate material are given and the cross-sectional geometry extends onto the floodplain, where sketchy information on vegetation in terms of deciduous and coniferous trees, bushes and grasses is provided. These attributes provide basic information on aquatic habitat and expected floodplain roughness for runoff simulations during floods.

From this cross-sectional profile, the following geometrical parameters can be determined as functions of stage: (1) top-channel width W, (2) wetted perimeter P, (3) cross-sectional area A, (4) mean flow depth $\bar{h} = A/W$, and (5) hydraulic radius $R_h = A/P$. These parameters describe the geometry of a cross section. It is important to consider that the mean flow depth is different from the stage. An increase/decrease in mean flow depth does not necessarily correspond to the same increase/decrease in stage.

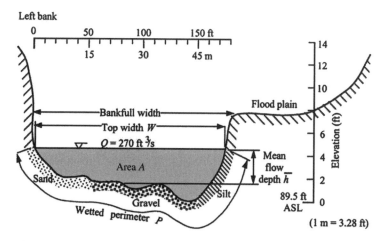

Figure 4.4. Cross section of the Matamek River (after Frenette and Julien, 1980).

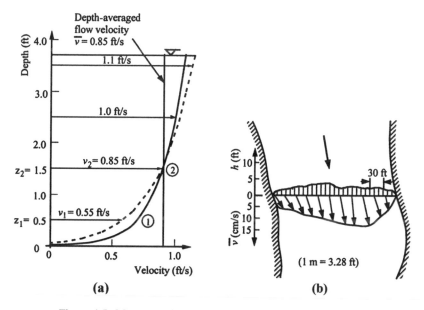

Figure 4.5. Matamek River: (a) vertical velocity profile and (b) transversal velocity profile (after Frenette and Julien, 1980).

The concept of at-a-station hydraulic geometry stems from velocity measurements along a cross section. An example of a velocity profile along one vertical of a cross section at a given stage provides sufficient measurements for the determination of the depth-averaged flow velocity $\bar{v}$ [Fig. 4.5(a)].

The depth-averaged velocity is normally obtained from a measured velocity profile.

Total discharge measurements are obtained from evenly spaced depth h_i and depth-averaged velocity $\bar{v}_i$ measurements along the cross section. The total cross-sectional area A is the sum of incremental areas a_i, $A = \sum_i a_i = \sum_i \Delta W_i h_i$, where ΔW is the spacing between verticals of a cross section. The total discharge is $Q = \sum_i a_i \bar{v}_i$, where $\bar{v}_i$ is the depth-averaged flow velocity normal to the incremental area. The cross-sectional average velocity is $V = Q/A$. Figure 4.5(b) shows a cross-sectional velocity profile with evenly spaced depth-averaged flow velocities along the cross-sectional profiles. At this cross section, the following hydraulic geometry is found: top width $W = 280$ ft, wetted perimeter $P = 280.5$ ft, cross-sectional area $A = 684$ ft^2, mean flow depth $\bar{h} = 2.44$ ft, and hydraulic radius $R_h = A/P = 2.43$ ft. The discharge is $Q = 210$ ft^3/s, and the cross-sectional average velocity $V = 0.31$ ft/s $= 9.4$ cm/s. Note that it is common to find that the wetted perimeter is close to the top width, $P \cong W$, and that the hydraulic radius is close to the mean flow depth, $R_h \cong \bar{h}$.

A flow-rating curve, or stage–discharge relationship, displays the change in stage with discharge. In channels with bedrock control, the stage–discharge relationship is unique and well defined, as shown in Fig. 4.6(a). In large alluvial rivers, the stage–discharge relationship may shift over time because of a combination of processes including (1) bed aggradation or degradation, (2) changes in bedform configuration, or (3) loop-rating effects that are due to dynamic flood routing. The example of the Mississippi River is shown in Fig. 4.6(b).

Specific-gauge records indicate the water level at a given flow discharge. The analysis of specific-gauge records can determine long-term aggradation or degradation trends of a river. For instance, Fig. 4.7(a) shows the specific-gauge records of the Atchafalaya River between Simmesport and Morgan City from 1950 to 1997. It clearly shows that the river has been gradually degrading between Simmesport and Chicot Pass and aggrading downstream of Chicot Pass.

In some alluvial rivers, water temperature can affect the specific-gauge records. For the example of the Mississippi River, in Fig. 4.7(b), the water temperature at a fixed discharge of 10^6 ft^3/s can alter the stage by as much as 5 ft. Higher stages correspond to warm water temperature.

The analysis of hydraulic-geometry relationships becomes simple once the flow-rating curve is known. At a given discharge, the mean flow depth is obtained from the flow-rating curve and the corresponding width, hydraulic radius, and cross-sectional area are obtained from the cross-sectional profiles.

This process can be repeated at various stages or discharges to define the at-a-station hydraulic-geometry relationships. The rate at which the

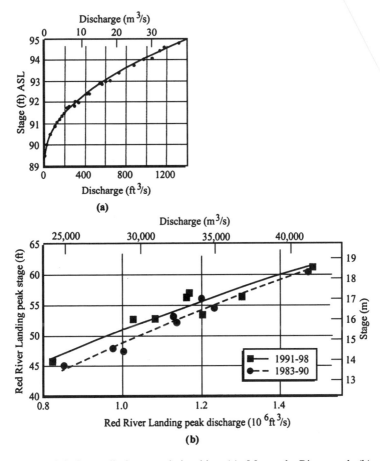

Figure 4.6. Stage–discharge relationship: (a) Matamek River and (b) Atchafalaya River.

hydraulic-geometry parameters change with discharge is quite important in defining at-a-station hydraulic geometry relationships. The example of the Matamek River is illustrated in Figs. 4.8. The analysis of 10 cross sections over a 1.4-km reach displays the local variability in hydraulic-geometry relationships of a nearly straight river. The wetted perimeter, or river width, varies by a factor of 2 within this reach; the rate of increase with discharge is fairly constant. When the hydraulic-geometry parameters are plotted vs. discharge on a log–log scale, the gradient of the profiles gives the exponent b of the relationship $P = aQ^b$.

For instance, the wetted perimeter in Fig. 4.8(a) can be approximated by $P \cong 126\, Q^{0.1}$, with Q in cubic feet per second and P in feet. Note that the

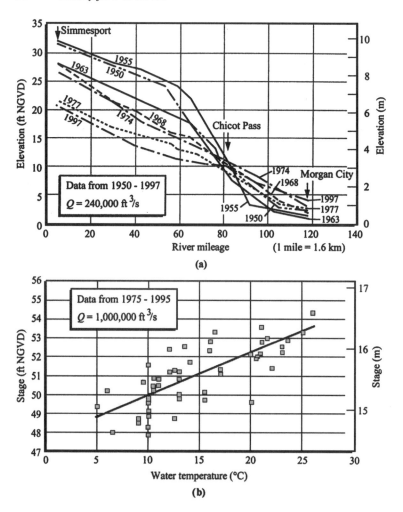

Figure 4.7. Specific-gauge records: (a) Atchafalaya River and (b) temperature effect (U.S. Army Corps of Engineers, 1999).

exponent is very small, which means that both the wetted perimeter and the river-surface width do not vary significantly with discharge. In fact, for many practical applications, the use of a rectangular cross section is a fairly good approximation. Beyond bankfull flows, the channel width suddenly increases, which alters calculations of mean flow depth and hydraulic radius.

Figure 4.8(b) shows the cross-section average flow depth vs. discharge at several cross sections of the same reach. Except at very low flows, the slope of the lines on this diagram indicates that the mean flow depth increases approximately as $\bar{h} = 0.35 \, Q^{0.36}$ with Q in cubic feet per second and $\bar{h}$ in feet.

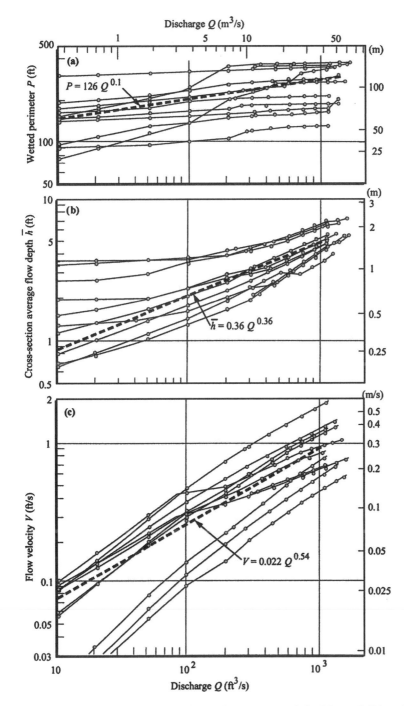

Figure 4.8. At-a-station hydraulic geometry of the Matamek River (after Frenette and Julien, 1980).

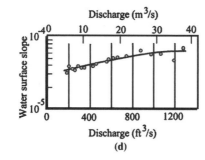

Figure 4.8. (*cont.*)

The cross-sectional averaged velocity can be plotted against discharge on a log–log plot, as shown in Fig. 4.8(c) for the same reach. At a station, the velocity varies largely with discharge, in this case, $V \cong 0.022 \, Q^{0.54}$.

The definition of volumetric flux requires that the product of width, depth, and velocity be equal to the discharge. Hence the relationships must satisfy $Q = W \bar{h} V \cong 126 \, Q^{0.1} \times 0.36 \, Q^{0.36} \times 0.022 \, Q^{0.54}$; thus the product of coefficients is $126 \times 0.36 \times 0.022 = 1$ and the sum of exponents is $0.1 + 0.36 + 0.54 = 1$.

Typical at-a-station hydraulic-geometry relationships show that changes in discharge affect primarily the flow velocity and, to a lesser extent, the flow depth. Unless the discharge exceeds bankfull conditions, the river surface width is fairly constant except at very low discharges. In some cases, the water-surface slope of a river may also change with discharge. The example shown in Fig. 4.8(d) illustrates a slight increase in slope with discharge.

4.1.2   *Steady-uniform river flow*

Steady-uniform flow implies no change in hydraulic conditions with either space or time. We easily obtain that $\partial Q / \partial x = 0$, or $Q = AV = \text{constant}$ and $\partial A / \partial t = 0$, or A is constant in space and time. Equation of motion (2.27) reduces to $S_f = S_0 = \tan \theta \cong \sin \theta$ because of steady flow $[(\partial V / \partial t) = 0]$ and uniform flow $[(\partial h / \partial x) = 0$ and $(\partial V / \partial x) = 0]$. The shear-stress relationship τ_0 for uniform flow reduces to

$$\tau_0 = \gamma R_h S_0. \tag{4.1}$$

For steady-uniform flow, shear-stress and resistance relationships can be directly written as a function of bed slope. With reference to Fig. 4.9, with identical pressure distributions at the upstream and the downstream cross sections, the weight component in the downstream x direction is balanced by only the

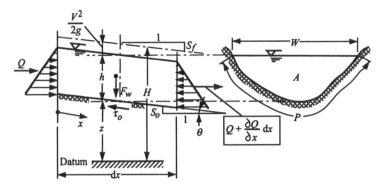

Figure 4.9. Equilibrium sketch for a river reach.

bed shear force $F_w \sin\theta = \tau_0 W dx$, which, for wide-rectangular channels with $F_w = \gamma W h dx$, and $\sin\theta \cong \tan\theta = S_0$ at small angles, simply reduces to $\tau_0 = \gamma h S_0$ because $S_f = S_0$.

Shear velocity u_* defines a kinematic substitute for the dynamic bed shear stress τ_0. The identity stems from $\tau_0 = \rho u_*^2$ or

$$u_* = \sqrt{\frac{\tau_0}{\rho}} = \sqrt{g R_h S_f}. \qquad (4.2) \; \blacklozenge$$

The shear velocity is not a measurable quantity but serves as a scaling parameter for kinematic-velocity profiles in turbulent boundary layers. The shear velocity is also often used in sediment-transport studies.

Resistance to flow is evaluated from steady-uniform flow conditions, $S_f = S_0$. Because no shear exists under hydrostatic conditions, the shear stress τ_0 is assumed to vary with fluid velocity and, after dimensional considerations, a dimensionless friction factor f has been defined after the work of Darcy and Weisbach as

$$f = \frac{8\tau_0}{\rho V^2} = \frac{8g R_h S_f}{V^2}. \qquad (4.3) \; \blacklozenge\blacklozenge$$

Resistance to flow can thus be described from the Darcy–Weisbach friction factor f. For wide-rectangular channels, $R_h = h$, we also easily demonstrate that $S_f = f \mathrm{Fr}^2/8$. We may also consider that $\sqrt{8/f} = V/u_*$. The depth-averaged flow velocity is then simply obtained as

$$V = \sqrt{\frac{8g}{f}} R_h^{1/2} S_f^{1/2} = C R_h^{1/2} S_f^{1/2}, \qquad (4.4) \; \blacklozenge\blacklozenge$$

where C is the Chézy coefficient. The identity $C^2 = 8g/f$ is always valid, and it is important to note that f describes flow resistance whereas C describes flow conveyance. The Chézy coefficient C is a constant as long as f is a constant.

It has been observed over the years that the Chézy coefficient C or the Darcy–Weisbach factor f varies with relative submergence and the Manning equation is a convenient approximation. Total resistance to flow can be described in terms of the Chézy coefficient C, the Darcy–Weisbach friction factor f, or the Manning coefficient n. The following identity among these factors has been established:
$$C \equiv \sqrt{(8g/f)} \equiv (R_h^{1/6}/n) \text{ (in S.I. units)} \equiv (1.49/n)\, R_h^{1/6} \text{ (in English units)}.$$

For steady-uniform flow, $S_f = S_0$, in a wide-rectangular channel, $R_h = h$, the normal depth h_n is obtained after $q = Vh$ is substituted into Eq. (4.3):

$$h_n = \left(\frac{fq^2}{8gS_0} \right)^{1/3} \qquad\qquad (4.5a) \;\blacklozenge$$

or

$$h_n = \left(\frac{nq}{S_0^{1/2}} \right)^{3/5} \text{ in S.I.} \qquad\qquad (4.5b) \;\blacklozenge$$

The normal depth thus increases with discharge and friction factor but decreases with increasing slope.

Evaluation of the Darcy–Weisbach friction factor for pipe flows yields three flow regimes clearly shown in Fig. 4.10. The flow is laminar on smooth surfaces when the Reynolds number $\mathrm{Re} < 2000$ and the Darcy–Weisbach friction factor is inversely proportional to the Reynolds number, $f = (k_t/\mathrm{Re})$. The flow becomes turbulent when $\mathrm{Re} > 2000$ and the surface boundary roughness exerts

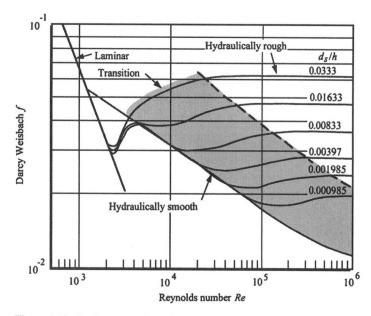

Figure 4.10. Resistance-to-flow diagram.

an influence on resistance to flow. When the flow is hydraulically smooth, the Darcy–Weisbach friction factor f gradually decreases with the Reynolds number. When the flow is hydraulically rough, resistance to flow depends on only the ratio of pipe diameter to surface roughness. For turbulent flows in rough pipes, the Darcy–Weisbach friction factor f and the Chézy coefficient C remain constant at any value of the Reynolds number.

The evaluation of the Darcy–Weisbach friction factor in rivers is a complex matter. The friction factor f varies with fluid viscosity, flow depth, grain size, bedform configuration, and vegetation. Exact values can be obtained experimentally for field conditions or laboratory conditions representative thereof.

River flows are turbulent, and we essentially need to differentiate between hydraulically rough versus hydraulically smooth boundaries. We can do this by comparing the bed-material grain size d_{50} with the laminar sublayer thickness $\delta = 11.6\,\nu/u_*$, where ν is the kinematic viscosity of water and u_* is the shear velocity from Eq. (4.2). The order of magnitude for δ corresponds to tenths of millimeters, that corresponds to sand sizes. When $\delta > 3\,d_s$, the flow is hydraulically smooth; the flow is hydraulically rough when $\delta < d_s/6$.

Resistance to flow in plane bed rivers with bed material finer than sand is said to be hydraulically smooth:

$$\frac{V}{u_*} = \sqrt{\frac{8}{f}} = \frac{C}{\sqrt{g}} = 5.75 \log\left(\frac{3.67u_*h}{\nu}\right), \tag{4.6}$$

where $u_* = \sqrt{\tau_0/\rho}$ is the shear velocity and ν is the kinematic viscosity of the fluid.

For hydraulically rough boundaries without bedforms, resistance to flow in rivers with bed material coarser than sand can be approximated by

$$\frac{V}{u_*} = \frac{C}{\sqrt{g}} = \sqrt{\frac{8}{f}} = 5.75 \log\left(\frac{12.2R_h}{k_s'}\right), \tag{4.7} \blacklozenge$$

where $k_s' \cong 3\,d_{90}$ or $k_s' \cong 6.8\,d_{50}$ can serve as first approximations.

Three equivalent flow-resistance formulations remain commonly used in river engineering practice: (1) the Chézy coefficient C, (2) the Manning coefficient n, and (3) the Darcy–Weisbach friction factor f. Both Chézy and Manning coefficients are dimensional, and the equivalence among the three is simply stated as

$$C \equiv \sqrt{\frac{8g}{f}} \equiv \frac{R_h^{1/6}}{n} \text{ (S.I. units)} \equiv \frac{1.49\,R_h^{1/6}}{n} \text{ (English units).} \tag{4.8} \blacklozenge\blacklozenge$$

Typical values for grain resistance are shown in Table 4.1. Both f and n increase with surface roughness. The Chézy coefficient describes flow conveyance and decreases with surface roughness.

Table 4.1. *Grain resistance and velocity formulations for turbulent flow over hydraulically rough plane boundaries* $(C = C'$ *and* $f = f')$

Formulation	Range	Resistance parameter	Velocity[a]
Chézy	$h/d_s \to \infty$	$C = \sqrt{\dfrac{8g}{f}}$ constant	$V = C R_h^{1/2} S_f^{1/2}$
Manning	$h/d_s > 100$	$\dfrac{C}{\sqrt{g}} \cong a\left(\dfrac{R_h}{d_s}\right)^{1/6} \cong \dfrac{R_h^{1/6}}{n}$ (S.I.)	$V = \dfrac{1}{n} R_h^{2/3} S_f^{1/2}$ (S.I.)
			$n \cong 0.062\, d_{50}^{1/6}$ (d_{50} in meters)
			$n \cong 0.046\, d_{75}^{1/6}$ (d_{75} in meters)
			$n \cong 0.038\, d_{90}^{1/6}$ (d_{90} in meters)
Logarithmic		$\dfrac{C}{\sqrt{g}} = \sqrt{\dfrac{8}{f}} = 5.75 \, \log\left(\dfrac{12.2\, R_h}{k_s'}\right)$	$V = \left(5.75 \, \log\dfrac{12.2\, R_h}{k_s'}\right)\sqrt{g R_h S_f}$
			$k_s' \cong 3\, d_{90}$
			$k_s' \cong 3.5\, d_{84}$
			$k_s' \cong 5.2\, d_{65}$
			$k_s' \cong 6.8\, d_{50}$

[a] The hydraulic radius $R_h = A/P$ is used, where A is the cross-sectional area, and P is the wetted perimeter; the friction slope S_f is the slope of the energy grade line.

For practical purposes, it is clear from Fig. 4.11 that the logarithmic equation applies over a wide range of h/d_{50}. As a first approximation, the following can be used:

$$\sqrt{\frac{8g}{f}} = 5.75 \log \frac{2h}{d_{50}}. \tag{4.9}$$

Manning's equation is also found to be applicable. For instance, the relationship in which $n = 0.064\, d_{50}^{1/6}$, with d_{50} in meters, should be in reasonable agreement with the field measurement when $h/d_s > 100$ and $h/d_s < 10,000$. It is concluded that Manning's equation may not applicable in shallow mountain streams ($h < 10\, d_s$) and in very deep sand-bed rivers ($h > 10,000\, d_s$).

The logarithmic form of grain resistance in Eq. (4.9) can be transformed into an equivalent-power form in which the exponent m varies with relative submergence h/d_s:

$$\frac{V}{u_*} = \sqrt{\frac{8}{f'}} = a\left(\frac{h}{d_s}\right)^m \equiv \hat{a} \ln\left(\frac{\hat{b}h}{d_s}\right) \tag{4.10} \blacklozenge$$

under the transformation that imposes the constraint that the value and the first

Table 4.2. *Typical values of resistance coefficients*

Boundary type	Darcy–Weisbach f	Manning n	Chézy C (m$^{1/2}$/s)
Smooth	0.0056	0.01	118
Plane sand bed	0.0046–0.0078	0.010–0.013	100–130
Sand antidunes	0.0078–0.015	0.013–0.018	72–100
Ripples	0.015–0.042	0.018–0.030	43–72
Sand dunes	0.018–0.076	0.020–0.040	32–65
Gravel bed	0.011–0.042	0.015–0.030	43–86
Cobble bed	0.018–0.057	0.020–0.035	37–65
Boulder bed	0.029–0.076	0.025–0.04	32–52
Vegetation	0.042–0.24	0.03–0.07	18–43

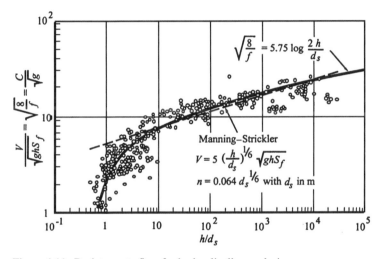

Figure 4.11. Resistance to flow for hydraulically rough rivers.

derivative be identical:

$$a = \frac{\hat{a}}{m}\left(\frac{d_s}{h}\right)^m, \tag{4.11a}$$

$$m = \frac{1}{\ln(\hat{b}h/d_s)}. \tag{4.11b}$$

Vegetation increases resistance to flow, and only crude empirical resistance coefficients can be obtained. Tables 4.2 provides typical values of resistance coefficients for various conditions.

Ranges of values for the Manning coefficient n are summarized in Tables 4.2 and 4.3. In the case of hydraulically smooth channels, values of $0.01 < n < 0.02$

Table 4.3. *Typical bedform characteristics*

Bedform	Manning n	Sediment concentration (mg/l)	Dominant type of roughness	Bedform surface profiles
Lower flow regime				
Plane bed	0.014	0	Grain	—
Ripples	0.018–0.028	10–200	Form	—
Dunes	0.020–0.040	200–3000	Form	Out of phase
Washed-out dunes	0.014–0.025	1000–4000	Variable	Out of phase
Upper flow regime				
Plane bed	0.010–0.013	2000–4000	Grain	—
Antidunes	0.010–0.020	2000–5000	Grain	In phase
Chutes and pools	0.018–0.035	5000–50000	Variable	In phase

are appropriate. In sand-bed channels, the presence of bedforms increases resistance and values of the Manning coefficient n can be as high as 0.05. In gravel-bed and cobble-bed streams, grain resistance is predominant, with $n \sim d_s^{1/6}$, as given in Tables 4.1 and 4.2. Other values of the Manning coefficient n are given in Table 4.3. Example 4.1 explains how to calculate the normal depth and shear stress in a river. Example 4.2 illustrates how to determine the main characteristics of channel flows, including shear stress, resistance parameters, and friction slope, from field measurements of a velocity profile.

Example 4.1   Application to smooth plane bed. Consider a steady-uniform flow in a 10-m-wide smooth rectangular channel; if the discharge is 10 m³/s and the slope is 26 cm/km, estimate the friction factor in terms of f and equivalent C and n and calculate the normal flow depth and the corresponding applied bed shear stress.

For a very smooth surface, we can consider $f \cong 0.01$ from Table 4.1, the calculated normal flow depth given the unit discharge

$$q = \frac{Q}{W} = \frac{10 \text{ m}^3}{\text{s } 10 \text{ m}} = \frac{1 \text{ m}^2}{\text{s}}$$

is

$$h_n = \left(\frac{0.01 \times 1 \text{ m}^4\text{s}^2}{\text{s}^2 \times 8 \times 9.81 \text{ m} \times 26 \times 10^{-5}} \right)^{1/3} = 0.788 \text{ m}.$$

The equivalent resistance parameters C and n are calculated, with the hydraulic radius given in S.I. units, as

$$R_h = \frac{A}{P} = \frac{W h_n}{W + 2h_n} = \frac{10\,\text{m} \times 0.788\,\text{m}}{10\,\text{m} + 2 \times 0.788\,\text{m}} = 0.681\,\text{m},$$

$$C = \sqrt{\frac{8 \times 9.81\,\text{m}}{\text{s}^2 \times 0.01}} = 88\,\text{m}^{1/2}/\text{s},$$

$$n = \frac{R_h^{1/6}}{C} = \frac{(0.681\,\text{m})^{1/6}}{88\,\text{m}^{1/2}\,\text{s}} = 0.01\,\text{s/m}^{1/3}.$$

The applied bed shear stress for steady-uniform flow $S_f = S_0$ is obtained from Eq. (4.1):

$$\tau_0 = \gamma R_h S_f = \gamma R_h S_0 = \frac{9810\,\text{N}}{\text{m}^3} \times 0.681\,\text{m} \times 26 \times 10^{-5}$$

$$= \frac{1.73\,\text{N}}{\text{m}^2} = 1.73\,\text{Pa}.$$

Example 4.2 Application to a turbulent-velocity profile. Consider the given measured velocity profile in a 200-ft-wide river in Fig. 4.5(a).

Consider two points, 1 and 2, near the bed,

$$v_1 = \frac{u_*}{\kappa} \ln \frac{z_1}{k_s'},$$

$$v_2 = \frac{u_*}{\kappa} \ln \frac{z_2}{k_s'},$$

and estimate the following parameters:

(a) shear velocity:

$$u_* = \frac{\kappa (v_2 - v_1)}{\ln\left(\dfrac{z_2}{z_1}\right)} = \frac{0.4(0.85 - 0.55)\text{ft}}{\ln\left(\dfrac{1.5}{0.5}\right)\text{s}} = 0.11\,\text{ft/s} = 0.0335\,\text{m/s}.$$

(b) boundary shear stress:

$$\tau_0 = \rho u_*^2 = \frac{1.92\,\text{slug}}{\text{ft}^3}(0.11)^2 \frac{\text{ft}^2}{\text{s}^2} = 0.023\,\frac{\text{lb}}{\text{ft}^2} = 1.1\,\text{Pa}.$$

(c) depth-averaged flow velocity:

$$V \cong 0.85\,\text{ft/s} = 0.26\,\text{m/s}.$$

(d) unit discharge:

$$q = Vh = 0.85\,\text{ft/s} \times 3.7\,\text{ft} = 3.1\,\text{ft}^2/\text{s} = 0.292\,\text{m}^2/\text{s}.$$

(e) hydraulic radius:

$$R_h = \frac{A}{P} = \frac{Wh}{W + 2h} = \frac{200 \times 3.7 \text{ ft}^2}{200 + 2 \times 3.7 \text{ ft}} = 3.56 \text{ ft} = 1.09 \text{ m};$$

the hydraulic radius $R_h = 3.56$ ft is close to flow depth 3.7 ft.

(f) Froude number:

$$\text{Fr} = \frac{V}{\sqrt{g R_h}} \cong \frac{V}{\sqrt{gh}} = \frac{0.85 \text{ ft/s}}{\sqrt{32.2 \times 3.7 \frac{\text{ft}^2}{\text{s}^2}}} = 0.078.$$

(g) friction slope from Eq. 4.1:

$$S_f = \frac{\tau_0}{\gamma R_h} \cong \frac{\tau_0}{\gamma h} = \frac{0.023 \text{ lb ft}^3}{\text{ft}^2 \, 62.4 \text{ lb} \times 3.7 \text{ ft}} = 9.96 \times 10^{-5} \cong \frac{10 \text{ cm}}{\text{km}}.$$

(h) Darcy–Weisbach friction factor from Eq. 4.3:

$$f = \frac{8 S_f}{\text{Fr}^2} = \frac{8 \times 9.96 \times 10^{-5}}{0.078^2} = 0.13.$$

(i) Manning coefficient from Table 4.1:

$$n = \frac{1.49}{V} R_h^{2/3} S_f^{1/2} = \frac{1.49 \text{ ft}^{1/3} \text{s}}{0.85 \text{ ft m}^{1/3}} (3.7 \text{ ft})^{2/3} (9.96 \times 10^{-5})^{1/2}$$

$$= 0.042 \text{ s/m}^{1/3}.$$

Note that, because of the conversion factor 1.49 ft$^{1/3}$/m$^{1/3}$, the value of n is the same in both S.I. and English units.

(j) Chézy coefficient from Eq. 4.8:

$$C = \sqrt{\frac{8g}{f}} = \sqrt{\frac{8 \times 32.2}{0.13}} = 44.5 \text{ ft}^{1/2}/\text{s} = 24.6 \text{ m}^{1/2}/\text{s}.$$

Note that the metric value of C is commonly used.

(k) momentum correction factor [Eq. (4.17)] in Subsection 4.2.1:

$$\beta_m = \frac{1}{A V_x^2} \int_A v_x^2 \, dA \cong \frac{1}{h V_x^2} \sum_i v_{xi}^2 \, dh_i,$$

Note that $dh_i = 1$ ft except for the uppermost velocity measurement

$$\beta_m \cong \frac{1}{3.7 \text{ ft}} \frac{\text{s}^2}{(0.85)^2 \text{ ft}^2} [0.55^2 + 0.85^2 + 1.0^2 + (1.1^2 \times 0.7)] \frac{\text{ft}^3}{\text{s}^2}$$

$$= 1.074.$$

(1) energy correction factor [Eq. (4.23b)] in Subsection 4.2.2:

$$\alpha_e = \frac{1}{AV_x^3} \int_A v_x^3 \, dA \cong \frac{1}{hV_x^3} \sum_i v_{xi}^3 \, dh_i,$$

$$\alpha_e \cong \frac{1}{3.7 \, \text{ft}} \frac{s^3}{(0.85)^3 \, \text{ft}^3} [0.55^3 + 0.85^3 + 1.0^3 + (1.1^3 \times 0.7)] \frac{\text{ft}^4}{s^3}$$

$$= 1.194.$$

Note that $\alpha_e > \beta_m$ and that both are greater than unity.

4.2   Steady-nonuniform river flow

Steady-nonuniform flow in rivers implies that the total discharge does not change with time but can vary in the downstream direction. Mathematically, steady flow implies that $(\partial h/\partial t) = 0$, $(\partial V/\partial t) = 0$, $(\partial W/\partial t) = 0$, and $(\partial Q/\partial t) = 0$. In a 1D channel without rainfall, infiltration, and lateral inflow, discharge also remains constant in the downstream direction. Nonuniform flow is possible under steady discharge when the mean flow velocity, channel width, and flow depth change in the downstream direction, or $(\partial V/\partial x) \neq 0$, $(\partial W/\partial x) \neq 0$, and $(\partial h/\partial x) \neq 0$. In the following discussion, momentum equations are derived (Subsection 4.2.1), followed by rapidly varied converging flow in rivers (Subsection 4.2.2) and gradually varied flow (Subsection 4.2.3).

4.2.1   *Momentum equations for steady flow*

Momentum equations define the hydrodynamic forces exerted by surface flows. After the equations of motion [Eqs. (2.18)] are multiplied by the mass density of the water ρ, the terms on the left-hand side of the equations represent the rate of momentum change per unit volume and the rate of impulse per unit volume is found on the right-hand side. Integration over the total volume $\forall$ shows that the rate-of-momentum change equals the impulse per unit time. For example, the x component in the Cartesian coordinates for steady flow is

$$\int_\forall \rho \left(v_x \frac{\partial v_x}{\partial x} + v_y \frac{\partial v_x}{\partial y} + v_z \frac{\partial v_x}{\partial z} \right) d\forall$$

$$= \int_\forall \rho g_x d\forall - \int_\forall \frac{\partial p}{\partial x} d\forall + \int_\forall \left(\frac{\partial \tau_{xx}}{\partial x} + \frac{\tau_{yx}}{\partial y} + \frac{\partial \tau_{zx}}{\partial z} \right) d\forall. \quad (4.12)$$

The integrand on the left-hand side can be rewritten as

$$\frac{\partial \rho v_x^2}{\partial x} + \frac{\partial \rho v_x v_y}{\partial y} + \frac{\partial \rho v_x v_z}{\partial z} - v_x \left(\frac{\partial \rho v_x}{\partial x} + \frac{\partial \rho v_y}{\partial y} + \frac{\partial \rho v_z}{\partial z} \right). \quad (4.13)$$

By virtue of the continuity Eq. (2.17), the terms in parentheses of expression (4.13) can be dropped. The volume integral of the remaining momentum and stress terms can be transformed into surface integrals by means of the divergence theorem Eq. 5.2. The result is the general impulse–momentum relationship.

x component

$$\int_A \rho v_x \left(v_x \frac{\partial x}{\partial n} + v_y \frac{\partial y}{\partial n} + v_z \frac{\partial z}{\partial n} \right) dA = \int_\forall \rho g_x d\forall - \int_A p \frac{\partial x}{\partial n} dA$$
$$+ \int_A \left(\tau_{xx} \frac{\partial x}{\partial n} + \tau_{yx} \frac{\partial y}{\partial n} + \tau_{zx} \frac{\partial z}{\partial n} \right) dA. \tag{4.14a}$$

y component

$$\int_A \rho v_y \left(v_x \frac{\partial x}{\partial n} + v_y \frac{\partial y}{\partial n} + v_z \frac{\partial z}{\partial n} \right) dA = \int_\forall \rho g_y \, d\forall - \int_A p \frac{\partial y}{\partial n} dA$$
$$+ \int_A \left(\tau_{xy} \frac{\partial x}{\partial n} + \tau_{yy} \frac{\partial y}{\partial n} + \tau_{zy} \frac{\partial z}{\partial n} \right) dA. \tag{4.14b}$$

z component

$$\int_A \rho v_z \left(v_x \frac{\partial x}{\partial n} + v_y \frac{\partial y}{\partial n} + v_z \frac{\partial z}{\partial n} \right) dA = \int_\forall \rho g_z d\forall - \int_A p \frac{\partial z}{\partial n} dA$$
$$+ \int_A \left(\tau_{xz} \frac{\partial x}{\partial n} + \tau_{yz} \frac{\partial y}{\partial n} + \tau_{zz} \frac{\partial z}{\partial n} \right) dA. \tag{4.14c}$$

It is observed that momentum is a vector quantity, the momentum change that as due to convection is embodied in the surface integral on the left-hand side of Eqs. (4.14), and all the stresses are expressed in terms of surface integrals.

Consider a detailed application of the momentum equations to open-channel flows. With reference to the rectangular channel sketched in Fig. 4.12, the momentum relationship [Eqs. (4.14)] in the downstream x direction is applied to this channel, now subjected to rainfall at an angle θ_r, velocity V_r over an area A_r, wind shear τ_w, bank shear $\tau_s = \tau_{yx}$, and bed shear $\tau_b = \tau_0 = \tau_{zx}$:

$$\int_A \rho v_x \left(v_x \frac{\partial x}{\partial n} + v_y \frac{\partial y}{\partial n} + v_z \frac{\partial z}{\partial n} \right) dA = \int_\forall \rho g_x d\forall - \int_A p \frac{\partial x}{\partial n} dA$$
$$+ \int_A \left(\tau_{xx} \frac{\partial x}{\partial n} + \tau_{yx} \frac{\partial y}{\partial n} + \tau_{zx} \frac{\partial z}{\partial n} \right) dA. \tag{4.15}$$

Several integrals vanish for 1D flow in impervious channels, $v_y = v_z = \tau_{xx} = 0$, leaving

$$\int_A \rho v_x^2 \frac{\partial x}{\partial n} dA + \int_A p \frac{\partial x}{\partial n} dA = \int_\forall \rho g_x d\forall + \int_A \tau_{zx} \frac{\partial z}{\partial n} dA + \int_A \tau_{yx} \frac{\partial y}{\partial n} dA$$
$$\tag{4.16}$$

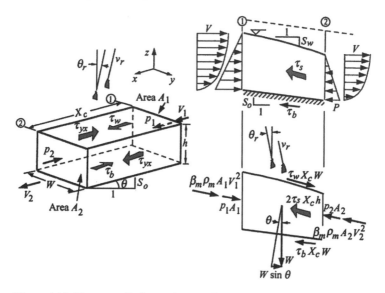

Figure 4.12. Forces applied on a river reach.

Consider an incompressible homogeneous fluid, $\rho_m = ct$, and define the momentum correction factor β_m, given the cross-sectional averaged velocity V_x:

$$\beta_m = \frac{1}{AV_x^2} \int_A v_x^2 \, dA. \tag{4.17}$$

The value of β_m is generally close to unity; the reader can refer to Example 4.2 for a detailed calculation from a measured velocity profile. With average values of pressure p, velocity V, and area A at upstream cross section 1 and downstream cross section 2, the integration of the momentum equation for this control volume $\forall$ of length X_c, width W, and height h yields

$$\beta_m \rho A_2 V_2^2 + p_2 A_2 - \beta_m \rho A_1 V_1^2 - p_1 A_1 - \rho A_r V_r^2 \sin(\theta + \theta_r)$$
$$= \gamma \forall \sin \theta - \tau_b W X_c - \tau_s 2h X_c + \tau_w W X_c. \tag{4.18}$$

Assuming that the boundary shear stress τ_0 equals the bank shear stress τ_s and the bed shear stress τ_b, the equation with negligible rainfall, $A_r \to 0$, without wind shear, $\tau_w \to 0$, can be rewritten when the channel inclination θ is small ($\sin \theta \cong S_0$, the bed slope) as

$$\underbrace{p_2 A_2 + \beta_m \rho A_2 V_2^2}_{\text{downstream force}} = \underbrace{p_1 A_1 + \beta_m \rho A_1 V_1^2}_{\text{upstream force}} + \underbrace{\gamma \left(\frac{A_1 + A_2}{2} \right) X_c S_0}_{\text{weight force}} - \underbrace{\tau_0 (W + 2h) X_c}_{\text{shear force}}.$$

$$(4.19) \; \blacklozenge$$

This equation indicates equilibrium of forces in the downstream direction. The downstream and the upstream forces equal the sum of a pressure force and a momentum force. Further reduction is possible when the weight-force component cancels the shear force. In the case of steady flow, $Q_1 = Q_2$, in a rectangular channel of width W, the cross-section area is $A = Wh$ and the average pressure is $p = 0.5\,\rho\,gh$. Further assuming that the Boussinesq correction factor $\beta_m \cong 1$, we find that the hydrodynamic-force component F_h reduces to

$$F_h = pA + \beta_m \rho A V^2 = 0.5\rho g h W h + \rho W h \left(\frac{Q}{Wh}\right)^2. \qquad (4.20)$$

The specific-momentum function M is obtained after the hydrodynamic force F_h is divided by the channel width W and the specific weight of the fluid ρg; thus

$$M = \frac{F_h}{\rho g W} = \frac{h^2}{2} + \frac{q^2}{gh}, \qquad (4.21) \; \blacklozenge$$

where the unit discharge $q = Q/W$.

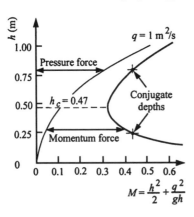

Figure 4.13. Specific-momentum diagram.

This specific-momentum function M can be plotted as a function of flow depth at a given unit discharge q. For instance, the specific-momentum functions for $q = 1\,\mathrm{m^2/s}$ are shown in Fig. 4.13. Two flow depths with identical specific momenta are called conjugate depths. The lowest value of the specific-momentum function is obtained when $\partial M/\partial h = 0$, which corresponds to the critical flow depth $h_c = (q^2/g)^{1/3}$. Given the property that $q = Vh$, the critical flow depth corresponds to $(q^2/gh_c^3) = (V^2/gh_c) = \mathrm{Fr}_c^2 = 1$. A critical value of the Froude number $\mathrm{Fr} = 1$ relates to the minimum value of the specific-momentum function. Flow is supercritical when $\mathrm{Fr} > 1$ or $h < h_c$ and subcritical when $\mathrm{Fr} < 1$ or $h > h_c$.

Example 4.3 Application of momentum to hydraulic jump. Steady flow, $Q = 10\,\mathrm{m^3/s}$, in a rectangular channel, $W = 10$ m, is such that the upstream flow velocity $V = 4\,\mathrm{m/s}$ is rapidly reduced over a short distance to form a

hydraulic jump. If the channel slope counterbalances the shear force, determine
the flow depth, velocity, and forces downstream of the hydraulic jump.

The increased turbulence in the aerated portion of the flow near the surface
is not accounted for by the bed-resistance equations. For this reason, energy
will be dissipated at a faster rate than it would through bed-resistance. Because
flow depth changes over a short distance, it can be shown that the bed friction
force on a smooth surface without baffle blocks is very small compared with
pressure forces. We can therefore consider that equilibrium will be approximately dictated by the balance of pressure and hydrodynamic forces upstream
and downstream of the hydraulic jump.

The answer can be found with the use of conjugate depths on the specific-
momentum diagram, Fig. 4.13. Exact values of the conjugate flow depth can
be calculated from $M_1 = M_2$ in the form

$$\frac{q^2}{g}\left(\frac{1}{h_1} - \frac{1}{h_2}\right) = \frac{1}{2}\left(h_2^2 - h_1^2\right).$$ (E.4.3.1)

The relationship $\text{Fr}_1^2 = (q^2/gh_1^3)$ allows to rewrite Eq. E.4.3.1 as

$$\text{Fr}_1^2 = \frac{1}{2}\frac{h_2}{h_1}\left(\frac{h_2}{h_1} + 1\right).$$ (E.4.3.2)

The solution of the quadratic equation in h_2/h_1 is called the Belanger
equation:

$$\frac{h_2}{h_1} = \frac{1}{2}\left(\sqrt{1 + 8\,\text{Fr}_1^2} - 1\right).$$ (E.4.3.3) ◆

The conjugate downstream flow depth h_2 is calculated directly from the upstream flow depth h_1 and the upstream Froude number Fr_1.

For the case in point, $Q = 10 \text{ m}^3/\text{s}$ and $W = 10 \text{ m}$, or $q = Q/W = 1 \text{ m}^2/\text{s}$,
the upstream flow depth is $(q/V_1) = [(1 \text{ m}^2\text{s}])/(\text{s} \times 4 \text{ m})] = 0.25 \text{ m}$ and the
upstream Froude number is

$$\text{Fr}_1^2 = \frac{q^2}{gh_1^3} = \frac{1 \text{ m}^4\text{s}^2}{\text{s}^2 \times 9.81 \text{ m} \times (0.25)^3 \text{ m}^3} = 6.52.$$

The corresponding downstream conditions are calculated from the Belanger
equation flow depth $h_2 = (0.25/2)(\sqrt{1 + 8 \times 6.52} - 1) = 0.788 \text{ m}$ and downstream velocity $V_2 = (q/h_2) = (1.27 \text{ m/s})$ (see Fig. E.4.3.1).

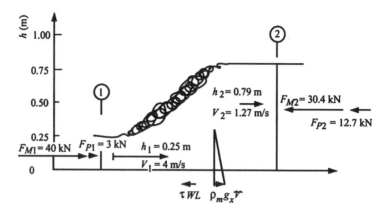

Figure E.4.3.1. Hydraulic jump.

Hydrodynamic forces are calculated by multiplying the specific-momentum value by $\rho g W$. For instance, the downstream pressure force is

$$F_{p2} = \rho g\, Wh^2/2 = \frac{9.81\,\text{kN}}{\text{m}^3} \times 10\,\text{m} \times \frac{(0.788\,\text{m})^2}{2} = 30.4\,\text{kN}.$$

Simplified solutions are possible for two types of nonuniform flow: (1) rapidly varied converging flow and (2) gradually varied flow. In both cases, convective changes in flow depth, width, and velocity head are significant and cannot be neglected.

Rapidly varied flows are usually induced by structures and other perturbations to the flow. Because flow conditions change over a short distance, it can generally be assumed that the energy loss and the change in bed elevation are small compared with the convective terms of the St. Venant equation. We can assume as a first approximation that

$$\frac{\mathrm{d}}{\mathrm{d}x}\left(h + \frac{V^2}{2g}\right) = 0, \qquad (4.22a)$$

which can be integrated over x to yield conservation of specific energy over this short reach. The specific-energy diagram hence becomes extremely useful in the analysis of rapidly varied steady flows.

Gradually varied flows are those in which changes in width, depth, and velocity take place over a reasonably long distance; the convective-acceleration terms can be neglected but friction losses remain important over a long reach. Gradually varied flows refer to changing flow depth in the downstream direction such that, with a change in flow depth h in the downstream x direction, all four terms of the St.-Venant equation are rearranged in the following manner:

$$\frac{dE}{dx} \cong \frac{d}{dx}\left(h + \frac{V^2}{2g}\right) = S_0 - S_f. \tag{4.22b}$$

4.2.2 Rapidly varied converging river flow

Rapidly varied flow conditions refer to large spatial derivatives over a short channel reach. In converging flows, it is often considered that the energy level will be maintained constant because the bed friction does not take place over a long distance. In wide-rectangular channels, the gradient of the integral form of the Bernoulli sum can be rewritten as

$$\frac{d\tilde{E}}{dx} = \frac{d}{dx}\left(\frac{p}{\gamma_m} + \alpha_e \frac{V^2}{2g}\right) = \frac{-dz_b}{dx} - S_f = S_0 - S_f \cong 0; \tag{4.23a}$$

this implies that the integral form of specific energy $\tilde{E} = (p/\gamma_m) + [\alpha_e(V^2/2g)]$ remains constant. The energy correction factor α_e is defined as

$$\alpha_e = \frac{1}{AV_x^3}\int_A v_x^3 \, dA. \tag{4.23b}$$

After considering that $\alpha_e \cong 1$, hydrostatic distribution $p = \gamma h$, and that the unit discharge $q = Vh$, we find that the specific energy E corresponds to the sum of pressure and velocity head above the channel-bed elevation:

$$E = h + \frac{q^2}{2gh^2}. \tag{4.24} \blacklozenge$$

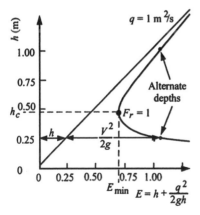

Figure 4.14. Specific-energy diagram.

The specific-energy function E can be plotted as a function of flow depth at a given unit discharge q, e.g., $q = 1 \text{ m}^2/\text{s}$ in Fig. 4.14.

Two flow depths with identical specific energies are called alternate depths. The lowest value of the specific energy corresponds to $(\partial E/\partial h) = 1 - (q^2/gh_c^3) = 0$, which defines the critical flow depth when the Froude number Fr $= 1$:

$$h_c = \left(\frac{q^2}{g}\right)^{1/3}. \tag{4.25} \blacklozenge\blacklozenge$$

We also easily demonstrate that the minimum value of specific energy $E_{min} = 1.5\,h_c$.

Once the critical flow depth is known, the Froude number can be directly calculated from $q^2 = gh_c^3$ as

$$\text{Fr} = \frac{q}{h\sqrt{gh}} = \left(\frac{h_c}{h}\right)^{3/2}. \qquad (4.26)$$

Applications of rapidly varied flows are shown for river flow contraction in Example 4.4 and for flow under a sluice gate in Example 4.5.

Example 4.4   Application to open-channel flow contractions. Consider steady flow at $Q = 10$ m³/s in a 10-m wide-rectangular channel from Example 4.3. Determine (a) the maximum possible elevation of a sill Δz_{max} in section A that will not cause backwater and (b) the maximum lateral contraction ΔW_{max} of the channel in section A that will not cause backwater.

In rapidly varied flow, it is expected that the energy losses through friction are equivalent to friction losses for steady-uniform flow. The accelerating flow is analyzed with the specific-energy diagram, Fig. E.4.4.1.

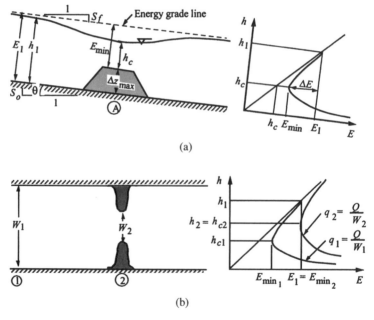

(a)

(b)

Figure E.4.4.1. Sill and river contraction.

Solution: (a) The maximum elevation of the sill Δz_{max} in section A is such that the flow will be critical on top of the sill and $\Delta z_{max} + E_{min} = E_1$ or $\Delta z_{max} = E_1 - E_{min}$. In this example, the unit discharge $q = (Q/W) = [(10 \, m^3/s)/10 \, m] = 1 \, m^2/s$, the critical flow depth

$$h_c = 3\sqrt{\frac{q^2}{g}} = \left(\frac{1 \, m^4 s^4}{s^2 \times 9.81 \, m}\right)^{1/3} = 0.467 \, m,$$

and the minimum specific energy $E_{min} = 3/2 \, h_c = 0.70 \, m$.

The approaching steady-uniform flow depth from Example 4.3 is $h_1 = 0.788 \, m$. The corresponding velocity is $V_1 = (q/h_1) = [1 \, m^2/(s \times 0.788 \, m)] = (1.27 \, m/s)$. The specific-energy level $E_1 = h_1 + V_1^2/2g = 0.788 \, m + (1.27 \, m/s)^2 \, 2 \times 9.81 \, m = 0.870 \, m$. The maximum sill elevation $\Delta z_{max} = E_1 - E_{min} = 0.870 \, m - 0.70 \, m = 0.17 \, m$.

(b) The minimum channel width W_2 in section A that does not cause backwater is such that the total discharge remains constant, $Q = W_1 q_1 = W_2 q_2$, and the flow is critical in the contracted section A_2, or $h_{c2} = 0.67 \, E_1 = 0.67 \, E_{min2}$, with $E_1 = 0.87 \, m$ and $1 = q_2^2/gh_{c2}^3$, or

$$W_2 = \frac{Q}{\sqrt{g(0.67E_1)^3}}.$$

In this example,

$$W_2 = \frac{10 \, m^3 s}{s\sqrt{9.81 \, m \, (0.667 \times 0.87 \, m)^3}} = 7.22 \, m.$$

The maximum lateral contraction $\Delta W_{max} = W_1 - W_2 = 10 \, m - 7.22 \, m = 2.78 \, m$.

Example 4.5 Application to rapidly varied flow under a sluice gate. A sluice gate controls the 10-m^3/s flow discharge in a 10-m wide-rectangular channel. If the flow depth downstream of the sluice gate is at $h_1 = 0.25 \, m$ and rapidly increases to the normal flow depth $h_2 = 0.788 \, m$ in a hydraulic jump located at the toe of the sluice gate, determine the following:

(a) What is the water level upstream of the sluice gate?
(b) What is the force applied onto the sluice gate?
(c) How much energy is lost in the hydraulic jump?

Solution: (a) The energy losses through friction along the sluice gate can be neglected because the flow is rapidly varied and converging. It is assumed that the specific-energy level on both sides of the gate are identical. At a unit flow discharge $q = (Q/W) = 1 \text{ m}^2/\text{s}$, the specific-energy diagram in Fig. 4.14 shows that the downstream specific energy corresponding to a downstream flow depth $h_1 = 0.25$ m is

$$E_1 = h_1 + \frac{q^2}{2gh_1^2} = 0.25 \text{ m} + \frac{1 \text{ m}^4 \text{ s}^2}{\text{s}^2 \times 2 \times 9.81 \text{ m} (0.25 \text{ m})^2} = 1.06 \text{ m}.$$

The alternate depth at the same specific-energy level corresponds to the flow depth $h_2 \cong 1$ m upstream of the gate.

(b) The force applied on the gate can be calculated from the specific-momentum function plotted in Fig. 4.13. We obtain graphically, at an upstream flow depth of 1 m, $M = 0.6 \text{ m}^2$, and, at the downstream flow depth of 0.25 m, $M \cong 0.45 \text{ m}^2$. The corresponding force applied to the gate corresponds to the difference in specific-momentum times γW, or $\Delta F = \gamma W \Delta M = (9{,}810 \text{ N/m}^3) \times 10 \text{ m} \times (0.6 - 0.45) \text{ m}^2 = 14.7 \text{ kN}$.

(c) The hydraulic jump with $q = 1 \text{ m}^2/\text{s}$ and upstream flow depth $h_1 = 0.25$ m was previously examined in Example 4.3. The specific-energy level at the upstream end is given by

$$E_1 = h_1 + \frac{q^2}{2gh_1^2} = 0.25 \text{ m} + \frac{1 \text{ m}^4 \text{s}^2}{\text{s}^2 \times 2 \times 9.81 \text{ m} \times (0.25 \text{ m})^2} = 1.06 \text{ m},$$

as shown in Fig. 4.14. At the downstream end of the hydraulic jump, the previously calculated conjugate depth $h_2 = 0.788$ m corresponds to the specific energy

$$E_2 = h_2 + \frac{q^2}{2gh_2^2} = 0.788 \text{ m} + \frac{1 \text{ m}^4 \text{s}^2}{\text{s}^2 \times 2 \times 9.81 \text{ m} \times (0.788 \text{ m})^2}$$
$$= 0.87 \text{ m}.$$

The specific energy lost in the jump is $\Delta E = 1.06 \text{ m} - 0.87 \text{ m} = 0.19 \text{ m}$. The specific energy lost in a hydraulic jump can be calculated directly after the following transformation:

$$\Delta E = \left(h_2 + \frac{q^2}{2gh_2^2} \right) - \left(h_1 + \frac{q^2}{2gh_1^2} \right)$$
$$= h_2 - h_1 + \frac{q^2}{2g} \left(\frac{1}{h_2^2} - \frac{1}{h_1^2} \right), \qquad \text{(E.4.5.1)}$$

given the constraint of conjugate depths

$$\frac{q^2}{g} \left(\frac{1}{h_1} - \frac{1}{h_2} \right) = \frac{1}{2}(h_2^2 - h_1^2)$$

from Eq. (E.4.3.1).

After q^2/g is substituted with a function of h_1 and h_2, ΔE in Eq. (E.4.5.1) can be solved as a function of h_1 and h_2 after algebraic manipulations as

$$\Delta E = \frac{(h_2 - h_1)^3}{4h_1 h_2}. \qquad \text{(E.4.5.2)} \blacklozenge$$

In our practical example,

$$\Delta E = \frac{(0.788 - 0.25)^3}{4 \times 0.788 \times 0.25} = 0.197 \, \text{m}.$$

The power loss ΔP in the hydraulic jump is then calculated from

$$\Delta P = \gamma Q \Delta E = \frac{9{,}810 \, \text{N}}{\text{m}^3} \times \frac{10 \, \text{m}^3}{\text{s}} \times 0.197 \, \text{m} = 19.3 \, \text{kW}.$$

The larger the Froude number upstream of the jump, the larger the power lost through turbulence in the hydraulic jump.

4.2.3   Gradually varied river flow

The term gradually-varied flow refers to flow conditions that change over long distances. Simplifications of the momentum equations are possible for 1D flows. As a first approximation, it is often assumed that resistance to flow in gradually varied flow can be calculated as for steady-uniform flow. In wide-rectangular channels, $h = R_h$ with steady-uniform flow, $S_f = S_0$, the normal flow depth h_n from Eq. (4.5a) compares with the critical flow depth h_c from Eqs. (4.25) and (4.26) according to

$$\frac{h_n}{h_c} = \left(\frac{f}{8S_0} \right)^{1/3} = \left(\frac{1}{\text{Fr}} \right)^{2/3}_{\text{at normal depth}}. \qquad (4.27)$$

Gradually-varied water-surface elevation profiles are commonly called back-water curves. Their analysis results from a direct application of the quasi-steady dynamic wave of the St. Venant equation. For a steady 1D flow, in wide-rectangular channels, relation 2.27 can be rewritten as

$$\frac{dE}{dx} = \frac{dE}{dh}\frac{dh}{dx} = S_0 - S_f = \frac{\partial h}{\partial x} + \frac{V}{g}\frac{\partial V}{\partial x} = \frac{dh}{dx}(1 - \text{Fr}^2). \qquad (4.28)$$

Note that, in this derivation, it is assumed that $V = q/h$ and $\text{Fr}^2 = q^2/gh^3$ and an equation for resistance to flow is not required because q is considered constant.

The relationship describing water-surface elevation for a steady 1D flow of an incompressible sediment-laden fluid is

$$\frac{dh}{dx} = \frac{S_0 - S_f}{1 - \text{Fr}^2}. \tag{4.29} \blacklozenge\blacklozenge$$

Using the properties of critical flow depth h_c from Eq. (4.25) and normal depth h_n from Eq. (4.5), in wide-rectangular channels, $R_h = h$, the governing equation for steady flow, with constant q and f, becomes

$$\frac{dh}{dx} = \frac{S_0 \left[1 - \left(\frac{h_n}{h} \right)^3 \right]}{\left[1 - \left(\frac{h_c}{h} \right)^3 \right]}. \tag{4.30}$$

Note that $dh/dx \to 0$ as the flow depth h approaches the normal depth h_n. Also, $dh/dx \to \infty$ near critical flow as $h \to h_c$.

The sign of dh/dx depends on the relative magnitude of h, h_n, and h_c. Five types of backwater profiles are possible:

1. *H* profiles for horizontal surfaces with $h_n \to \infty$,
2. *M* profiles for mild slopes when $h_n > h_c$ or $S_0 < f/8$,
3. *C* profiles for critical slopes when $h_n = h_c$ or $S_0 = f/8$,
4. *S* profiles for steep slopes when $h_n < h_c$, or $S_0 > f/8$,
5. *A* profiles or adverse slopes when $S_0 < 0$.

A subcritical normal depth $h_n > h_c$ corresponds to $f > 8 S_0$ and a supercritical normal depth $h_n < h_c$ is obtained when $f < 8 S_0$. Accordingly, a stream slope is said to be mild when $S_0 < f/8$ and steep when $S_0 > f/8$. The ratio of friction slope S_f to bed slope S_0 can be obtained from $S_f = (fq^2/8\, gh^3)$ and $S_0 = (fq^2/8\, gh_n^3)$; thus

$$\frac{S_f}{S_0} = \left(\frac{h_n}{h} \right)^3, \tag{4.31} \blacklozenge$$

and, finally, the ratio of applied bed shear stress τ_b to the bed shear stress applied at normal depth τ_{bn} for wide-rectangular channels with constant q and f is

$$\frac{\tau_b}{\tau_{bn}} \cong \frac{\gamma h S_f}{\gamma h_n S_0} = \frac{h}{h_n} \left(\frac{h_n}{h} \right)^3 = \left(\frac{h_n}{h} \right)^2. \tag{4.32} \blacklozenge$$

Typical water-surface profiles in open channels are shown in Fig. 4.15.

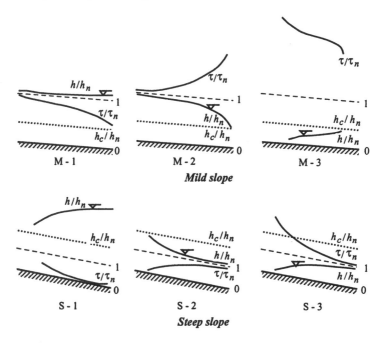

4.15 Backwater profiles for mild and steep slopes.

Numerical calculations can be initiated from a given flow depth h_1, and the distance increment Δx at which $h_2 = h_1 \pm \Delta h$ is approximated by

$$\Delta x \cong \frac{\Delta h \left[1 - \left(\frac{h_c}{h_1} \right)^3 \right]}{S_0 \left[1 - \left(\frac{h_n}{h_1} \right)^3 \right]} \qquad (4.33a)$$

or

$$\Delta h \cong \frac{S_0 \Delta x \left[1 - \left(\frac{h_n}{h} \right)^3 \right]}{\left[1 - \left(\frac{h_c}{h} \right)^3 \right]}. \qquad (4.33b)$$

Note that it is also possible to iterate on Δh until a predetermined value of Δx is obtained. The friction slope S_f in gradually varied flows with constant q and f can be approximated by Eq. (4.31), which shows that $S_f < S_0$ when the flow depth exceeds the normal depth and increases very rapidly when $h < h_n$.

This shows that the bed shear stress increases ($\tau > \tau_n$) at flow depths less than normal depth ($h < h_n$). Bed shear-stress distributions for 1D mild M and steep S backwater curves are sketched in Figs. 4.15 and 4.16. Shear stress

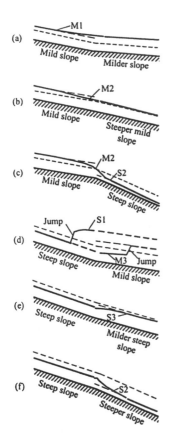

4.16 Typical backwater profiles.

increases in the downstream direction for converging flows (M-2 and S-2 backwater curves), and decreases for diverging flows (M-1, M-3, S-1, and S-3 backwater curves). Consequently, bed sediment transport is expected to increase in the downstream direction for converging flows and decrease for diverging flows.

Example 4.6 Application to gradually varied flow. Consider a 10-m-wide canal discharging 10 m³/s at a slope of 26 cm/km along the reach sketched in Fig. E.4.6.1. Given the sluice-gate opening at 0.2 m, a sill height at 0.17 m, and free downstream overfall conditions, draw the water-surface profile, also called the hydraulic grade line (HGL), and the energy grade line (EGL). The flow depth at C is 0.2 m and the bed surface is smooth ($f \cong 0.01$). Identify the backwater curves and sketch the shear-stress distribution along the reach.

Given the specific-energy diagram for the same unit discharge $q = 1$ m²/s in Fig. 2.10, the alternate depth to $h = 0.2$ m is approximately 1.5 m upstream of the gate at B. Critical flow depth, $h_c = (q^2/g)^{1/3} = (1/9.81)^{1/3} = 0.467$ m, controls the flow depth on top of the sill at G and near the free overfall at J. The sill controls subcritical flow upstream of G and the reach H–I is sufficiently long to sustain steady-uniform flow between EF and HI at a normal depth $h_n = 0.788$ m, from Example 4.1. Rapidly varied flow is obtained between BC and FH, as is the hydraulic jump DE previously calculated in Example 4.3. Gradually-varied flow is observed among AB, CD, and IJ. The bed slope $S_0 = 26$ cm/km corresponds to a mild slope because $S_0 < f/8 = 125$ cm/km. In the reach AB, $h > h_n > h_c$ and the backwater type is classified as M-1. The flow is accelerating between IJ with $h_n > h > h_c$ to form an M-2 backwater curve, and the reach CD, with $h_n > h_c > h$, corresponds to an M-3 backwater curve.

The EGL is drawn with conservation of energy in rapidly-varied flow between BC and FH taken into consideration. The specific energy for steady-uniform

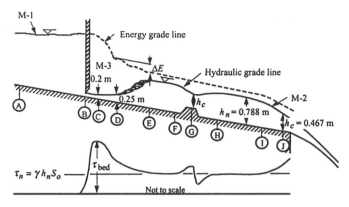

E.4.6.1 Sketch of HGL and EGL.

flow was previously calculated in Example 4.5 at 0.87 m, which remains constant between E and I. From Example 4.5, the energy lost in the hydraulic jump is $\Delta E = 0.197$ m and, from Eq. (4.47), the friction slope S_f is much greater than S_0 when $h < h_n$, as observed between CD and IJ. Finally, the velocity head is very small in the reservoir and the EGL practically corresponds to the free surface.

Bed shear-stress distribution is also sketched along the reach as calculated from Eq. 4.32, shear stress increases largely when $h < h_n$ and should be largest downstream of the sluice gate in the M-3 backwater curve. Energy losses in the hydraulic jump are caused by near-surface turbulence. Bed shear stress should be highly variable but larger than that for steady-uniform flow. Around the sill, bed shear stress increases in converging flow between F and G and decreases in diverging flow between G and H. Bed shear stress increases again near the free overfall as $h < h_n$.

4.3 Sediment transport in rivers

This section presents a brief summary of sediment transport in rivers. Equilibrium conditions are covered in Subsection 4.3.1 and aggradation–degradation characteristics are discussed in Subsection 4.3.2. More specific details are available in Julien (1995).

4.3.1 Equilibrium sediment transport

The dimensionless particle diameter d_* is defined from the specific gravity G of sediment, the kinematic viscosity of the fluid ν, and the gravitational acceleration g as

$$d_* = d_s \left[\frac{(G-1)g}{\nu^2} \right]^{1/3}. \tag{4.34}$$

The settling velocity ω of a sediment particle in still water is defined as

$$\omega = \frac{8\nu}{d_s} \left[\left(1 + \frac{d_*^3}{72} \right)^{0.5} - 1 \right]. \tag{4.35} \blacklozenge$$

The ratio of shear force to bed particle weight defines the Shields parameter τ_* as

$$\tau_* = \frac{\tau_0}{(\gamma_s - \gamma) d_s} = \frac{u_*^2}{(G-1)g\, d_s}, \tag{4.36} \blacklozenge\blacklozenge$$

where τ_0 is the bed boundary shear stress, u_* is the shear velocity, γ_s is the specific weight of a sediment particle, γ_m is the specific weight of water, d_s is the particle size, and g is the gravitational acceleration.

The critical value of the Shields parameter τ_{*c} corresponding to the beginning of motion ($\tau_0 = \tau_c$) depends on d_*, as shown in Fig. 4.17. Critical values of the Shields parameter τ_{*c} and shear stress τ_c for different particle sizes are listed in Table 2.4.

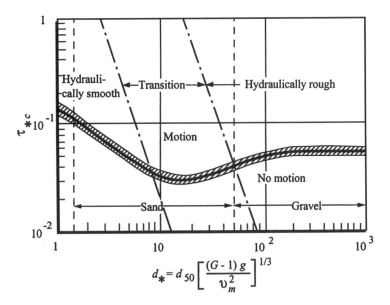

4.17 Particle-motion diagram (after Julien, 1995).

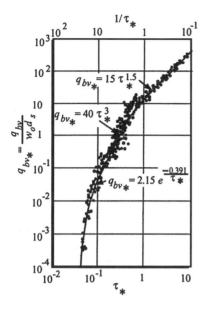

4.18 Sediment-transport diagram (after Julien, 1995).

The unit sediment discharge by volume can be defined from the Meyer–Peter and the Müller formulas as

$$q_{bv} = 8\,(\tau_* - \tau_{*c})^{1.5}$$
$$\times \big[(G-1)g\,d_s^3\big]^{0.5}. \qquad (4.37a)$$

An alternative formulation based on the Shields parameter and the settling velocity is shown in Fig. 4.18. Also, a very crude approximation for sands, where $0.1 < \tau^* < 1$, is

$$q_{bv} \approx 18\sqrt{g}\,d_s^{3/2}\tau_*^2. \qquad (4.37b)$$

Sediment transport can be subdivided into three zones that describe the dominant mode of transport: bedload, mixed load, and suspended load. It is interesting that, for turbulent flow over rough boundaries, incipient motion corresponds to $u_*/\omega \cong 0.2$. Figure 4.19 shows the ratio of suspended to total load as a function of u_*/ω and h/d_s. In most rivers, bedload is dominant at values of u_*/ω less than $\sim$0.4. A transition zone, called a mixed load, is found

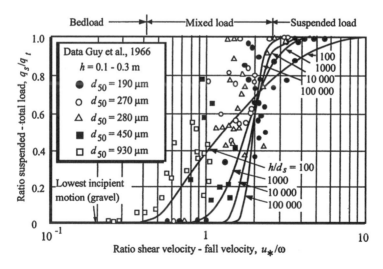

4.19 Ratio of suspended to total sediment load (after Julien, 1995).

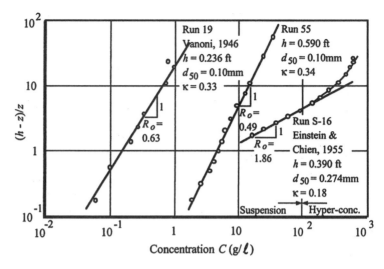

4.20 Sediment concentration profiles (after Woo et al., 1988).

where $0.4 < u_*/\omega < 2.5$ in which both the bedload and the suspended load contribute to the total load.

The sediment concentration C at an elevation z above the bed for the suspended load can be calculated from the Rouse equation as

$$C = C_a \left[\left(\frac{h-z}{z} \right) \left(\frac{a}{h-a} \right) \right]^{\frac{\omega}{\kappa u_*}}, \tag{4.38}$$

where C_a is the concentration at an elevation a above the bed, h is the flow depth, and κ is the von Kármán constant ($\kappa \cong 0.4$). An example of a concentration profile is shown in Fig. 4.20.

4.3.2 Riverbed aggradation and degradation

Owing to continuity of sediment, part of the total load deposits on the channel bed as the sediment-transport capacity decreases in the downstream direction. The sediment continuity relationship for advective fluxes is

$$\frac{\partial C_v}{\partial t} + \frac{\partial q_{tx}}{\partial x} + \frac{\partial q_{ty}}{\partial y} + \frac{\partial q_{tz}}{\partial z} = 0, \tag{4.39} \blacklozenge$$

where the mass fluxes q_{tx}, q_{ty}, and q_{tz} account for the total unit sediment discharge by volume in the x, y, and z directions, respectively.

Assuming a steady supply of sediment ($\partial C_v / \partial t = 0$), Eq. (4.39) for a wide channel without lateral sediment inflow ($\partial q_{ty} / \partial y = 0$) reduces to

$$\frac{\partial q_{tx}}{\partial x} + \frac{\partial q_{tz}}{\partial z} = 0, \tag{4.40}$$

which reduces further after it is assumed that the advective fluxes $q_{tx} = v_x C_v$ and $q_{tz} = -\omega C_v$ are dominant:

$$\frac{\partial v_x C_v}{\partial x} - \frac{\partial \omega C_v}{\partial z} = 0. \tag{4.41}$$

A practical approximation is obtained for gradually varied flow ($\partial v_x / \partial x \to 0$), constant fall velocity ω, and $\partial C_v / \partial z \cong -C_v / h$; thus

$$\frac{v_x \partial C_v}{\partial x} + \frac{\omega C_v}{h} = 0. \tag{4.42}$$

The solution for grain sizes of a given fraction i (constant fall velocity) at a constant unit discharge $q = Vh$, given $v_x = V$, is a function of the upstream sediment concentration C_{0i} of fraction i at $x = 0$:

$$C_i = C_{0i} e^{-\frac{X \omega_i}{hV}}. \tag{4.43}$$

This shows that the concentration left in suspension is negligible ($C_i / C_{0i} = 0.01$) at a distance X_{C_i}:

$$X_{C_i} = 4.6 \frac{hV}{\omega_i}. \tag{4.44}$$

This relationship is very useful in the design of settling basins. For the analysis of reservoir sedimentation, the percentage of sediment fraction i that settles within a given distance X defines the trap efficiency T_{E_i} as:

$$T_{E_i} = \frac{C_{0i} - C_i}{C_{0i}} = 1 - e^{-\frac{X \omega_i}{hV}} = 1 - e^{\frac{-W X \omega_i}{Q}}. \tag{4.45} \blacklozenge$$

It is interesting to note that the trap efficiency for particles of given settling velocity ω_i, basin length x, and discharge $Q = WhV$ increases with basin width W. We thus conclude that, for a given discharge and sediment discharge Q_{tx}, increasing the channel width induces aggradation. When the trap efficiency of silt and clay particles is calculated, careful consideration must also be given to density currents and possible flocculation, in which case the flocculated settling velocity must be used instead of ω_i:

$$T_{Ei} \frac{\partial Q_{txi}}{\partial x} + (1 - p_0) \frac{W \partial z_i}{\partial t} = 0 \tag{4.46a}$$

or

$$\frac{\partial z_i}{\partial t} = -\frac{T_{Ei}}{(1 - p_0)} \frac{\partial Q_{txi}}{W \partial x}.$$ (4.46b)

Values of porosity p_0 depend of the specific weight of sediment deposits. Dry specific weight of sediment deposits varies with the proportion of sand, silt, and clay and changes with time. The conversion of the incoming weight of sediment to volume necessitates knowledge of the average dry specific weight of a mixture γ_{md}, defined in Chap. 2 as the dry weight of sediment per unit total volume including voids. For material coarser than 0.1 mm, the dry specific weight of the mixture remains practically constant around $\gamma_{md} = 14.75$ kN/m^3 or 93 lb/ft^3. As a rough first approximation, 1 ton of sediment corresponds to 20 ft^3 of bed material. The corresponding dry mass density of the mixture, ρ_{md}, is given by $\rho_{md} = \gamma_{md}/g = 1500$ kg/m^3 or 2.9 slug/ft^3. The porosity p_0 of sand material is then obtained from $p_0 = 1 - \gamma_{md}/\gamma_s = 0.43$. The volumetric sediment concentration is $C_v = 1 - p_0$ and the void ratio is $e = p_0/(1 - p_0)$.

For distances separating successive cross sections ΔX larger than X_C, the trap efficiency is essentially unity and aggradation responds directly to changes in the sediment-transport capacity of the stream. For ΔX smaller than X_C, only part of the sediment load in suspension will settle within the given reach. The sediment load at the downstream end will then exceed the sediment-transport capacity of the stream. The reader is referred to Julien (1995) for a detailed analysis of erosion and sedimentation.

Exercise 4.1

Demonstrate the conjugate flow-depth relationship, Eq. (E.4.3.3), from the specific-momentum identity $M_1 = M_2$ for flow in a wide-rectangular channel.

Exercise 4.2

Carry out the algebraic transformations from Eq. (E.4.5.1) to demonstrate that the specific-energy loss in a hydraulic jump is

$$\Delta E = \frac{(h_2 - h_1)^3}{4h_1 h_2}.$$

Exercise 4.3

Apply the law of conservation of volume to an incompressible fluid flowing through a 1D control volume of discharge $Q = \int_0^A v_x dA$ given the cross-sectional area A and top width W. Consider the lateral inflow of unit discharge

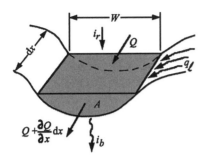

Ex.4.3.1 Conservation of mass.

q_l, rainfall intensity i_r, and infiltration rate i_b leaving through the wetted perimeter P (see Fig. Ex.4.3.1).

[*Hint*: calculate the volume fluxes entering the control volume $\forall = A dx$ and compare to the rate of change in control volume $(\partial \forall / \partial t = (\partial A / \partial t) dx.]$

♦Exercise 4.4

Demonstrate the 1D formulation of the equation of motion (4.15). Consider the momentum flux $F = \beta_m \rho A V_x^2 = \rho \int_0^A v_x^2 dA$ passing through a cross-sectional area A and top width W. Neglect momentum contributions from rainfall, lateral inflow, and infiltration. The shear stress τ_0 is applied over the wetted perimeter P and the bed slope is $S_0 = \tan \theta$ (see Fig. Ex.4.4.1).

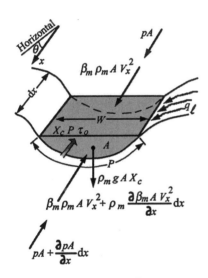

Ex.4.4.1 Conservation of momentum.

♦Problem 4.1

The cross-sectional and depth-averaged flow-velocity profiles are given in digital form with the orientation reference along the cross section. Complete Table P.4.1.1 to calculate the cross-section area $A = \sum_i a_i$ and the flow discharge $Q = \sum_i Q_i = \sum a_i v_i \sin \theta$. Note that a large eddy formed near the right bank and reverse flow is measured. Draw the cross section profile and the velocity profile as per Fig. 4.5(b).

♦Problem 4.2

Consider the cross section of a 10-m wide-rectangular minor channel with a bankfull depth of 2 m. Given a floodplain width that extends 50 m on each side of the minor channel bounded with a near-vertical escarpment (see Fig. P.4.2.1). Calculate the following parameters as functions of flow depth h up to 5 m: (1) top channel width W, (2) wetter perimeter P, (3) cross-sectional area

Table P.4.1.1. *Cross-section data*

Distance from left bank (ft)	Flow depth (ft)	Depth-averaged flow velocity (ft/s)	Deviation angle (°)	a_i (ft²)	Q_i (ft³/s)
0	0	0	—	0	
10	1.3	+0.6	260		−7.68
20	1.7	+0.2	280	17	
30	1.8	0.4	110		6.76
40	2.1	1.1	90	31.5	
60	2.4	1.4	89		67.19
80	2.9	2.1	86	58	
100	3.6	2.6	85		186.48
120	4.1	1.9	88	82	
140	2.2	0.8	90		35.2
160	0	0	—	0	

Answer: $A = 383.5$ ft², $Q = 596.5$ ft³/s.

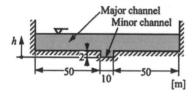

P.4.2.1 Cross section.

A, (4) near flow depth $\bar{h}$, and (5) hydraulic radius R_h. Discuss the effect of the width discontinuity at $h = 2$ m.

♦Problem 4.3

The cross section in Fig. P.4.3.1 is located 1.25 km upstream of the control section on the Matamek River. Given the water-surface slope [Fig. 4.8(d)] and the stage–discharge relationship

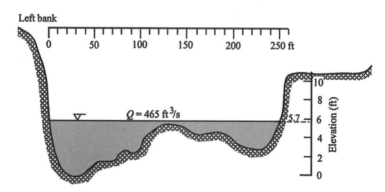

P.4.3.1 Cross section.

[Fig. 4.6(a)], determine the following:

(a) A pool is formed because of downstream bedrock control. Draw the water surface elevation when $Q = 0$.

(b) At the given discharge, $Q = 465$ ft^3/s, calculate surface width, wetted perimeter, cross-section area, mean flow depth, mean flow velocity, and hydraulic radius. Plot these points on Figs. 4.8.

(c) Estimate the bankfull discharge at that cross section, assuming a constant Manning coefficient n at all stages.

♦♦Problem 4.4

Consider a steady-uniform flow discharge of 15 m^3/s in a 70-m-wide 2–4-mm gravel-bed stream. Given the bed slope of 5 cm/km, determine the following:

(a) the critical flow depth;
(b) the resistance coefficients f, n, and C;
(c) the normal flow depth and compare it with the hydraulic radius;
(d) the Froude number at normal flow depth;
(e) the type of slope for backwater profile calculations;
(f) the applied bed shear stress at normal flow depth.

♦Problem 4.5

Draw the position of the free surface, the HGL, and the EGL in a 10-m wide-rectangular channel discharging 10 m^3/s in the smooth canal at a bed slope of 26 cm/km (see Fig. P.4.5.1).

Answer: Combine the information from Examples 4.1 to 4.6. Energy is conserved between A–B and D–E. Friction losses occur between C and D, to

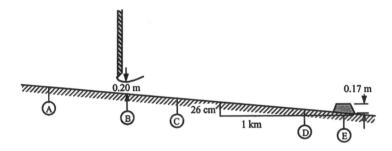

P.4.5.1 Definition sketch.

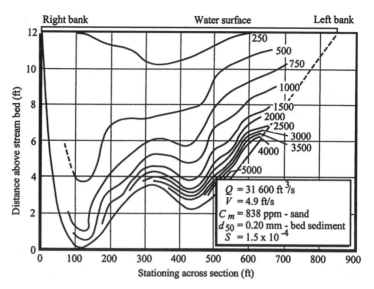

P.4.6.1 Concentration profile of the Missouri River (after Bondurant, 1963).

which turbulence losses are added between B and C. Momentum is conserved between B and D, and friction forces can be neglected between B and C, but are dominant between C and D. Added forces are provided by the sill and the sluice gate, which can be calculated as per Example 4.3.

♦♦Problem 4.6

From the cross section in Fig. P.4.6.1, do the following:

(a) Plot the surface width vs. stage and cross-sectional areas vs. stage.
(b) Estimate the Manning coefficient n of the river if the slope is 1.5×10^{-4}.

♦♦Problem 4.7

Use the values of $\hat{a}$ and $\hat{b}$ from Eqs. (4.10) and use the transforms for Eqs. (4.11) to define a and m for a power relationship applicable when $h \cong 5\,d_{50}$. Once a and m are defined, plot the straight-line resistance relationship on Fig. 4.11 and compare with the field measurements. Determine what the range of applicability is of this equation.

   Answer: From $\hat{a} = 2.5$, $\hat{b} = 2$, and $h/d_{50} = 5$, we obtain $m = 0.434$ and $a = 2.86$. Plotting $\sqrt{(8/f)} = 2.86\,(h/d_{50})^{0.434}$ in Fig. 4.11 is in very good agreement with the field measurements when $h/d_{50} < 50$.

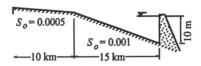

CP.4.1.1 Definition sketch.

♦♦Computer Problem 4.1
Backwater behind a reservoir
Consider steady flow ($q = 3.72$ m²/s) in the impervious rigid boundary channel sketched in Fig. CP.4.1.1. Assume that the channel width remains large and constant regardless of flow depth and that $f = 0.03$. Determine the distribution of the following parameters along the 25-km reach of the channel when the water-surface elevation at the dam is 10 m above the bed elevation: (a) flow depth in meters, (b) mean flow velocity in meters per second, and (c) bed shear stress in newtons per square meter.

5

Unsteady flow in rivers

This chapter is concerned with unsteady flow in rivers. The governing equations describing unsteady flow in open channels include continuity in Section 5.1 and momentum in Section 5.2 for 1D flow. Governing equations are followed by the concept of floodwave propagation (Section 5.3), loop-rating effects (Section 5.4), flood routing (Section 5.5), and flow and sediment-duration curves in Section 5.6.

5.1   River continuity equation

The continuity equations simply express conservation of mass. In most rivers, it is assumed that the mass density ρ is constant, and the continuity equations imply conservation of volume. A general formulation is presented in Sub-section 5.1.1, followed by the commonly used two-dimensional (2D) form in Subsection 5.1.2. A simplified 1D form is then given in Subsection 5.1.3.

5.1.1   General continuity formulation

The three-dimensional (3D) form of the continuity equation is simply the integral over a control volume $\forall$ of the differential form [Eq. 2.16a]:

$$\int_\forall \frac{\partial \rho}{\partial t} \, d\forall + \int_\forall \left(\frac{\partial \rho v_x}{\partial x} + \frac{\partial \rho v_y}{\partial y} + \frac{\partial \rho v_z}{\partial z} \right) \, d\forall = 0. \qquad (5.1)$$

This volume integral of partial derivatives can be transformed into surface integrals owing to the divergence theorem, e.g., application to a vector F,

$$\int_\forall \frac{\partial F}{\partial x} \, d\forall = \int_A F \frac{\partial x}{\partial n} \, dA, \qquad (5.2)$$

in which $\partial x / \partial n$ is the cosine of the angle between the coordinate x and the normal vector $\mathbf{n}$ pointing outside the control volume. The integral form of the continuity equation simply states that the difference between inflow and outflow results in volumetric storage, as shown in Example 5.1.

Example 5.1 Application of the continuity equation. Consider the fairly straight river reach of length X_c, constant width W, and flow depth h, as sketched in Fig. E.5.1.1.

The first integral of (5.1) is zero in this case because the fluid is incompressible and the control volume $V = WhX_c$ is constant. The divergence theorem (Eq. 5.2) is applied to the second term:

$$\int_A \left(\rho v_x \frac{\partial x}{\partial n} + \rho v_y \frac{\partial y}{\partial n} + \rho v_z \frac{\partial z}{\partial n} \right) dA = 0. \qquad (\text{E.5.1.1})$$

The values of $\partial x/\partial n$, $\partial y/\partial n$, and $\partial z/\partial n$ are the cosines of the angle between the vector normal to the surface **n** pointing outside the control volume and the Cartesian coordinates x, y, and z, respectively. Figure E.5.1.1 illustrates the direction cosines on the downstream face. In this example, the direction cosine $\partial x/\partial n$ vanishes on all faces except the upstream ($\partial x/\partial n = -1$) and downstream ($\partial x/\partial n = +1$) cross sections. Similarly, the direction cosine $\partial y/\partial n$ vanishes except on the left bank ($\partial y/\partial n = +1$) and right river bank ($\partial y/\partial n = -1$). The direction cosine $\partial z/\partial n$ vanishes except at the free surface ($\partial z/\partial n = +1$) and the riverbed ($\partial z/\partial n = -1$). Moreover, for an incompressible homogeneous suspension $\rho_{m1} = \rho_{m2} = \rho_m$ with the vertical velocity, $v_z = dh/dt$ at the free surface, the infiltration rate through the banks $v_y = i$, and $v_z = -i$ at the bed. The surface integral of the first term in Eq. (E.5.1.1) yields the downstream flux $A_2 V_2 \partial x/\partial n$ (note that $\partial x/\partial n = +1$) and the upstream flux $-A_1 V_1$ (note that V_1 is positive in the x direction and $\partial x/\partial n = -1$ upstream).

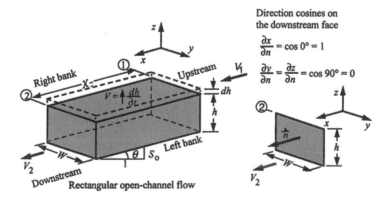

Direction cosines on the downstream face

$$\frac{\partial x}{\partial n} = \cos 0° = 1$$

$$\frac{\partial y}{\partial n} = \frac{\partial z}{\partial n} = \cos 90° = 0$$

Rectangular open-channel flow

Figure E.5.1.1. Definition sketch.

The second term gives the infiltration through the channel banks $2ihX_c$ (note that on the right bank the flux is in the negative y direction and $\partial y/\partial n = -1$). The third term gives the flux through the of free-surface area $WX_c \, dh/dt$ and the infiltration through the bed iWX_c. The sum of all components results in

$$\underbrace{A_2V_2 \; - \; A_1V_1}_{\text{outflow}} \; \underbrace{+ \; 2ihX_c + iWX_c}_{\text{infiltration}} \; \underbrace{+ \; WX_c\frac{\partial h}{\partial t}}_{\text{storage}} = 0. \qquad \text{(E.5.1.2)}$$

In general, continuity simply formulates that the storage is the inflow minus the outflow. In the case of an impervious channel, $i = 0$, the integral form of the continuity equation reduces to

$$\underbrace{A_2V_2 \; - \; A_1V_1}_{\substack{\text{outflow} \quad \text{inflow}}} \; + \; \underbrace{WX_c\frac{\partial h}{\partial t}}_{\text{storage}} = 0. \qquad \text{(E.5.1.3)}$$

This form is equivalent to 1D equation (5.8) written as

$$\frac{A_2V_2 \; - \; A_1V_1}{X_c} = \frac{-W\partial h}{\partial t}. \qquad \text{(E.5.1.4)}$$

5.1.2   *Two-dimensional continuity for rivers*

Depth-integrated formulations of the governing equations are useful in describing most river systems with typical variability in planform geometry such as changes in lakes, reservoirs, and estuaries. The system of coordinates typically sets the x axis in the main downstream direction at a small slope $\bar{S}_{0x} = \tan\theta$ from the horizontal, a horizontal y axis toward the left bank, and the upward z axis deviating at a slope $\bar{S}_{0x}$ from the vertical. The depth-integrated continuity relationship is obtained for homogeneous suspensions (constant ρ) from the integration of Eq. (2.16a) along the upward z axis.

The surface integration is carried out over a control volume of fixed grid dx and dy with bed elevation z_b and water-surface elevation z_w that vary in space, as sketched in Fig. 5.1.

Given the flow depth $h = z_w - z_b$, the depth-integrated velocity components V_x and V_y are

$$V_x = \frac{1}{h}\int_{z_b}^{z_w} v_x \, dz, \quad \text{and} \quad V_y = \frac{1}{h}\int_{z_b}^{z_w} v_y \, dz. \qquad (5.3)$$

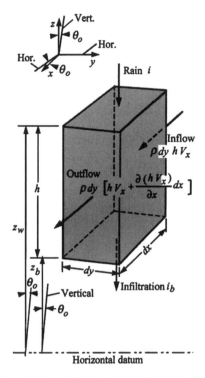

Figure 5.1. Depth-integrated control volume.

The net volume flux leaving the control volume in the x direction is

$$dy \left[hV_x + \frac{\partial (hV_x)}{\partial x} dx \right] - dy h V_x$$

$$= dx \, dy \frac{\partial (hV_x)}{\partial x}. \qquad (5.4)$$

Repeating the process in the y direction gives $dx \, dy \, \partial(hV_y)/\partial y$ and, considering the infiltration flux $i_b \, dx \, dy$ and the entering rainfall flux $i \, dx \, dy$, the net volume change within the control volume is

$$dx \, dy \frac{\partial h}{\partial t} + dx \, dy \frac{\partial hV_x}{\partial x} + dx \, dy \frac{\partial hV_y}{\partial y}$$

$$+ \, (i_b - i) \, dx \, dy = 0. \qquad (5.5)$$

The depth-integrated form of the continuity equation is obtained after Eq. (5.5) is divided by $dx \, dy$:

$$\frac{\partial h}{\partial t} + \frac{\partial hV_x}{\partial x} + \frac{\partial hV_y}{\partial y} + (i_b - i) = 0.$$

$$(5.6) \; \blacklozenge\blacklozenge$$

The rate of change in flow depth, dh/dt, relates to the net downstream flux $\partial(hV_x)/\partial x$, the net lateral flux $\partial(hV_y)/\partial y$, the vertical infiltration rate i_b defined as positive outward, and the vertical rainfall intensity i defined as positive inward. In this form, $h = z_w - z_b$, where both water-surface elevation z_w and bed elevation z_b can vary and V_x and V_y are depth-integrated velocities. All parameters, h, V_x, V_y, i_b, and i, can vary in space $x - y$ and time. Equation (5.6) is applicable to homogeneous suspensions (constant ρ_m) and to the orthogonal system of coordinates in which the z axis is near vertical (small angle θ_0).

5.1.3   One-dimensional continuity for rivers

The definition sketch in Fig. 5.2 describes a river reach with a top width W, cross-section area A, wetted perimeter P, hydraulic radius $R_h = A/P$, and mean flow depth $h = A/W$. The total discharge Q is given from the product of the mean

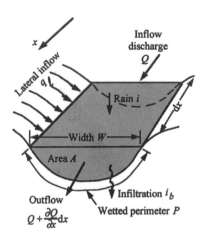

Figure 5.2. Continuity for a river reach.

flow velocity V and area A; the unit discharge of the lateral flow is q_l. The rainfall intensity is i, and the infiltration rate through the wetted perimeter is i_b.

The net volumetric flux leaving the control volume is $(\partial Q/\partial x)dx + i_b P dx$. The net volumetric flux entering the control volume is $q_l dx + i W dx$. The difference between entering and leaving volumetric fluxes corresponds to volumetric storage $\partial A dx = \partial(Wh)dx$ per unit time ∂t.

After dividing by dx, we easily demonstrate that

$$\frac{\partial A}{\partial t} + \frac{\partial Q}{\partial x} + i_b P - i W - q_l = 0, \tag{5.7}$$

where i_b is the rate of infiltration through the wetted perimeter P, i is the rainfall intensity through the reach-averaged river width W, A is the reach-averaged cross-sectional area, and q_l is the unit discharge of lateral inflow.

For an impervious channel ($i_b = 0$) without rainfall ($i = 0$) and without lateral inflow ($q_l = 0$), the 1D equation of continuity simply reduces to

$$\frac{\partial Q}{\partial x} + \frac{\partial A}{\partial t} = 0. \tag{5.8} \blacklozenge\blacklozenge$$

This simple differential equation that expresses conservation of mass is widely used in the analysis of floodwave propagation.

5.2   River momentum equations

In this section, the momentum equations are presented for 2D flows in Subsection 5.2.1, followed by a 1D formulation in Subsection 5.2.2.

5.2.1   *Two-dimensional momentum for rivers*

For most rivers, depth-integrated formulations of the equation of motion are sufficiently accurate as long as the acceleration in the near-vertical z direction can be neglected. The resulting depth-integrated formulation is 2D in x and y

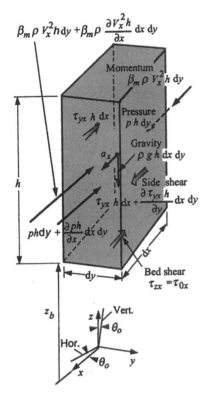

$$\beta_m \rho V_x^2 h dy + \beta_m \rho \frac{\partial V_x^2 h}{\partial x} dx\, dy$$

Momentum
$\beta_m \rho V_x^2 h\, dy$

$\tau_{yx} h\, dx$ | Pressure
$p\, h\, dy$

Gravity
$\rho g h\, dx\, dy$

a_x

Side shear
$\tau_{yx} h\, dx + \frac{\partial \tau_{yx} h}{\partial y} dx\, dy$

$\tau_{yx} h\, dx$

$p h dy + \frac{\partial p h}{\partial x} dx\, dy$

dy

Bed shear
$\tau_{zx} = \tau_{0x}$

z_b

Vert.
z θ_o

Hor.
θ_o y
x

Figure 5.3. Stresses on a depth-integrated volume.

along with a hydrostatic pressure distribution. For homogeneous suspensions (constant ρ_m), the integration over depth is applied to a control volume, as shown in Fig. 5.3.

It is assumed that the wind and the rainfall forces applied through the free surface are negligible. The bed shear stress $\tau_{zx} = \tau_{0x}$ and the pressure is hydrostatic such that the depth-averaged pressure p is given by $p = 0.5\, \rho g h$. The sum of forces in the x direction gives

$$\rho a_x h\, dx\, dy = \rho g h\, dx\, dy \sin\theta$$

$$-\frac{\rho g}{2} \frac{\partial h^2}{\partial x} dx\, dy$$

$$+\frac{\partial \tau_{yx} h}{\partial y} dx\, dy$$

$$-\tau_{0x} dx\, dy. \tag{5.9}$$

Dividing through by the mass $\rho h\, dx\, dy$ with small θ_0, such that $\sin\theta_0 \cong \tan\theta_0 = \bar{S}_{0x}$ (and $\sin\theta = \bar{S}_{0x} - \partial z_b/\partial x$), and using Eq. (2.14a), results in

$$a_x = \underbrace{\frac{\partial V_x}{\partial t}}_{|\text{local}|} + \underbrace{V_x \frac{\partial V_x}{\partial x} + V_y \frac{\partial V_x}{\partial y} + V_z \frac{\partial V_x}{\partial z}}_{\text{convective}}$$

$$= \underbrace{g\bar{S}_{0x}}_{|\text{gravity}|} - \underbrace{g\frac{\partial z_b}{\partial x}}_{|\text{bed elevation}|} - \underbrace{g\frac{\partial h}{\partial x}}_{|\text{pressure}|} - \underbrace{\frac{\tau_{0x}}{\rho h}}_{|\text{bed shear}|} + \underbrace{\frac{\partial \tau_{yx}}{\rho \partial y}}_{|\text{side shear}|}. \tag{5.10}$$

It is assumed that the momentum correction factors are close to unity, i.e., that the flow velocity is fairly uniform over the depth. It can be further assumed that V_z is very small and that the variability of riverbank shear stress is negligible compared with that of the bed shear stress. The following approximation for the depth-integrated equation of motion is

$$\frac{\partial V_x}{\partial t} + V_x \frac{\partial V_x}{\partial x} + V_y \frac{\partial V_x}{\partial y} = g \bar{S}_{0x} + g S_{0x} - g \frac{\partial h}{\partial x} - \frac{\tau_{0x}}{\rho h}. \qquad (5.11)$$

With a similar analysis applied in the y direction we obtain

$$\frac{\partial V_y}{\partial t} + V_x \frac{\partial V_y}{\partial x} + V_y \frac{\partial V_y}{\partial y} = g S_{0y} - g \frac{\partial h}{\partial y} - \frac{\tau_{0y}}{\rho h}, \qquad (5.12)$$

where $\bar{S}_{0x}$ is the average downstream bed slope and the average lateral slope is $\bar{S}_{0y} = 0$. The local bed slopes are $S_{0x} = -(\partial z_b/\partial x)$ and $S_{0y} = -(\partial z_b/\partial y)$, respectively.

These simplified equations are applicable to 2D raster-based models of rivers with large width–depth ratios. The main difficulty lies in the evaluation of the bed and the side shear-stress components.

For flow in bends, the relative magnitude of radial acceleration terms in cylindrical coordinates indicates that the centrifugal acceleration is counterbalanced by pressure gradient and radial shear stress, as suggested by Rozovskii (1957):

$$\frac{V_\theta^2}{r} = g S_{wr} - \frac{1}{\rho} \frac{\partial \tau_r}{\partial z}, \qquad (5.13)$$

where the local downstream velocity V_θ, the radial shear stress τ_r, and the radial water-surface slope S_{wr} vary with the vertical elevation z and/or the radius of curvature r.

5.2.2   One-dimensional momentum for rivers

When the channel width remains fairly constant, the lateral velocity component V_y from Eq. (5.12) can be neglected. The velocity reduces to $V = V_x$, the bed slope to $S = S_{0x}$, and the bed shear stress to $\tau_0 = \tau_{0x}$:

$$\frac{\partial V_x}{\partial t} + \frac{V_x \partial V_x}{\partial x} = g \bar{S}_{0x} + g S_{0x} - g \frac{\partial h}{\partial x} - \frac{\tau_{0x}}{\rho_m h}, \qquad (5.14)$$

where $\bar{S}_{0x}$ is the reach-averaged positive slope and $S_{0x} = -\partial z_b/\partial x$. It is left to the reader in Exercise 5.1 to demonstrate with the continuity equation [Eq. (5.8)] that the following formulation reduces to Eq. (5.14) when A is constant and $\beta_m = 1$:

$$\frac{\partial Q}{\partial t} + \frac{\partial \beta_m Q V}{\partial x} = g A S_0 - \frac{A \partial h}{\partial x} - \frac{\tau_0 P}{\rho}. \qquad (5.15)$$

Assume a hydrostatic pressure distribution, $p = \rho g h$, and assume that the bed shear stress τ_0 in wide-rectangular channels ($R_h \cong h$ and $P \simeq W$) approximately equals $\tau_0 = \rho g h S_f$ and that g_x is a function of bed slope $g_x = g S_0$, where $S_0 = \bar{S}_{0x} + S_{0x}$. The 1D equation of motion [Eq. (5.14)] with a single bed slope S_0 simply reduces to St. Venant equation (2.27) following dimensionless form after being rearranged and divided throughout by gravitational acceleration g and A:

$$S_f \cong \underset{(1)}{S_0} - \underset{(2)}{} \underset{(3)}{\frac{\partial h}{\partial x}} - \underset{(4)}{\frac{V \partial V}{g \partial x}} - \underset{(5)}{\frac{1}{g} \frac{\partial V}{\partial t}}. \qquad (5.16)$$

| kinematic |

| diffusive |

| quasi-steady |

| dynamic |

The terms of the dimensionless equation of motion are sketched in Fig. 5.4.

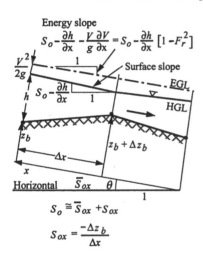

Figure 5.4. Slope definition sketch.

They describe (1) the friction slope or the slope of the EGL, (2) the bed slope of the channel, (3) the pressure gradient or downstream change in flow depth that is due to backwater, (4) the velocity head gradient or downstream change in velocity head that is due to backwater and/or changes in channel width, and (5) the local-acceleration term for unsteady flow.

The full use of Eq. (5.16) is referred to as the dynamic-wave approximation of the St. Venant equation. In many applications, however, the last term (5) is very small and can be neglected. For all steady flows, as well as unsteady flows in which (5) is negligible, the quasi-steady dynamic-wave approximation includes the first four terms of Eq. (5.16); this corresponds to the backwater equations for gradually varied flows used in Chap. 4. In most rivers, the flow is subcritical, Fr < 1, and the velocity head gradient (4) can be neglected before the pressure gradient (3). The diffusive-wave approximation, $S_f = S_0 - \partial h/\partial x$, is commonly used in river mechanics. Finally, the kinematic-wave approximation is obtained in which all but the first two terms vanish. This is the case when $\partial h/\partial x \ll S_0$. Example 5.2 illustrates how the continuity and the momentum equations can be combined to determine the celerity of surface waves.

Example 5.2   Application to surface perturbations. Consider a solitary wave traveling in a frictionless channel without change of shape or velocity. For instance, a solitary wave can be produced by a sudden horizontal displacement of a vertical gate in a laboratory flume.

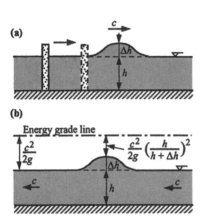

Figure E.5.2.1. Propagation of a small perturbation.

As sketched in Fig. E.5.2.1(a), the solitary wave travels to the right with celerity c in a stationary fluid. An observer moving along the wave crest at a velocity equal to the celerity will perceive steady flow [Fig. E.5.2.1(b)] in which the wave appears to stand still while the flow moves at a velocity equal to c in magnitude.

When continuity relationship (5.8) is applied to the steady-flow case, constant discharge implies that the relative velocity under the crest is $ch/(h + \Delta h)$. It is interesting to note that in relative motion [Fig. E.5.2.1(b)], the relative velocity decreases with flow depth. The equation of motion is then applied to the steady relative motion. When friction is neglected and a small slope is assumed, the energy equation between the normal section of the flow and the section at the wave crest simply describes conservation of specific energy, or

$$h + \frac{c^2}{2g} = h + \Delta h + \frac{c^2}{2g} \left(\frac{h}{h + \Delta h} \right)^2 . \tag{E.5.2.1}$$

Solving for c, we obtain

$$c = \sqrt{\frac{2g\,(h + \Delta h)^2}{2h + \Delta h}} . \tag{E.5.2.2}$$

This is commonly known as the Lagrangian celerity equation for 1D propagation of small waves in still water. For waves of moderate amplitude, the wave celerity increases with wave height.

5.3   River floodwaves

This section treats the subject of floodwave propagation in 1D channels. Unsteady flow describes changes in flow discharge Q with time t, and downstream distance x. Therefore the exact differential is

$$dQ = \frac{\partial Q dx}{\partial x} + \frac{\partial Q dt}{\partial t}. \tag{5.17}$$

By definition, the celerity $c = dx/dt$ defines the location where the flow is steady, i.e., $dQ = 0$. The celerity c at which space–time changes take place is simply given by the solution of Eq. (5.17) for dx/dt, or

$$c \equiv \frac{dx}{dt} \equiv \frac{-\frac{\partial Q}{\partial t}}{\frac{\partial Q}{\partial x}}. \tag{5.18}$$

When considering conservation of mass in a one-dimensional impervious channel, we obtain from Eq. (5.8) that $(\partial Q/\partial x) = (-\partial A/\partial t)$. Substituting this into Eq. (5.18) gives

$$c = \frac{\partial Q}{\partial A}. \tag{5.19} \blacklozenge\blacklozenge$$

This relationship for floodwave celerity is referred to as the Kleitz–Seddon law. It is interesting to point out that it is essentially the result of the conservation of mass with the restrictions pertaining to Eq. (5.8).

For a wide-rectangular channel, the unit discharge q varies with depth h according to resistance relationships, such as $q = Vh = \alpha h^\beta$. The equation of wave propagation and the celerity relationship in flowing water then reduce to

$$\frac{\partial h}{\partial t} + c\frac{\partial h}{\partial x} = 0, \tag{5.20}$$

$$c = \frac{\partial q}{\partial h} = \beta\alpha h^{\beta-1} = \beta V. \tag{5.21} \blacklozenge$$

The floodwave celerity c is always faster than the flow velocity when $\beta > 1$. For instance, $\beta = 3$ for laminar flow and $\beta = 1.5$ for turbulent flow with constant Darcy–Weisbach factor f (or Chézy coefficient C); see Table 3.3. It is of foremost importance to understand that the celerity of floodwaves increases with flow depth. It implies that larger floodwaves (larger flow depth) propagate faster than small floodwaves. This causes nonlinearity in the downstream propagation of floodwaves, and linear techniques based on superposition, such as the unit hydrograph, fail to adequately simulate floodwave propagation in channels. It also indicates that the method of isochrons used in hydrology is not applicable to both small and large floodwaves.

The following treatment of equation of motion (5.16) considers the resistance and the continuity relationships. The resistance relationship $V = \alpha h^{\beta-1}$ yields the following derivatives $\partial V = (\beta-1)(V/h)\partial h$. The reader can demonstrate in Exercise 5.2 that equation of motion (5.16) can be rewritten as

$$S_f = S_0 - [1 + (\beta - 1)\mathrm{Fr}^2]\frac{\partial h}{\partial x} - \frac{1}{g}\frac{\partial V}{\partial t}. \tag{5.22}$$

Note that for steady flow this equation is slightly different from Eq. (4.28) because resistance to flow is considered and q is not constant.

A complete list of conversion factors considering $q = Vh = \alpha h^\beta$, $\partial h/\partial t = -\partial q/\partial x$, and $c = \beta V$ is presented in Table 5.1. Continuity relationship (5.8) then yields $\partial h/\partial t = -\partial q/\partial x = -V\partial h/\partial x - h\partial V/\partial x$, which can be combined with $\partial V = [(\beta - 1)V/h]\partial h$ to reduce Eq. (5.22) to

$$S_f = S_0 - \underbrace{[1 - (\beta - 1)^2\,\mathrm{Fr}^2]}_{\text{floodwave diffusivity}}\frac{\partial h}{\partial x}, \tag{5.23a} \blacklozenge$$

and, for Manning's equation,

$$S_f = S_0 - \left(1 - \frac{4}{9}\frac{Q^{0.2}\,S^{0.9}}{gn^{1.8}\,W^{0.2}}\right)\frac{\partial h}{\partial x}. \tag{5.23b}$$

The floodwave-diffusivity term depends on the value of β and Fr. For instance, Manning's equation is applicable in most rivers ($\beta = 5/3$), and, for Fr < 1.5, the floodwave-diffusivity term is positive. Conversely, the term is negative when Fr > 1.5. In rivers with low Froude number, the floodwave diffusivity becomes close to unity.

It is also quite important to examine the ratio of the second to the first term of the St. Venant equation. From Eq. (5.22), the ratio $(-\partial h/S_0\partial x)$ is a measure of the floodwave attenuation at low values of the Froude number. From Table 5.1, this diffusivity ratio can be written (see Exercise 5.5) as

$$\frac{-1}{S_0}\frac{\partial h}{\partial x} = \frac{1}{S_0 c^2}\frac{\partial q}{\partial t} = \frac{1}{\beta^2 S_0}\frac{W^{1-2/\beta}}{\alpha^{2/\beta}Q^{2-2/\beta}}\frac{\partial Q}{\partial t}. \tag{5.24}$$

In the case of the Manning equation, $\alpha = (S_0^{1/2}/n)$, $\beta = 5/3$, the diffusivity ratio reduces to

$$\frac{-1}{S_0}\frac{\partial h}{\partial x} = \frac{9}{25}\frac{n^{1.2}}{W^{0.2}S_0^{1.6}Q^{0.8}}\frac{\partial Q}{\partial t}. \tag{5.25} \blacklozenge$$

The reader will note that rapid changes in discharge (large $\partial Q/\partial t$) increase floodwave diffusivity. For a given floodwave, a given Q, and $\partial Q/\partial t$, diffusivity increases with Manning coefficient n and decreases with channel slope. It is

Table 5.1. *Conversion factors for floodwave propagation*

	$\dfrac{\partial h}{\partial x}$	$\dfrac{\partial h}{\partial t}$	$\dfrac{\partial V}{\partial x}$	$\dfrac{\partial V}{\partial t}$	$\dfrac{\partial q}{\partial x}$	$\dfrac{\partial q}{\partial t}$
$\dfrac{\partial h}{\partial x} = \dfrac{\partial h}{\partial x}$		$\dfrac{-1}{\beta V}\dfrac{\partial h}{\partial t}$	$\dfrac{h}{(\beta-1)V}\dfrac{\partial V}{\partial x}$	$\dfrac{-h}{\beta(\beta-1)V^2}\dfrac{\partial V}{\partial t}$	$\dfrac{h}{\beta q}\dfrac{\partial q}{\partial x}$	$\dfrac{-h^2}{\beta^2 q^2}\dfrac{\partial q}{\partial t}$
$\dfrac{\partial h}{\partial t} = -\beta V\dfrac{\partial h}{\partial x}$	$-\beta V\dfrac{\partial h}{\partial x}$	$\dfrac{\partial h}{\partial t}$	$\dfrac{-\beta h}{(\beta-1)}\dfrac{\partial V}{\partial x}$	$\dfrac{h}{(\beta-1)V}\dfrac{\partial V}{\partial t}$	$\dfrac{-\partial q}{\partial x}$	$\dfrac{h}{\beta q}\dfrac{\partial q}{\partial t}$
$\dfrac{\partial V}{\partial x} = (\beta-1)\dfrac{V}{h}\dfrac{\partial h}{\partial x}$	$(\beta-1)\dfrac{V}{h}\dfrac{\partial h}{\partial x}$	$\dfrac{-(\beta-1)}{\beta h}\dfrac{\partial h}{\partial t}$	$\dfrac{\partial V}{\partial x}$	$\dfrac{-1}{\beta V}\dfrac{\partial V}{\partial t}$	$\dfrac{(\beta-1)}{\beta}\dfrac{V}{q}\dfrac{\partial q}{\partial x}$	$\dfrac{-(\beta-1)}{\beta^2 q}\dfrac{\partial q}{\partial t}$
$\dfrac{\partial V}{\partial t} = -\beta(\beta-1)\dfrac{V^2}{h}\dfrac{\partial h}{\partial x}$	$-\beta(\beta-1)\dfrac{V^2}{h}\dfrac{\partial h}{\partial x}$	$(\beta-1)\dfrac{V}{h}\dfrac{\partial h}{\partial t}$	$-\beta V\dfrac{\partial V}{\partial x}$	$\dfrac{\partial V}{\partial t}$	$-(\beta-1)\dfrac{V^2}{q}\dfrac{\partial q}{\partial x}$	$\dfrac{(\beta-1)}{\beta}\dfrac{V}{q}\dfrac{\partial q}{\partial t}$
$\dfrac{\partial q}{\partial x} = \dfrac{\beta q}{h}\dfrac{\partial h}{\partial x}$	$\dfrac{\beta q}{h}\dfrac{\partial h}{\partial x}$	$\dfrac{-\partial h}{\partial t}$	$\dfrac{\beta q}{(\beta-1)V}\dfrac{\partial V}{\partial x}$	$\dfrac{-q}{(\beta-1)V^2}\dfrac{\partial V}{\partial t}$	$\dfrac{\partial q}{\partial x}$	$\dfrac{-1}{\beta V}\dfrac{\partial q}{\partial t}$
$\dfrac{\partial q}{\partial t} = \dfrac{-\beta^2 q^2}{h^2}\dfrac{\partial h}{\partial x}$	$\dfrac{-\beta^2 q^2}{h^2}\dfrac{\partial h}{\partial x}$	$\dfrac{\beta q}{h}\dfrac{\partial h}{\partial t}$	$\dfrac{-\beta^2 q}{(\beta-1)V}\dfrac{\partial V}{\partial x}$	$\dfrac{\beta q}{(\beta-1)V}\dfrac{\partial V}{\partial t}$	$-\beta V\dfrac{\partial q}{\partial x}$	$\dfrac{\partial q}{\partial t}$

Note: Calculated from $q = hV = \alpha h^\beta$, $(\partial h/\partial t) = (-\partial q/\partial x)$, and $c = (dx/dt) = \beta V$.

therefore clear that, in many cases, channel straightening (higher S_0) and channel lining (lower n) decrease floodwave diffusivity of natural channels.

The following relationships are important in determining the flow velocity V and the floodwave celerity c, assuming that the Manning equation is applicable:

$$V = \frac{1}{n} R_h^{2/3} \left\{ S_0 - [1 - (\beta - 1)^2 \, \text{Fr}^2] \frac{\partial h}{\partial x} \right\}^{1/2}, \tag{5.26}$$

$$c = \frac{\beta}{n} R_h^{2/3} \left\{ S_0 - [1 - (\beta - 1)^2 \, \text{Fr}^2] \frac{\partial h}{\partial x} \right\}^{1/2}. \tag{5.27} \blacklozenge$$

The floodwave-diffusivity term plays a dominant role in the alteration of floodwaves. This process is illustrated with the following example. Consider a triangular floodwave propagating in a wide-rectangular channel, as sketched in Fig. 5.5. Two points at equal flow depth are located upstream (A) and downstream (B) of the wave crest. The reference discharge Q_r corresponds to the kinematic-wave approximation, where, $S_0 = S_f$, thus assuming Manning's equation $Q_r = (A/n)R_h^{2/3} S_0^{1/2}$. Three types of floodwaves are recognized: (1) a dynamic wave when $\text{Fr} > 1/(\beta - 1)$; (2) a kinematic wave when $\text{Fr} = 1/(\beta - 1)$, and (3) a diffusive wave when $\text{Fr} < 1/(\beta - 1)$.

Dynamic waves in steep channels thus tend to form pulsating flows or surges, also called roll waves. In laminar flow ($\beta = 3$), roll waves can theoretically form when $\text{Fr} > 0.5$. Measurements in sheet flows are possible for subcritical flow at $\text{Fr} > 0.7$ (Julien and Hartley, 1985). In turbulent flows ($\beta = 5/3$), roll waves develop on very steep smooth channels under supercritical flows ($\text{Fr} > 1.5$). Roll waves and supercritical flows should be avoided when open channels are being designed because of surface instabilities and cross waves incurred by any perturbation of the bank and/or the bed. This can best be achieved by an increase in boundary roughness to the extent that the flow will remain subcritical.

Kinematic waves are obtained when $\text{Fr} = 1/(\beta - 1)$. As sketched in Fig. 5.5(b), the bed and the friction slopes are identical. This implies that wave celerity and discharge increase solely with flow depth. The wave celerities and discharges at points (A) and (B) are therefore identical, and the distance separating (A) and (B) remains constant as the floodwave travels downstream. The celerity of the wave crest at (C) is nevertheless larger than that of (A) or (B), and the crest gradually moves to form a well-defined wave front, also referred to as kinematic shock.

In most rivers, the flow is subcritical and flood routing is adequately described by the diffusive-wave approximation of the St. Venant equation. In such cases, the wave celerity and discharge do not vary solely with flow depth but also depend on the gradient of flow depth in the downstream direction. For instance,

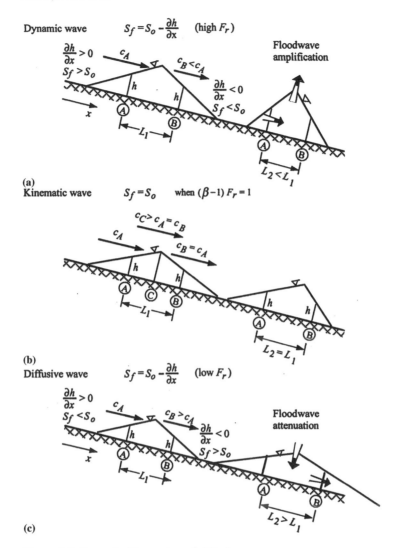

Figure 5.5. Dynamic, kinematic, and diffusive waves.

as sketched on Fig. 5.5(c), the initially triangular floodwave shows $\partial h / \partial x < 0$ on the downstream side and $\partial h / \partial x > 0$ on the upstream side of the crest.

This implies that the two points, (A) and (B), at identical flow depth will propagate at different celerities where the downstream point (B) moves faster downstream than point (A). This will result in floodwave attenuation, or stretching of the distance separating (A) and (B), as the flood propagates in the downstream direction. The floodwave will elongate and the peak discharge decrease as the wave travels downstream. Floodwave attenuation is most effective when the Froude number is low, and $\frac{\partial h}{\partial x}$ is large compared to S_0.

5.4   Loop-rating curves

Solutions to unsteady flow problems in most rivers require the use of the diffusive-wave formulation of the St. Venant equation. Based on the friction-slope relationship, total discharge Q in a river can be written as

$$Q = \frac{A}{n} R_h^{2/3} \left\{ S_0 - \left(\frac{\partial h}{\partial x} \right) [1 - (\beta - 1)^2 \, \text{Fr}^2] \right\}^{1/2}. \tag{5.28}$$

At a given flow depth, the friction slope S_f at low Froude numbers is slightly larger than the bed slope during the rising stage $(\partial h / \partial x) < 0$. Unless the kinematic-wave approximation is applicable, river floods will induce loops in the rating curve relationship because $S_f \neq S_0$ during the rising and the falling levels of flood hydrographs. Typically, at a given flow depth, the rising limb of a floodwave will display a higher discharge than the falling limb. A counterclockwise loop is obtained for stage–discharge relationship (h versus Q) of most channels. The example of the Mississippi River is shown in Fig. 5.6. The maximum discharge is reached before the maximum flow depth.

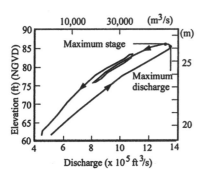

Figure 5.6. Loop-rating curve of the Mississippi River at Vicksburg (after Combs, 1994).

The loop-rating curve also has an impact on shear stress and sediment-transport calculations. Bed shear-stress τ_0 calculations based on the friction slope are given by

$$\tau_0 = \gamma R_h S_f = \gamma R_h \left\{ S_0 - [1 - (\beta - 1)^2 \, \text{Fr}^2] \frac{\partial h}{\partial x} \right\}. \tag{5.29}$$

Shear-stress calculations deviate from kinematic-wave considerations. It is particularly shown that at a given flow depth, shear stress is larger on the rising limb than on the falling limb. Bedload sediment transport also increases with bed shear stress, thus resulting in larger sediment transport during the rising limb at a given flow depth than for the falling limb. For instance, the Meyer–Peter Muller formula shows that $q_s \sim (\tau_0 - \tau_c)^{1.5}$.

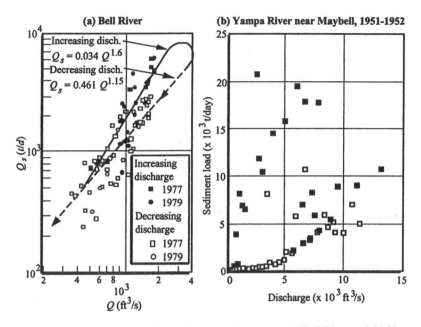

Figure 5.7. Loop of the sediment-rating curves: (a) Bell River and (b) Yampa River.

Sediment transport is described by the sediment-rating curve (sediment discharge Q_s versus water discharge Q). Although both Q_s and Q increase during the rising stages, Q_s increases faster than Q. The sediment-rating curve is characterized by a clockwise loop-rating effect, as shown in Fig. 5.7. This effect is predominant for streams transporting mostly bedload. Streams carrying predominantly washload may not respond to changes in bed shear stress.

In summary, Fig. 5.8 sketches the effects of the dynamic, kinematic, and diffusive waves on discharge and bedload sediment transport. The dynamic wave usually found in upland areas tends to cause riverbed degradation as the floodwave amplifies downstream. Conversely, the diffusive wave typically found in most subcritical rivers causes floodwave attenuation and riverbed aggradation.

Example 5.3   Application of the St. Venant equation. An observer measures the flow depth in a 50-m wide-rectangular channel inclined at $S_0 = 3 \times 10^{-3}$ with Manning coefficient $n = 0.03$. Initially, the flow depth is 1.0 m and the water level rises at a rate of 1 m/h. Calculate (1) the initial discharge at a distance of 1 km downstream, (2) calculate the relative magnitude of the acceleration terms in the St. Venant equation, and (3) determine whether the floodwave attenuates as it propagates downstream.

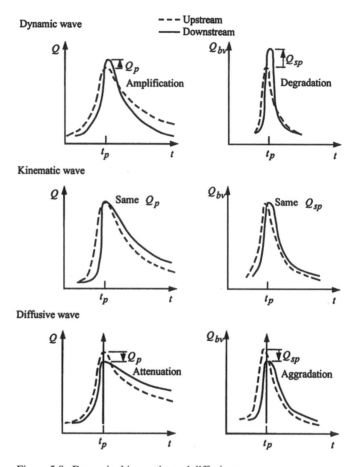

Figure 5.8. Dynamic, kinematic, and diffusive waves.

Step 1: The initial upstream discharge is

$$Q_u = \frac{W}{n} h_u^{5/3} S_0^{1/2} = \frac{50}{0.03} 1^{5/3} \sqrt{3 \times 10^{-3}} = 91.3 \text{ m}^3/\text{s}.$$

From continuity equation (5.8),

$$\Delta Q = \frac{-W \Delta h \Delta X}{\Delta t} = -50 \times \frac{1}{3,600 \text{ s}} \times 1,000 \text{ m}^3 = -13.9 \text{ m}^3/\text{s}.$$

The downstream discharge Q_d is $91.3 - 13.9 = 77.4 \text{ m}^3/\text{s}$.

Step 2: The downstream flow depth is

$$h_d = \left(\frac{nQ}{W S^{1/2}} \right)^{3/5} = \left(\frac{0.03 \times 77.4}{50 \times \sqrt{3 \times 10^{-3}}} \right)^{0.6} = 0.906 \text{ m}.$$

The upstream and the downstream velocities are, respectively,

$$V_u = \frac{Q_u}{W h_u} = \frac{91.3}{50 \times 1} = 1.826\,\text{m/s}, \quad V_d = \frac{Q_d}{W h_d} = \frac{77.4}{50 \times 0.906} = 1.708\,\text{m/s}.$$

Over a distance of 1 km, the flow depth changes by 0.094 m and $\Delta V = 0.118$ m/s. The terms of the St. Venant equation (Eq. 5.16) are

(2) $S_0 = 0.003 = 3 \times 10^{-3}$,

(3) $\dfrac{\partial h}{\partial x} = \dfrac{-0.094\,\text{m}}{1{,}000\,\text{m}} = -9.4 \times 10^{-5}$,

(4) $\dfrac{V}{g} \dfrac{\partial V}{\partial x} = \dfrac{1.77}{9.81} \dfrac{(-0.118)}{1{,}000} = -2.13 \times 10^{-5}$.

Given the floodwave celerity $c = \beta V = 5/3 \times 1.77\,\text{m/s} = 2.95\,\text{m/s}$, the 1-km distance is traveled in 390 s.

(5) $\dfrac{1}{g} \dfrac{\partial V}{\partial t} = \dfrac{s^2}{9.81\,\text{m}} \dfrac{0.118\,\text{m}}{340\,\text{s}^2} = 3.53 \times 10^{-5}$.

The friction slope is $S_f = S_0 - (\partial h / \partial x) - (V \partial V / g \partial x) - (1/g)(\partial V / \partial t) = 3.08 \times 10^{-3}$. Note that the Froude number,

$$\text{Fr} = \frac{V_u}{\sqrt{g h_u}} = \frac{1.826}{\sqrt{9.81}} = 0.58$$

is less than unity and term (4) is smaller than term (3).

Step 3: We can use Eq. (5.22) to calculate the floodwave diffusivity term as $1 - (\beta - 1)^2 \text{Fr}^2 = 1 - \frac{4}{9}(0.58)^2 = 0.85$. We determine that the floodwave is diffusive and attenuates as it propagates downstream.

5.5   River flood routing

We can calculate the propagation of floodwaves in rivers by solving the equations of conservation of mass and momentum. The basic approach for the numerical simulation of floodwave routing stems from the concept of celerity, which is sketched for a single wide-rectangular channel in Fig. 5.9. Any change in flow depth corresponds to a change in discharge, and the perturbation propagates at a celerity c. Because $c = dx/dt$, numerical models should attempt to set Δx and Δt such that $c \cong \Delta x / \Delta t$. The stability of several numerical

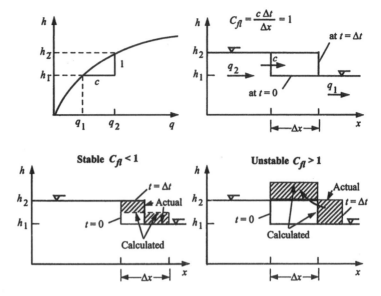

Figure 5.9. Sketch of stable and unstable numerical schemes.

schemes depends on the Courant–Friedrich–Levy condition, which can be written as

$$C_{fl} = \frac{c\Delta t}{\Delta x},$$ (5.30)

where C_{fl} is the Courant number. When $C_{fl} \le 1$, the numerical scheme is stable; the scheme is usually unstable when $C_{fl} > 1$ m, as sketched in Fig. 5.9.

Because the floodwave celerity is maximum near the peak discharge of the floodwave ($c = \beta V$), the time and space increments Δt and Δx, respectively, can be related as

$$\Delta t = \frac{C_{fl}\,\Delta x}{\beta V} = \frac{\Delta x\, C_{fl}}{\beta \alpha^{1/\beta} q^{\frac{\beta-1}{\beta}}},$$ (5.31)

and, for Manning's equation in S.I. units, we easily demonstrate in Exercise 5.6 that, for stability,

$$\Delta t < \frac{3}{5}\frac{\Delta x\, n}{h^{2/3}\, S_0^{1/2}} = \frac{3n^{0.6}\, W^{0.4}}{5Q^{0.4}\, S^{0.3}}\Delta x.$$ (5.32)

Alternatively, we can calculate C_{fl} at any node at any time step from fixed values of Δx and Δt as

$$C_{fl} = \frac{\Delta t}{\Delta x}\frac{5}{3}\frac{Q^{0.4}\, S^{0.3}}{n^{0.6}\, W^{0.4}} = \frac{\Delta t}{\Delta x}\frac{5}{3}V;$$ (5.33)

when C_{fl} exceeds 1, the highest value of the Courant number C_{fl} is indicative of where numerical instabilities are likely to originate in numerical models. It is also clear from Eq. (5.32) that numerical instabilities are caused by high discharges and steep slopes whereas larger resistance and channel width increase numerical stability.

Example 5.4   Application to flood routing. The continuity relationship is to be used to calculate floodwave propagation through a 325-km-long reach of a 260-m-wide river. The bed slope is 1.5×10^{-4} and the Manning coefficient n is 0.015. The hypothetical daily inflow discharge is Q_{in} of 1,000, 2,000, 3,000, 2,500, 2,000, 1,500, and 1,000 m^3/s. Calculate the outflow discharge and the daily change in flow depth during this period. We can first check the value of the Courant number from Eq. (5.33):

$$C_{fl} = \frac{86,400 \text{ s}}{325,000 \text{ m}} \frac{5}{3} \frac{(Q)^{0.4} (1.5 \times 10^{-4})^{0.3}}{(0.015)^{0.6}(260)^{0.4}} = 4.24 \times 10^{-2} \, Q^{0.4}.$$

For the maximum discharge $Q = 3,000$ m^3/s, $C_{fl} = 1.04$, which can be acceptable only because $C_{fl} > 1$ for a single time step. The Froude number is

$$\text{Fr} = \frac{Q^{0.1} \, S^{0.45}}{g^{0.5} \, n^{0.9} \, W^{0.1}} = \frac{(3,000)^{0.1} \, (1.5 \times 10^{-4})^{0.45}}{(9.81)^{0.5} \, (0.015)^{0.9} \, (260)^{0.1}} = 0.34.$$

The natural floodwave should attenuate because $\text{Fr} < 1.5$.

The Froude number is sufficiently small and the diffusivity term from Eq. (5.23), $1 - (\beta - 1)^2 \, \text{Fr}^2 = 1 - \frac{4}{9}(0.34)^2 = 0.95$, remains close to unity. The diffusive-wave approximation is therefore recommended over the full-dynamic-wave approximation.

The maximum diffusivity ratio from Eq. (5.25) is

$$\frac{-1}{S_0} \frac{\partial h}{\partial x} = \frac{9}{25} \frac{n^{1.2}}{W^{0.2} \, S_0^{1.6} \, Q^{0.8}} \frac{\Delta Q_{max}}{\Delta t}$$

$$= \frac{9}{25} \frac{(0.015)^{1.2} \, 1,000}{(260)^{0.2}(1.5 \times 10^{-4})^{1.6}(3,000)^{0.8} \, 86,400} = 0.019.$$

The diffusivity is so small that the solution should be very close to kinematic routing.

The initial water-surface elevation at $Q = 1,000$ m^3/s is $h = 2.534$ m. The equation of continuity for an impervious channel without rainfall and lateral inflow is rewritten as

$$\Delta h_i = \frac{(Q_{in} - Q_{out}) \, \Delta t}{W \, \Delta x}. \tag{E.5.4.1}$$

Table E.5.4.1. *Floodwave routing calculation*

Days (i)	Q_{in} (m³/s)	h_u (m)	Δh_i (m)	$h_d{}^a$ (m)	Kinematic Q_{out} (m³/s)	C_{fl}	Diffusive $Q_{out}{}^b$ (m³/s)	$h_d{}^c$ (m)
1	1,000	2.53	0	2.53	1,000	0.67	1,000	2.53
2	2,000	3.84	1.02	2.53	1,000	0.67	1,015	2.55
3	3,000	4.90	1.28	3.56	1,759	0.84	1,786	3.65
4	2,500	4.39	−0.43	4.82	2,925	1.03	2,912	4.71
5	2,000	3.84	−0.51	4.39	2,499	0.97	2,485	4.42
6	1,500	3.23	−0.54	3.88	2,033	0.89	2,021	3.90
7	1,000	2.53	−0.59	3.33	1,579	0.81	1,567	3.33

$^a h_{d_{i+1}} = h_{d_i} + \Delta h_i$

$^b Q_{out} = \left(\dfrac{W}{n}\right) h_d^{5/3} S_0^{1/2} \left(1 - \dfrac{\partial h}{S_0 \partial x}\right)^{1/2} \cong 212.3\, h_d^{5/3}\left[1 + \dfrac{\Delta h_i}{S_0 C_{fl}\Delta x}\right]^{1/2}.$

^{c}Floodstage calculated with $\Delta x = 32.5$ km and $\Delta t = 2$ h.

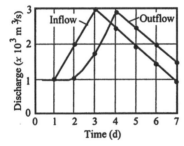

Figure E.5.4.1. Calculated flood routing.

On a daily basis ($\Delta t = 86{,}400$ s), the discharge Q_{out} at the downstream station is then updated from the new downstream flow depth $h_{d_{i+1}} = h_{d_i} + \Delta h_i$, and the downstream discharge is calculated with the diffusive-wave approximation in $S_f = S_0 - \partial h/\partial x$. The procedure is illustrated in Table E.5.4.1 and the results are shown in Fig. E.5.4.1. For the sake of comparison, the floodstage calculated on a finer grid, $\Delta x = 32.5$ km and $\Delta t = 2$ h, is shown in the last column of Table E.5.4.1. The results are quite similar to those of the diffusive-wave formulation.

5.6   River flow and sediment-duration curves

Duration curves describe the percentage of time that a certain water or sediment discharge is exceeded. The curves are usually based on daily records and are useful in estimating how many days per year an event will be exceeded. Let the uppercase letter X denote a random variable, and let the lowercase letter x denote a possible value of X. For a random variable X, its cumulative distribution function (cdf), denoted as $F(x)$, is the probability P that the random variable X will be less than or equal to x:

$$F(x) = P(X \leq x). \qquad (5.34)$$

$F(x)$ is also called the nonexceedance probability for the value x. The exceedance probability $E(x) = 1 - F(x)$. The probability density function (pdf), $p(x)$, is the derivative of the cdf:

$$p(x) = \frac{dF(x)}{dx}. \tag{5.35}$$

Exceedance probability transforms are useful in the analysis of duration curves. Properties of an exponential pdf (epdf) of variable x are such that after variable x is divided by the mean value $\bar{x}$, a reduced variable $\bar{\Psi}$ is defined as $\bar{\Psi} = x/\bar{x}$. The properties of the epdf $p(\bar{\Psi})$ and the exceedance probability $E(\bar{\Psi})$, defined as $E(\bar{\Psi}) = \int_{\bar{\Psi}}^{\infty} p(\bar{\Psi}) \, d\bar{\Psi}$, are

$$E(\bar{\Psi}) = p(\bar{\Psi}) = e^{-\bar{\Psi}}. \tag{5.36}$$

It is interesting to note that both the epdf and the exceedance probability are identical.

The purpose of the transform is to determine whether the pdf of a runoff or sediment variable x reduces to an epdf of variable $\bar{\Psi}$ after the following transform:

$$\bar{\Psi} = ax^b, \tag{5.36a}$$

where a and b are, respectively, the transform coefficient and exponent, hereby referred to as transform parameters. When successful, this transforms the unknown exceedance probability of variable x into a simple epdf of the reduced variable $\bar{\Psi}$.

The inverse transform is simply defined from Eq. (5.36a) as

$$x = \left(\frac{1}{a}\right)^{\frac{1}{b}} \bar{\Psi}^{\frac{1}{b}} = \hat{a}\bar{\Psi}^{\hat{b}}, \tag{5.36b}$$

where $\hat{a}$ and $\hat{b}$ are the inverse transform parameters simply calculated from a and b as $\hat{a} = (1/a)^{1/b}$ and $\hat{b} = (1/b)$. Useful conversions are $a = \hat{a}^{-1/\hat{b}} = \hat{a}^{-b}$ and $b = (1/\hat{b})$. The inverse transform exponent $\hat{b}$ is important in determining the degree of nonlinearity of the variable x and can be related to deterministic relationships. An example of range of values of $\hat{b}$ for various processes is summarized in Table 5.2. In increasing order of nonlinearity, we find point rainfall, surface runoff, chemical and sediment concentration, and chemical and sediment yield.

The properties of the transform are such that once the transform parameters a and b are known, the exceedance probability of variable x is calculated directly

Table 5.2. *Typical values of $\hat{b}$ for different variables*

Process	$\hat{b}$
Point rainfall	0.85–1.28
Upland snowmelt	1.15–1.37
Upland chemical concentration	0.46–1.40
Upland runoff	1.46–1.62
River flow discharge	1.25–1.77
River sediment concentration	1.13–2.14
Upland chemical yield	1.48–2.33
Upland sediment yield	2.20–2.35
River sediment discharge	1.70–2.84

from Eqs. (5.35) and (5.36a). The pdf $p(x)$ is

$$p(x) = abx^{b-1}e^{-ax^b} = ab\left(\frac{\bar{\Psi}}{a}\right)^{\frac{b-1}{b}}e^{-\bar{\Psi}}. \tag{5.37}$$

We can also demonstrate that $p(x)dx = p(\bar{\Psi})d\bar{\Psi}$ and $E(x) = E(\bar{\Psi})$.

The procedures to evaluate the parameters of this distribution are presented in Subsection 5.6.1, followed by practical considerations in Subsection 5.6.2.

5.6.1   *Parameter evaluation*

Two procedures are examined for the transform parameter evaluation: (1) a graphical method and (2) the method of moments.

Graphical method. The graphical method capitalizes on the properties of the epdf and the exceedance probability in Eq. (5.36). The natural logarithm of Eq. (5.36) is combined with Eq. (5.36a) to give

$$-\ln E(\bar{\Psi}) = \bar{\Psi} = ax^b. \tag{5.38} \blacklozenge$$

The transform parameters a and b are evaluated graphically after the natural logarithm of Eq. (5.38) is taken in the form

$$\Pi = \ln[-\ln E(\bar{\Psi})] = \ln\bar{\Psi} = \ln a + b\ln x, \tag{5.39} \blacklozenge$$

where the shorthand exceedance probability Π designates the double logarithm of the exceedance probability. Of course, this transform requires that all values of x be positive.

A straight line is fitted on the graphical presentation of Π as a function of $\ln x$ to yield estimates for a and b. It becomes clear from the properties of Eq. (5.39) that $\ln a$ corresponds to the value of Π when $x = 1$. Likewise, if the points on the graph Π vs. $\ln x$ assemble on a straight line, the slope of the line gives directly the transform exponent b. The graphical method is subjective but provides the user a qualitative appreciation of the goodness of fit. Ideally, all points should plot on a straight line for the power transform in Eq. (5.38) to be exactly applicable.

Method of moments. Parametric estimation from the method of moments takes advantage of the information contained in the first and the second moments of the sample. The transform parameters a and b can be evaluated after being equated with the first and the second moments of the transformed variables. Specifically, the first moment, or mean value, of the sample $\bar{x}$ is equated to the first moment M_1 of the transformed distribution, given $x = \left(\bar{\Psi}/a\right)^{1/b}$ and $p(x)\,dx = p(\bar{\Psi})d\bar{\Psi}$:

$$M_1 = \int_0^\infty xp\,(x)\,dx = \left(\frac{1}{a}\right)^{\frac{1}{b}} \int_0^\infty \bar{\Psi}^{\frac{1}{b}} e^{-\bar{\Psi}}\,d\bar{\Psi}$$

$$= \left(\frac{1}{a}\right)^{\frac{1}{b}} \Gamma\left(1 + \frac{1}{b}\right) = \hat{a}!\hat{b} = \bar{x}. \tag{5.40}$$

Accordingly, the expected value $\bar{x}$ can be simply evaluated from a simple gamma function of the inverse transform parameters as $\bar{x} = \hat{a}!\hat{b}$, where ! designates the factorial function of the argument $\hat{b}$. Likewise, the second moment of the sample $\overline{x^2}$ is equated to the second moment M_2 of the transformed distribution:

$$M_2 = \int_0^\infty x^2 p(x)\,dx = \left(\frac{1}{a}\right)^{\frac{2}{b}} \int_0^\infty \bar{\Psi}^{\frac{2}{b}} e^{-\bar{\Psi}}\,d\bar{\Psi}$$

$$= \left(\frac{1}{a}\right)^{\frac{2}{b}} \Gamma\left(1 + \frac{2}{b}\right) = \hat{a}^2!(2\hat{b}) = \overline{x^2}. \tag{5.41}$$

The evaluation of the transform parameters a and b follows after the transform coefficient a is eliminated from the ratio of Eq. (5.41) to the square of Eq. (5.40); thus

$$\frac{\Gamma\left(1 + \frac{2}{b}\right)}{\left[\Gamma\left(1 + \frac{1}{b}\right)\right]^2} = \frac{\overline{x^2}}{\bar{x}^2}. \tag{5.42}$$

From the calculated values of $\bar{x}$ and $\overline{x^2}$ of the sample, the value of b on the left-hand side of Eq. (5.42) can best be evaluated numerically. For instance, an interpolation procedure is based on the numerical values given in Table 5.3

Table 5.3. *Useful values of the transform parameter b*

b	$\overline{x^2}/\overline{x}^2$	$\Gamma(b)$	$!(b)$
0.1	184,756	9.51	0.951
0.15	2,213	6.22	0.933
0.2	252	4.59	0.918
0.25	70	3.62	0.906
0.3	30.2	2.99	0.897
0.35	16.77	2.54	0.891
0.4	10.86	2.21	0.887
0.45	7.79	1.97	0.886
0.5	6	1.77	0.886
0.55	4.861	1.61	0.888
0.6	4.090	1.49	0.893
0.65	3.543	1.38	0.900
0.7	3.138	1.30	0.908
0.75	2.830	1.22	0.919
0.8	2.588	1.16	0.931
0.85	2.395	1.11	0.945
0.9	2.238	1.07	0.962
0.95	2.108	1.02	0.980
1	2	1.00	1.00
1.05	1.907	0.973	1.02
1.1	1.828	0.951	1.046
1.2	1.700	0.918	1.10
1.3	1.601	0.897	1.17
1.4	1.523	0.887	1.24
1.5	1.460	0.886	1.33
1.6	1.409	0.893	1.43
1.7	1.366	0.909	1.54
1.8	1.330	0.931	1.68
1.9	1.299	0.962	1.83
2	1.273	1.00	2.0
2.5	1.183	1.33	3.32
3	1.132	2.0	6.0
3.5	1.100	3.32	11.6
4	1.078	6.0	24.0
4.5	1.063	11.6	52.3

proves to be sufficiently accurate. The value of the transform coefficient a then follows from $\hat{a}!\hat{b} = \bar{x}$ in Eq. (5.40) as

$$a = \left[\frac{\Gamma(1+\frac{1}{b})}{\bar{x}}\right]^b. \tag{5.43}$$

Parameter estimation from the method of moments is direct and not subjective but lacks the visual information inherent in the graphical method. The evaluation of the transform parameters with both the graphical method and the method of moments is illustrated in Example E.5.5.

Example 5.5   Application of the duration curve procedure. Consider the following sample of an unknown variable: $\bar{x}$ = 4.5, 1.0, 7.0, 2.0, 9.0, 0.5, 6.0, 11.0, 3.5. The first step consists of ranking the n = 9 numbers in decreasing order of x, as shown in the second column of Table E.5.5.1; the values of x are squared in the third column. The second step consists of calculating the exceedance probability by use of the Weibull plotting position. Accordingly, the numbers in decreasing order are ranked from 1 to n as shown in the fourth column, 1 being the largest and n being the smallest number. After the rank is divided by $1 + n$, the plotting position corresponds to the exceedance probability $E(\bar{\Psi})$ or $E(x)$ in the fifth column. The values of $\ln x$ and $\Pi = \ln[-\ln E(\bar{\Psi})]$ are tabulated in the sixth and the seventh columns, respectively, for the plot shown in Fig. E.5.5.1. Graphically, the parameter estimation gives $a \approx 0.1$ and $b \approx 1.3$. The line is usually fit through the higher values of $\ln x$ because, for sediment-transport parameters, the large values of x are usually those contributing to most of the sediment load.

With the method of moments, the average value, $\bar{x}$ = 4.94, is calculated at the bottom of the second column of Table E.5.5.1. The average value of $\overline{x^2}$ = 36.1 is compiled at the bottom of the third column. From Table E.5.5.1, the ratio $(\overline{x^2}/\bar{x}^2) = [36.1/(4.94)^2] = 1.479$ in the second column corresponds to a value of $b \approx 1.45$ from interpolation with the values given in the first column of Table 5.3. The value of $a \approx 0.085$ is thereafter calculated from Eq. (5.43).

Table E.5.5.1. *Example of transformation procedure*

Sample $\bar{x}$	Ranked x	x^2	Rank	$E(x)$	$\ln(x)$	$\ln[-\ln E(x)]$
4.5	11.0	121.00	1	0.1	2.3979	0.8340
1.0	9.0	81.00	2	0.2	2.1972	0.4759
7.0	7.0	49.00	3	0.3	1.9459	0.1856
2.0	6.0	36.00	4	0.4	1.7918	−0.0874
9.0	4.5	20.25	5	0.5	1.5041	−0.3665
0.5	3.5	12.25	6	0.6	1.2528	−0.6717
6.0	2.0	4.00	7	0.7	0.6931	−1.0309
11.0	1.0	1.00	8	0.8	0.0000	−1.4999
3.5	0.5	0.25	9	0.9	−0.6931	−2.2504
Average:	$\bar{x}$ = 4.94	$\overline{x^2}$ = 36.1				

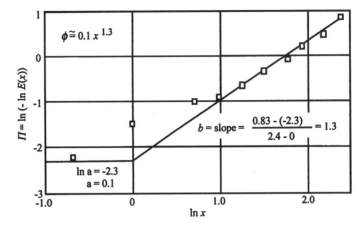

Figure E.5.5.1. Flow duration example.

The gamma function can also be approximated by Stirling's asymptotic series as

$$!(x) = \Gamma(x+1) = \sqrt{2\pi x}\, x^x e^{-x} \left\{ 1 + \frac{1}{12x} + \frac{1}{288x^2} - \frac{139}{51,840x^3} \cdots \right\},$$

$$(E.5.5.1)$$

which is readily available on most scientific calculators.

5.6.2   *Practical considerations*

The transforms provide useful information in the analysis of exceedance probability and flow duration curves. Indeed, the exceedance probability $E(x)$ of a variable x can be directly calculated from

$$E(x) = e^{-ax^b}, \qquad (5.44)$$

given the transform parameters. For instance, the flow discharge in the Chaudière river ($x = Q$) is given by $\phi = 0.048\,Q^{0.66}$, or $a = 0.048$ and $b = 0.66$. The exceedance probability $E(1{,}000\ \mathrm{m}^3/\mathrm{s})$ of a daily discharge $Q = 1{,}000\ \mathrm{m}^3/\mathrm{s}$ is calculated from $E(1{,}000) \cong e^{-0.048\times(1{,}000)0.66} = 0.01$. The transforms also enable the user to estimate what value of a parameter will be exceeded a certain fraction of the time directly from

$$x = [-\ln E(x)]^{1/b} \left(\frac{1}{a}\right)^{1/b} = \hat{a}\,[-\ln E(x)]^{\hat{b}}. \qquad (5.45)$$

For instance, we calculate the daily sediment discharge of the Colorado River at Lee's Ferry that is exceeded 1% of the time, or 3.65 days a year, from Eq. (5.45) after solving for $x = Q_s$. For the Colorado River, $\hat{a} = 120,230$ and $\hat{b} = 1.7$, a daily sediment discharge exceeded 1% of the time, $E(Q_s) = 0.01$, is simply given by $Q_s = 120,230(-\ln 0.01)^{1.7} = 1.6 \times 10^6$ tons/day.

Once the magnitude x_1 and exceedance probability E_1 of an event is known one can determine the unknown magnitude x_2 of another event of exceedance probability E_2 simply as a function of the inverse transform exponent $\hat{b}$. At a given $\hat{b}$ we define $\xi = E_2/E_1$ to obtain the unknown $\eta = x_2/x_1$. From Eq. (5.44), we obtain $E_1 = e^{-ax_1^b}$ and $\xi E_1 = e^{-a(\eta x_1)^b}$ to be solved for ξ as a function of η and b as

$$\xi = e^{-ax_1^b(\eta^b-1)} = E_1^{(\eta^b-1)}, \tag{5.46}$$

or conversely for η as a function of ξ and $\hat{b}$,

$$\eta = \left(1 + \frac{\ln \xi}{\ln E_1}\right)^b = \left(\frac{\ln E_2}{\ln E_1}\right)^b, \tag{5.47}$$

which shows that at any given value of ξ and E_1, the magnitude of η increases with the exponent $\hat{b}$.

As a final result, the transforms enable the user to calculate the magnitude of infrequent events from the mean value of a variable and the inverse transform parameter $\hat{b}$. From Equations 5.40 and 5.45, we demonstrate that the exceedance probability of the mean value is only a function of $\hat{b}$ as

$$-\ln E(\bar{x}) = \left(!\hat{b}\right)^{1/\hat{b}}. \tag{5.48}$$

The value of x that has an exceedance probability $E(x)$ is then directly calculated from $\bar{x}$ and $\hat{b}$ as

$$x = \left[\frac{\ln E(x)}{\ln E(\bar{x})}\right]^b \bar{x} = \left\{\frac{[-\ln E(x)]^b}{!\hat{b}}\right\} \bar{x}. \tag{5.49} \blacklozenge\blacklozenge$$

Values of the multiplication coefficient in the braces of Eq. (5.49) are given in Table 5.4. One of the practical applications of this method is the determination of the two-year flood, which usually corresponds to the dominant discharge of alluvial rivers.

Extreme rainfall events and the resulting floods can claim thousands of lives and cause billions of dollars in damage. Floodplain management and the design of flood-control works, reservoirs, bridges, and other investigations need to reflect the likelihood or probability of such events. Figure 5.10 shows extreme specific-discharge conditions as functions of drainage area.

A frequency distribution is precisely determined when a sufficiently long record is available, e.g., flood discharges, rainfall precipitation, or pollutant loadings. Frequency analyses are required when available data from past

Table 5.4. *Multiplication factor k for infrequent events* $x = k\bar{x}$

| | | Inverse transform coefficient $\hat{b}$ | | | | |
| | | rainfall | | runoff | | sediment |
Exceeded	$E(x)$	1	1.5	2	2.5	3
5% of the time	0.05	3.0	3.9	4.49	4.68	4.48
1% of the time	0.01	4.6	7.4	10.6	13.7	16.3
1 day per year	2.738×10^{-3}	5.9	10.7	17.4	25.5	34.2
1 day per 1.5 years	1.825×10^{-3}	6.3	11.9	19.9	30.1	41.8
1 day per 2 years	1.37×10^{-3}	6.6	12.7	21.7	33.6	47.8
1 day per 5 years	5.47×10^{-4}	7.51	15.5	28.2	46.6	70.6
1 day per 10 years	2.73×10^{-4}	8.2	17.6	33.6	58.0	92

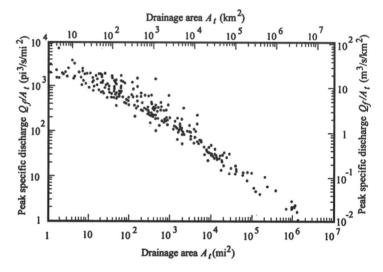

Figure 5.10. Specific discharge vs. drainage area (modified after Creager et al., 1945).

measurements are insufficient to define precisely the risk of large floods, extreme rainstorms, and pollutant loadings. Engineers typically have to work with a sample of 10 to 100 observations to estimate events with exceedance probabilities of 1 in 100, 1 in 1,000, and even the flood flows for spillway design exceeded with a chance of 1 in 10,000.

The return period (sometimes called the recurrence interval) is often specified rather than the exceedance probability. A return period T can be understood as follows: In a fixed T-year period the expected number of exceedances of the T-year event is exactly one if the distribution of floods does not change over that period. Thus an average of one flood greater than the T-year flood level occurs in a T-year period. A return period has been incorrectly understood to mean that one and only one T-year event can occur every T years. Actually, the probability of the T-year flood's being exceeded is $1/T$ every year; thus the 1% exceedance event can be described as a value with a 1 in 100 chance of being exceeded each year.

Case Study 5.1   Daily sediment discharge of the Colorado River, United States. The duration curve of 5 years of daily sediment-discharge measurements of the Colorado River at Taylor's Ferry before the construction of the dams is shown in Fig. CS.5.1.1. The magnitude of infrequent events can be estimated from the mean values $\bar{x}$ and $\hat{b}$. For instance, given the mean daily sediment discharge of 185,700 tons/day in the Colorado River and $\hat{b} = 1.7$, estimate (1) the parameters a, b, $\hat{a}$, and $E(\bar{Q}_s)$ and (2) the magnitude of the daily sediment discharge x_2 exceeded 1 day per year ($P_2 = 0.00274$).

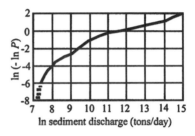

Figure CS.5.1.1. Sediment discharge of the Colorado River (after Julien, 1996).

First, $b = 1/\hat{b} = 0.588$, $!(1.7) = 1.54$ from Table 5.3, $\hat{a} = \bar{x}/1.54 = 120,230$ from Eq. (5.40), and $a = \hat{a}^{-b} = 1.028 \times 10^{-3}$ from Eq. (5.36b). The exceedance probability of the mean daily sediment discharge of the Colorado River, $\bar{Q}_s = 185,700$ tons/day, is calculated from Eq. (5.44) with $X = \bar{x}$. This gives $E_1 = E(\bar{Q}_s) = e^{-1.028 \times 10^{-3} \times 185,700^{0.588}} = 0.275$.

Second, the daily sediment discharge exceeded 1 day/yr is calculated from Eq. (5.49) as $Q_s = [(-\ln 0.00274)^{1.7}/1.54]185,700 = 2.46$ million tons per day.

As a last practical example, estimate the daily sediment discharge with a period of return of 10 yr from this 5-yr record. Using Eq. (5.49), we obtain $Q_{s_{10yr}} = [(-\ln 2.74 \times 10^{-4})^{1.7}/1.54]\,185,700 = 4.31$ million tons per day, which is 23 times the mean daily sediment discharge.

Case Study 5.2   Fluvial data set for the Rhine River, The Netherlands. This case study presents a summary of the 1998 Rhine River flood data in Table CS.5.2.1, followed with stage measurements along the Rhine-Waal River in The Netherlands in Table CS.5.2.2. Selected particle-size distributions of the bed material are presented in Table CS.5.2.3 and selected velocity and

Table CS.5.2.1. *Flood of the Rhine river near the Pannerdens Canal*

Date in 1998	Discharge (m³/s)	Stage (m)	Depth (m)	Velocity (m/s)	Slope (cm/km)	Large Dune length (m)	Dune height (m)	Small Dune length (m)	Dune height (m)
29 Oct.	4,077	11.71	8.51			7.43	0.34		0
30	4,783	12.33	9.13			8.31	0.41		0
31	6,180	13.42	10.22		10.42	11.12	0.48		0
1 Nov.	8,119	13.96	10.76		11.77				
2	9,045	14.48	11.28		12.35	16.24	0.74		0
3	9,464	14.88	11.52	1.82	13.12	19.78	0.87		0
4	9,149	15.01	11.49	1.75	13.70	22.57	0.97		0
5	8,267	14.91		1.71	13.32	23.90	1.08		0
6	7,273		11.17	1.71	11.58	26.02	1.13		0
7		14.14	10.94		10.61	28.84	1.15		0
8		13.76	10.56		9.84				
9		13.46	10.05	1.63	9.07				
10	5,640	13.11			8.49	32.42	0.93		0
11		12.85	9.82	1.53	7.91				
12	5,122	12.69	9.49			34.78	0.79	6.94	0.27
13	4,850	12.46	9.25			36.63	0.74	6.72	0.25
14									
15									
16	4,522	12.18	9.1			40.01	0.66	6.61	0.28
17									
18									
19	4,527	12.17	9.1			42.28	0.56	7.95	0.26

sediment-concentration profiles are shown in Table CS.5.2.4. The navigable width of the Rhine River is maintained at 260 m. The assistance of G. Klaassen at Delft Hydraulics and W. ten Brinke at the Rijkswaterstaat, is gratefully acknowledged.

Exercise 5.1

Demonstrate that the left-hand side of Eq. (5.15) reduces to Eq. (5.14) when $\beta_m = 1$ in wide rivers. *Hint*: use continuity equation (5.8).

♦Exercise 5.2

Demonstrate from the derivatives of the resistance relationship $V = \alpha h^{\beta-1}$ that equation of motion (5.16) reduces to Eq. (5.22).

> *Answer:* $h\partial V = (\beta - 1)\, V\partial h$,
> or $(V/g)\,(\partial V/\partial x) = (\beta - 1)\mathrm{Fr}^2\,(\partial h/\partial x)$.

Table CS.5.2.2. *Stages along the Rhine-Waal River*

Station kilometer section Distance Date	Time	Lobith 862.18 4,998.1 cm	Pannerdense Kop 867.22 10,180.858 cm	Nijmegen Haven 884.87 27,225.159 cm	Tiel 913.25 55,528.35 cm	Zaltbommel 934.78 76,652.202 cm	Vuren 951.78 93,204.14 cm
Oct. 1	09:00	1372	1318	1118	776	467	224
Nov. 1	12:00	1457	1395	1188	840	536	277
2	12:00	1514	1450	1231	890	585	315
3	12:00	1554	1486	1266	939	646	377
4	11:00	1573	1502	1283	955	670	409
4	16:00	1574	1502	1284	958	673	409
4	17:00	1573	1502	1284	958	673	412
4	18:00	1573	1502	1284	958	674	414
4	19:00	1572	1502	1284	958	674	415
4	20:00	1572		1283	959	675	416
5	07:00	1563	1494	1277	958	675	418
5	08:00	1562	1494	1278	957	675	419
5	09:00	1561	1493		957	674	418
5	11:00	1559	1490	1276	956	673	145
5	12:00	1558	1489	1274	956	673	414
6	12:00	1515	1455	1248	933	657	402
7	12:00	1468	1413	1215		626	370
8	13:00	1427	1376	1189	871	593	337
9	11:00	1392	1346	1165	852	568	321
10	10:00	1358	1314	1137	826	541	300
11	13:00	1327	1285	1110	799	514	276

Table CS.5.2.3. *Particle-size distribution of the bed material*

Position (m from axis)	d_{90} (mm)	d_{84} (mm)	d_{75} (mm)	d_{65} (mm)	d_{50} (mm)	d_{35} (mm)	d_{25} (mm)	d_{16} (mm)	d_{10} (mm)
−33	7.810	5.524	3.000	1.864	0.948	0.609	0.453	0.372	0.315
−33	11.480	8.768	5.296	2.539	0.980	0.651	0.469	0.378	0.319
−33	7.894	5.770	3.040	1.474	0.808	0.556	0.448	0.378	0.325
0	11.506	8.809	5.767	3.347	1.901	1.306	0.952	0.762	0.635
0	12.190	9.905	6.400	3.040	1.182	0.710	0.495	0.401	0.338
0	12.000	9.600	4.000	1.851	0.942	0.698	0.536	0.414	0.339
33	11.789	9.263	6.800	4.800	2.686	1.109	0.684	0.459	0.396
33	11.767	9.227	6.109	3.232	0.901	0.472	0.397	0.334	0.297
33	5.642	3.717	2.622	1.925	1.435	0.967	0.775	0.602	0.486
Average (mm)	10.231	7.843	4.782	2.675	1.310	0.786	0.579	0.456	0.383

Table CS.5.2.4. *Velocity and concentration profiles*

Height z (m)	Concentration c (mg/1)	Velocity v (m/s)	Height z (m)	Concentration c (mg/1)	Velocity v (m/s)
October 31, 1998			1.9	71	1.66
0.2	233	0.97	3.1	40	1.81
0.2	246	1.02	3.7	41	1.82
0.2	159	1.00	5.1	29	2.06
0.3	83	1.28	5.5	23	2.00
0.3	75	1.23	5.9	28	1.94
0.4	97	1.25	6.8	20	1.99
1.1	39	1.50	7.9	18	2.03
1.1	38	1.48	9.0	8	2.07
1.2	9	1.45	Surface 10.5		
2.5	4	1.73	November 7, 1998		
3.0	5	1.71	0.2	108	0.84
4.4	4	1.72	0.2	140	0.89
6.5	3	1.90	0.2	99	0.85
7.3	2	1.87	0.7	53	1.02
7.3	3	1.85	0.6	96	1.11
8.3	2	1.98	0.2	303	0.93
8.2	2	1.96	1.1	31	1.35
Surface 8.9			1.3	24	1.46
November 3, 1998			1.3	26	1.36
0.3	494	0.74	2.1	20	1.45
0.3	488	0.81	2.5	23	1.46
0.3	498	0.72	3.3	21	1.43
0.5	398	0.84	4.7	11	1.76
0.8	293	1.22	5.1	8	1.77
0.4	432	0.47	5.9	8	1.79
1.2	132	1.38	6.8	7	1.79
0.9	185	1.34	7.9	5	1.77
1.3	134	1.47	9.0	4	1.84
2.2	83	1.63	Surface 9.4		
3.5	49	1.92	November 10, 1998		
4.1	43	1.86	0.3	143	0.85
4.0	43	1.85	0.3	153	0.95
5.3	35	1.99	0.3	103	0.91
6.0	33	1.98	0.4	124	0.95
7.3	26	2.08	0.2	181	0.73
8.0	25	2.04	0.4	87	0.96
9.0	23	1.90			
Surface 9.9			1.3	15	1.30
November 5, 1998			0.9	42	1.11
0.2	216	1.27	1.1	28	1.24
0.2	212	1.33	2.0	11	1.38
0.3	250	1.30	3.2	7	1.51
0.5	227	1.47	3.8	8	1.53
0.5	218	1.36	4.5	7	1.44
0.5	230	1.47	5.1	7	1.52
0.8	157	1.41	6.3	6	1.62
0.9	170	1.46	6.9	5	1.70
0.9	160	1.41	8.1	4	1.74
			Surface 9.1		

♦♦Exercise 5.3

Demonstrate Eq. (5.23a) from the derivatives of the resistance relationship $V = \alpha h^{\beta-1}$ combined with the continuity relationship [Eq. (5.8)]. *Hint:*

$$\frac{1}{g}\frac{\partial V}{\partial t} = (\beta - 1)\frac{V}{gh}\frac{\partial h}{\partial t}, \quad \frac{\partial h}{\partial t} = -\frac{\partial q}{\partial x} = -\left(\frac{V\partial h}{\partial x} + \frac{h\partial V}{\partial x}\right).$$

♦Exercise 5.4

As per Exercises 5.2 and 5.3, demonstrate that, for floodwave propagation, term (5) of St. Venant equation (5.16) is always $-\beta$ times larger than term (4).

♦Exercise 5.5

Demonstrate Eqs. (5.23a) and (5.23b) from Eq. 5.16, Table 5.1, $c = \beta V$, $Q = Wq$, and $q = \alpha h^\beta$. For Manning's equation, substitute $\alpha = S^{1/2}/n$ and $\beta = 5/3$.

Exercise 5.6

Demonstrate Eqs. (5.24) and (5.25) from $q = \alpha h^\beta$, $\alpha = S^{1/2}/n$ and $\beta = 5/3$ for Manning's equation.

♦Problem 5.1

Determine the celerity of the floodwave in Example 5.3. What is the time required for the upstream perturbation to reach the downstream station?

Answer:

$$c = \beta V = 5 \times 1.82/3 = 3.03 \text{ m/s},$$
$$T = \frac{1,000 \text{ m}}{c} = 5.5 \text{ min}.$$

♦♦Problem 5.2

Repeat the calculations from Example 5.4 for different values of Δx. Describe and explain the influence of the Courant number on flood routing calculations.

Problem 5.3

Estimate the peak discharge of a 3-km² watershed under extreme rainfall conditions. What is the equivalent rainfall intensity?

◆**Problem 5.4**

With reference to Case Study CS.5.1, for the daily sediment discharge of the Colorado River, $\hat{b} = 1.7$, the daily sediment discharge of $x_1 = 1 \times 10^6$ tons/day is exceeded 10 days/yr ($E_1 = 0.0274$). Calculate the exceedance probability E_2 of a daily sediment discharge $x_2 = 2 \times 10^6$ tons/day.

Answer: In this case, ξ is calculated, given $\eta = x_2/x_1 = 2$ and $b = 1/\hat{b} = 0.59$ for the Colorado River; thus $\xi = (0.0274)^{(2^{0.59}-1)} = 0.162$, or $E_2 = \xi E_1 = 4.45 \times 10^{-3}$ or 1.62 days/yr.

◆**Problem 5.5**

From the cross section of the Missouri River (Fig. P.4.6.1) determine the following:

(a) If the Manning coefficient n is constant at any stage, determine the celerity relationship from the relation between the cross-section area and discharge.
(b) When the discharge is $Q = 31,600$ ft^3/s and the water level rises at a rate of 1 ft per day, what is the flow discharge at a cross section located 10 miles upstream (assume that the cross section is unchanged and without lateral inflow along the reach).

◆◆**Problem 5.6**

Consider that the Missouri River flow in Fig. P.4.6.1 is controlled by clearwater releases from a reservoir and determine the following:

(a) Determine the celerity of the wave generated from suddenly increasing the discharge from a steady 10,000 ft^3/s to a discharge fixed at 20,000 ft^3/s.
(b) Assume simple approximations for partial derivatives over a reach on a daily basis and estimate the relative magnitude of the various terms of the St. Venant equation when the discharge increases from 20,000 to 30,000 ft^3/s in 1 day. What approximation of the St. Venant equation is best suited for floodwave routing?

Answer: (a) c ≃ 2.5 m/s, (b) diffusive-wave approximation.

♦♦**Problem 5.7**

Take a sample, at least 100 values, containing daily measurements of either discharge, sediment concentration, sediment discharge, or nutrient or chemical concentration. Determine the following:

(a) exceedance probability curve
(b) inverse transform parameters $\hat{a}$ and $\hat{b}$
(c) average value of the sample and value exceeded 1 day/yr

♦♦**Problem 5.8**

From the data section in Case Study 5.2, determine the floodwave celerity. Determine which approximation of the St. Venant equation is best suited for flood routing.

6

River equilibrium

In a strict sense, a channel is stable when all particles along the wetted perimeter are not moving. This implies that, without transport of bed material, a cross-sectional geometry cannot change with time. The geometry of stable channels, also termed nonalluvial channels, depends on rock outcrops and artificial riprap and does not depend on sediment transport. Many rivers, however, flow in their own deposits, called alluvium, to form alluvial channels. Equilibrium of alluvial channels implies a balance between incoming and outgoing water discharge and sediment load. Poised and graded streams are synonyms describing equilibrium conditions. Whenever a balance is obtained between incoming and outgoing sediment discharges, the cross-sectional geometry may locally change as long as the deposition volume with a river reach is equal to the erosion volume. For instance, river bends may reach equilibrium condition between the rate of erosion on the outside bank and the rate of sedimentation on the point bar. In a broad sense, the cross-sectional geometry of a meandering channel is in equilibrium. However, lateral migration of the bend implies that the planform geometry of the stream is not stable.

This chapter deals with stable and equilibrium river conditions. Section 6.1 details particle stability in a strict sense, and the concept of stable channel is used in Section 6.2 to define the ideal at-a-station cross-sectional geometry of a straight channel. Empirical regime relationships in Section 6.3 are followed by flow conditions in river bends in Section 6.4. The downstream hydraulic geometry of alluvial channels is derived in Section 6.5. Planform geometry and bars are discussed in Section 6.6, followed by a discussion of river meandering in Section 6.7. Lateral migration rates are discussed in Section 6.8, followed by a case study of a meandering river.

6.1   Particle stability

Figure 6.1 illustrates the forces acting on a cohesionless particle resting on an embankment inclined at a sideslope angle Θ_1 and a downstream bed-slope

158

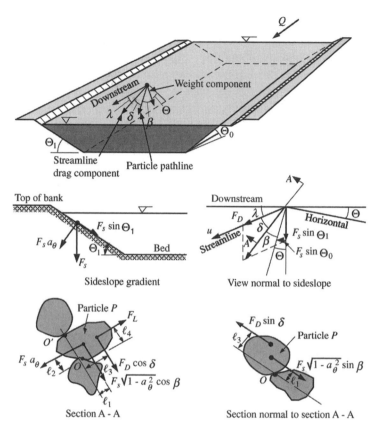

Figure 6.1. Particle-stability analysis.

angle Θ_0. These are the lift force F_L, the drag force F_D, the buoyancy force F_B, and the weight of the particle F_W. As long as the water-surface slope in the downstream direction is small, the buoyancy force can be subtracted from the particle weight to give the submerged weight, $F_S = F_W - F_B$. The lift force is defined as the fluid force normal to the embankment plane whereas the drag force is acting along the plane in the same direction as that of the velocity field surrounding the particle.

For notational convenience, we define two geometrical parameters, a_Θ and $\tan \Theta$, from the sideslope angle Θ_1 and the downstream bed-slope angle Θ_0. These two parameters describe the projection of the submerged weight vector along the embankment plane. The angle Θ is obtained from the ratio of the two projection components of F_S in the embankment plane as $\tan \Theta = [(\cos \Theta_1 \sin \Theta_0)/(\cos \Theta_0 \sin \Theta_1)]$, which can be approximated by $\tan \Theta \cong (\sin \Theta_0/ \sin \Theta_1)$ as long as both angles are fairly small (less than $\sim 20°$). The fraction of the submerged weight that is normal to the embankment plane is given by $a_\Theta = \sqrt{1 - \cos^2\Theta_0 \sin^2\Theta_1 - \cos^2\Theta_1 \sin^2\Theta_0}$, which is approximated by $a_\Theta \cong \sqrt{\cos^2\Theta_1 - \sin^2\Theta_0}$ when both angles are small (less than $\sim 20°$). As a realistic approximation, the submerged weight has one sideslope component, $F_S \sin\Theta_1$, one downslope component, $F_S \sin\Theta_0$, and a component normal to the plane, $F_S a_\Theta$, as shown in Fig. 6.1. The streamline deviates from the downstream direction at an angle λ along the embankment plane (λ is defined positive downward). When unstable, a particle moves at an angle β from the direction of steepest descent. A particle moves along the horizontal when $\beta = 90°$ and moves in the downstream direction when $\beta + \Theta = 90°$. When $\beta + \Theta > 90°$, a particle moves up the sideslope toward the free surface. Conversely, when $\beta + \Theta < 90°$, a particle moves down the sideslope toward the thalweg. In most streams, the downstream slope Θ_0 will be sufficiently small to consider that $\Theta \cong 0$ and $a_\Theta \cong \cos\Theta_1$. The downstream direction thus practically corresponds to $\beta = 90°$, values of $\beta < 90°$ indicate particles moving toward the thalweg, and $\beta > 90°$ indicates motion toward the free surface.

Stability against rotation of a particle determines incipient motion when the equilibrium of moments about the point of rotation is satisfied. The deviation angle δ is measured between the particle direction and the streamline. Considering the angles δ and β and the moment arms l_1, l_2, l_3, and l_4 that are shown in Fig. 6.1, we find that stability about point 0 corresponds to

$$l_2 F_S a_\Theta = l_1 F_S \sqrt{1 - a_\Theta^2} \cos\beta + l_3 F_D \cos\delta + l_4 F_L. \qquad (6.1)$$

The left-hand side of Eq. (6.1) defines the stabilizing moment that is due to the particle weight. Clearly the last term on the right-hand side denotes the lift moment, which always destabilizes the particle. The first two terms on the right-hand side determine about which pivot point particle P is to rotate. In most cases, their net sum is positive, and moments about 0 are considered. Should their net sum be negative, when $\lambda < 0$, the particle will then rotate about point $0'$ instead of about 0.

The stability factor SF_0, for rotation about point 0, is defined as the ratio of the resisting moments to the moment generating motion. In the case shown in Fig. 6.1, in which both $\cos\delta$ and $\cos\beta$ are positive, the stability factor SF_0 is the ratio of the sum of counterclockwise moments about 0 to the sum of clockwise moments about 0; thus

$$SF_0 = \frac{l_2 F_S a_\Theta}{l_1 F_S \sqrt{1 - a_\Theta^2} \cos\beta + l_3 F_D \cos\delta + l_4 F_L}. \qquad (6.2) \blacklozenge\blacklozenge$$

Note that each term in Eq. (6.2) must be positive; otherwise the formulation is changed to express the ratio of positive stabilizing moments to positive destabilizing moments.

Because the stability factor SF_0 equals unity when the angle Θ_0 or Θ_1 equals the angle of repose ϕ under static fluid conditions ($F_D = F_L = 0$), it is found that $\tan \phi = l_2/l_1$. Dividing both the numerator and the denominator by $l_1 F_S$ transforms Eq. (6.2) into

$$SF_0 = \frac{a_\Theta \tan \phi}{\eta_1 \tan \phi + \sqrt{1 - a_\Theta^2} \cos \beta}, \qquad (6.3) \blacklozenge$$

in which $\eta_1 = M + N \cos \delta$ after we define

$$M = \frac{l_4 F_L}{l_2 F_S}, \qquad N = \frac{l_3 F_D}{l_2 F_S}. \qquad (6.4)$$

The parameter M/N represents the ratio of lift to drag moments of force. The case of no lift is given by $M/N = 0$ and equal moments are described by $M = N$ or $(M + N)/N = 2$. The variable η_1 is called the sideslope stability number for the particle on the embankment. The variable η_1 relates to the stability number $\eta_0 = M + N$ for particles on a plane-horizontal surface ($\Theta_0 = \Theta_1 = \delta = 0$) after we consider $\lambda + \delta + \beta + \Theta = 90°$:

$$\eta_1 = \eta_0 \left\{ \frac{(M/N) + \sin(\lambda + \beta + \Theta)}{1 + (M/N)} \right\}, \qquad (6.5) \blacklozenge$$

$$\eta_0 = \frac{\tau_0}{\tau_c} = \frac{\tau_0}{(G - 1)\rho \, g \, d_s \, \tau_{*_c}}. \qquad (6.6) \blacklozenge$$

The stability number η_0 is calculated from the applied shear stress τ_0, the critical shear stress on a plane-horizontal surface τ_c, the particle diameter d_s, the mass density of the particle $\rho_s = G\rho$, the mass density of the fluid ρ, the gravitational acceleration g, and the critical value of the Shields parameter τ_{*_c}.

This normalized form of the Shields parameter shows that $\eta_0 = 1$ describes the incipient motion of particles on a plane-horizontal bed. When the flow is fully turbulent over a hydraulically rough horizontal surface, incipient motion approximately corresponds to $\tau_{*_c} = 0.047$ and $SF_0 = \eta_0 = 1$.

The second equilibrium condition indicates the direction of a moving particle from equilibrium conditions along the section normal to A-A in Fig. 6.1:

$$l_3 F_D \sin \delta = l_1 F_S \sqrt{1 - a_\Theta^2} \sin \beta. \qquad (6.7)$$

After δ is written as a function of λ, Θ, and β, solving for β gives

$$\beta = \tan^{-1}\left[\frac{\cos(\lambda + \Theta)}{\dfrac{(M + N)\sqrt{1 - a_\Theta^2}}{N\eta_0 \tan\phi} + \sin(\lambda + \Theta)}\right]. \qquad (6.8) \blacklozenge$$

The angle β determines the direction of motion of a sediment particle in contact with the inclined plane. The particle-orientation angle depends on (1) surface topography by means Θ and a_Θ, (2) streamflow direction at an angle λ, and (3) particle characteristics such as angle of repose ϕ and excess shear $\eta_0 = \tau_0/\tau_c$. A complete calculation example is given in Example 6.1 .

Example 6.1   Application to particle-stability analysis. The following example details the calculations of particle stability for typical river-bend conditions by use of Eqs. (6.3), (6.5), (6.6), and (6.8). The overall agreement with field data being slightly better with $M/N = 0$, this value was given preference for all calculations.

A 16-mm quartz particle stands on the bed of a channel; if the downstream channel-slope angle is $\Theta_0 = 0.074°$ and the sideslope angle is $\Theta_1 = 15°$, calculate the particle-direction angle under an applied bed shear $\tau_0 = 12$ Pa when the streamlines are deflected upward at $\lambda = -10°$. The calculation procedure is as follows:

(a) The particle size is $d_s = 16$ mm.
(b) The angle of repose is approximately $\phi = 37°$, and the specific gravity $G = 2.65$ is assumed.
(c) The sideslope angle is $\Theta_1 = 15°$.
(d) The downstream slope angle is $\Theta_0 = 0.074°$.
(e) The angle $\Theta = \tan^{-1}(\sin\Theta_0/\sin\Theta_1) = 0.28°$.
(f) The geometric factor $a_\Theta = \sqrt{\cos^2\Theta_1 - \sin^2\Theta_0} = 0.965$.
(g) The applied bed shear stress is $\tau_0 = 12$ Pa; this corresponds to a flow depth of ~ 1 m.
(h) The streamline deviation angle $\lambda = -10°$ means that the downstream shear-stress component is $\tau_0 \cos\lambda = 11.8$ Pa and the transverse shear stress is $\tau_0 \sin\lambda = -2.08$ Pa toward the free surface (negative).
(i) From Eq. (6.6) and $\tau_{*_c} = 0.047$, the plane-bed-stability number is

$$\eta_0 = \frac{12\,\mathrm{N m^3 s^2}}{\mathrm{m^2}\,(1.65)\,1{,}000\,\mathrm{kg} \times 9.81\,\mathrm{m} \times 0.016\,\mathrm{m} \times 0.047} = 0.985.$$

(j) From Eq. (6.8), assuming $M/N = 0$, the 16-mm particle-direction angle is

$$\beta = \tan^{-1} \left[\frac{\cos(-10° + 0.28°)}{\dfrac{\sqrt{1 - (0.965)^2}}{0.985 \tan 37°} + \sin(-10° + 0.28°)} \right] = 79.4°.$$

(k) The 16-mm particle moves toward the thalweg because $\beta + \Theta < 90°$.

(l) The stability factor is calculated from Eq. (6.3).

$$\text{SF}_\circ = \frac{0.965 \tan 37°}{0.985 \sin(-10° + 79.4 + 0.28°) \tan 37° + \sqrt{1 - 0.965^2} \cos 79.4°}$$
$$= 0.97.$$

The particle is unstable. We easily repeat the calculations with a 32-m particle to find out that it is stable under the given hydraulic conditions.

6.2   Channel stability

Consider the cross-sectional geometry of a straight channel for which all particles of weight F_s along the wetted perimeter are at incipient motion. The critical shear stress τ_{sc} on a sideslope θ_1 is determined in analogy to the critical shear stress τ_c that corresponds to the angle of repose ϕ. A very simplified but elegant equilibrium relationship was proposed by Lane (1953) who, after assuming that $\tau_c \sim F_s \tan \phi$ and the resultant on the sideslope $R/F_s \cos \theta_1 = \tan \phi$, obtained

$$\frac{\tau_{sc}}{\tau_c} = \sqrt{1 - \frac{\sin^2 \theta_1}{\sin^2 \phi}} = \cos \theta_1 \sqrt{1 - \frac{\tan^2 \theta_1}{\tan^2 \phi}}. \tag{6.9} \blacklozenge$$

A detailed derivation can also be found in Julien (1995). As sketched in Fig. 6.2, the thalweg shear stress is $\tau_c = \gamma h_0 S$ and $\tau_0 = \gamma h S \cos \theta_1$ on the sideslope. Isolating γS gives the following identity:

$$\tau_0 h_0 = \tau_c h \cos \theta_1. \tag{6.10}$$

The differential equation for the ideal cross-section geometry is obtained from $\tan \theta_1 = -dh/dy$ after τ_0/τ_c is canceled from Eqs. (6.9) and (6.10), thus

$$\left(\frac{dh}{dy} \right)^2 + \left(\frac{h}{h_0} \right)^2 \tan^2 \phi = \tan^2 \phi, \tag{6.11}$$

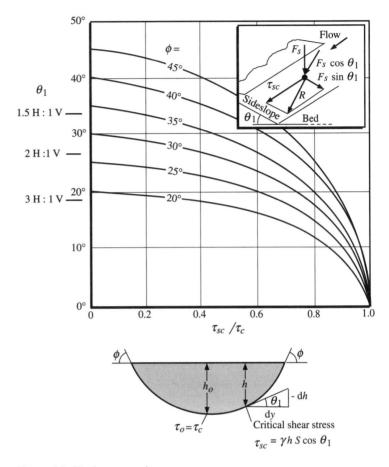

Figure 6.2. Ideal cross-section geometry.

where h_0 is the flow depth at the centerline of the channel. The ideal cross-section geometry in which all particles are at incipient motion has a cosinusoidal shape, with y measured laterally from the centerline:

$$\frac{h}{h_0} = \cos\left(\frac{y \tan\phi}{h_0}\right). \tag{6.12} \blacklozenge$$

The flow depth h_0 corresponds to incipient motion on a plane surface $h_0 \approx 0.047\,(G-1)d_s/S$. The surface width, $h_0\pi/\tan\phi$, is obtained from the values of y in Eq. (6.12) where $h = 0$. The width–depth ratio is constant and equal to $\pi \cot\phi$. The cross-section area is $A = 2\,h_0^2\cot\phi$, and the hydraulic radius is

$$R_h = \frac{8h_0 \cos \phi}{\pi (4 - \sin^2 \phi)}.$$

The mean flow velocity V is then obtained from a resistance relationship, e.g., the Manning equation. The flow discharge is then obtained from A and V (see Problem 6.2). In practice, only coarse bed channels may reach incipient motion at high discharge. Sediment transport in all sand-bed channels cannot be ignored, and the width–depth ratio of most alluvial streams far exceeds the ideal geometry conditions.

6.3 Regime relationships

The construction of irrigation canals in India and Pakistan fostered investigations of the design of canals under regime conditions, i.e., a canal that is nonsilting and nonscouring. Empirical relationships have been proposed by Kennedy (1895), Lacey (1929), and Blench (1969), among many others. The Lacey silt factor f_e was defined to designate the properties of "Kennedy's standard silt" of the Upper Bari Doab canal. The geometry of canals with different bed material were compared in terms of different values of f_e. The silt factor was shown to increase with grain size; the approximate relationship $f_l = 1.59 \, d_{mm}^{1/2}$ can be used. The key relationships from Lacey determined the mean flow velocity V in feet per second, the hydraulic radius R_h in feet, the cross-section area A in square feet, the wetted perimeter P in feet, and the dimensionless slope as a function of the design discharge Q in cubic feet per second, and the Lacey silt factor:

$$V = 0.794 \, Q^{1/6} f_l^{1/3}, \tag{6.13}$$

$$R = 0.47 \, Q^{1/3} f_l^{-1/3}, \tag{6.14}$$

$$A = 1.26 \, Q^{5/6} f_l^{-1/3}, \tag{6.15}$$

$$P = 2.66 \, Q^{1/2}, \tag{6.16}$$

$$S = 0.00053 \, f_l^{5/3} Q^{-1/6}. \tag{6.17}$$

In wide-shallow channels, the hydraulic radius is approximately equal to the flow depth, and the channel width can often be approximated by the wetted perimeter. The width–depth ratio thus increases slightly with discharge and decreases slightly with grain size. Another characteristic of the regime equation is that once the discharge and grain size are determined a unique value of bed slope is calculated from Eq. (6.17). Note that the method does not allow

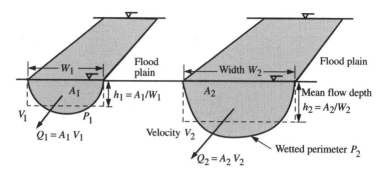

Figure 6.3. Downstream hydraulic geometry.

changes in slope at a given discharge and grain size. Consequently, the velocity in Eq. (6.13) does not depend on slope.

The regime equations refer to downstream channel geometry at a given bankfull discharge. As shown in Fig. 6.3, different discharges in Eqs. (6.13)–(6.17) correspond to bankfull conditions in different channels. Because these channels can be located at different positions in the same fluvial system, channel properties are referred to as downstream hydraulic geometry. Downstream hydraulic geometry describes bankfull conditions for different cross sections, as opposed to the at-a-station hydraulic geometry that describes channel properties at different discharges at a given cross section.

6.4   Equilibrium in river bends

Secondary circulation in curved channels is generated through a change in downstream channel orientation. The streamlines near the surface are deflected toward the outer bank whereas those near the bed are deviated toward the inner bank. The near-bed velocity, the tangential bed shear stress, and the drag on the bed particles are commonly directed toward the inner bank.

Flow in bends is analyzed in cylindrical coordinates. The relative magnitude of radial-acceleration terms indicates that the centrifugal acceleration is counterbalanced by pressure gradient and radial shear stress, as suggested by Rozovskii (1957):

$$\frac{v^2}{r} = g S_r - \frac{1}{\rho} \frac{\partial \tau_r}{\partial z}, \tag{6.18}$$

where the local downstream velocity v, the radial shear stress τ_r, and the radial water-surface slope S_r vary with the vertical elevation z and/or the radius of curvature r. In Fig. 6.4, the transverse boundary shear stress τ_{rR} at point R_A, the

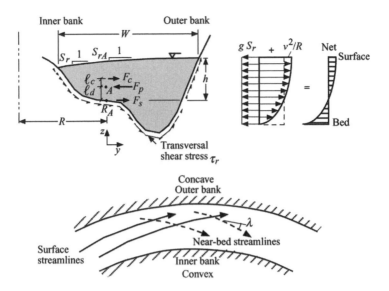

Figure 6.4. Flow in river bends.

radial water-surface slope S_{rA} at point A, the radius of curvature R at the same point, the average flow depth h, and the top channel width W serve as scaling factors.

These scaling factors define dimensionless parameters for channel width $w^* = w/W$, flow depth $z^* = z/h$, radius of curvature $r^* = r/R$, velocity $v^* = v/\bar{V}$, radial shear stress $\tau_r^* = \tau_r/\tau_{rR}$, and the radial surface slope $S_r^* = S_r/S_{rA}$. The element of fluid volume $d\forall = dx\, dy\, dz$ for a reach of given length $dx = R\,d\theta$ is reduced to a dimensionless volume $d\forall^* = d\forall/WRh$. Radial equation of motion (6.18) is multiplied by ρ and $d\forall$, reduced in dimensionless form, and then integrated over the dimensionless volume $\forall^*$ of the reach. The resulting dimensionless momentum equation in the radial direction is

$$\rho W h \bar{V}^2 \int_{\forall^*} \frac{v^{*2}}{r^*} d\forall^* = \rho g R W h S_{rA} \int_{\forall^*} S_r^* d\forall^* - R W \tau_{rR} \int_{\forall^*} \frac{\partial \tau_r^*}{\partial z^*} d\forall^*. \quad (6.19)$$

The corresponding force diagram is sketched in Fig. 6.4, after the centrifugal force is denoted on the left-hand side of Eq. (6.19) by F_c, the pressure force F_p describes the first term on the right-hand side, and the last integral represents the shear force F_s. The pressure force is found to balance the sum of the centrifugal force exerted at a distance l_c above point A whereas the shear force is exerted at a distance l_d below point A. Moment equilibrium around point A

gives

$$\Omega_R = \frac{\rho h \bar{V}^2}{R \tau_{rR}} = \frac{l_d \int_{\forall^*} \frac{\partial \tau_r^*}{\partial z^*} d\forall^*}{l_c \int_{\forall^*} \frac{v^{*2}}{r^*} d\forall^*}. \tag{6.20}$$

The dimensionless parameter Ω_R denotes the ratio of the centrifugal force generating secondary motion to the shear force abating the motion and dissipating energy. The resulting ratio of radial shear stress τ_{rR} to the downstream bed shear stress τ_θ defines the deviation angle λ of the streamlines near the bed. Therefore, combining Eqs. (6.20) and (4.10), we obtain

$$\tan \lambda = \frac{\tau_{rR}}{\tau_\theta} = \left[\frac{a^2}{\Omega_R} \left(\frac{h}{d_s} \right)^{2m} \right] \frac{h}{R}. \tag{6.21} \blacklozenge$$

Rozovskii (1957) found that the value of the term in brackets of Eq. (6.21) is approximately equal to 11. Slightly different values have been proposed by Engelund (1974), de Vriend (1977), Odgaard (1981), and Hussein and Smith (1986). We conclude that deviation angle λ depends primarily on the ratio of flow depth to radius of curvature. Sharp bends will exhibit stronger secondary flows.

In curved channels, secondary-flow effects on particle stability are examined through the influence of the deviation angle λ on the values of the stability factor SF_0. Stability calculations based on Eqs. (6.6), (6.8), (6.5), and (6.2) with $\lambda \neq 0$ demonstrate that slight downward deviations decrease particle stability, whereas small upward deviations increase particle stability. On the other hand, large upward deviations will also decrease particle stability.

Quantitative dimensionless results are summarized in Fig. 6.5 for typical 2D flow conditions in a curved alluvial channel. The local cross-sectional shape is described as the ratio of the embankment slope θ to the angle of repose ϕ. This figure illustrates the relative particle-stability ratio $\Psi_\lambda = SF_0 (\lambda \neq 0)/ SF_0(\lambda = 0)$ of the stability factor with secondary circulation ($\lambda \neq 0$) over the stability factor without secondary circulation ($\lambda = 0$). It is shown in Fig. 6.5 that, when the deviation angle λ is relatively small, for example less than $15°$, the downward deflected streamlines near the outer bank ($\theta/\phi > 0$) induce particle motion, as expected from $\Psi_\lambda < 1$. Conversely, opposite effects are observed near the inner bank ($\theta/\phi < 0$); the upward deviations of the streamlines increase the stability of particles, as shown by the values of $\Psi_\lambda > 1$. The conditions induced by secondary circulation at angles $\lambda < 15°$ decrease particle stability near the outer bank and increase particle stability near the inner bank of curved alluvial channels.

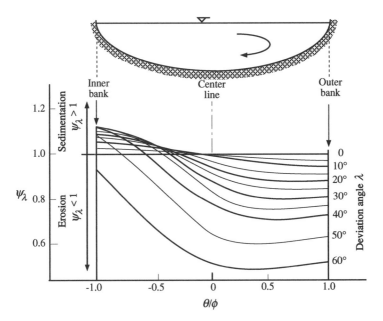

Figure 6.5. Relative particle stability in curved channels.

When the strength of secondary circulation increases, $15° < \lambda < 55°$, asymmetry in the particle-stability curves shown in Fig. 6.5 develop and a larger proportion of the channel becomes unstable. For extreme conditions ($\lambda > 55°$), the entire cross section becomes unstable and scour occurs for all particles on the wetted perimeter of the alluvial channel, thus widening the channel.

This analysis of secondary-circulation effects on particle mobility in curved alluvial channels highlights a continuum of conditions between the following two extremes: (1) at small deviation angles, around $\lambda < 15°$, equilibrium prevails between outer-bank erosion and inner-bank deposition; and (2) at large deviation angles, typically when $\lambda > 55°$, the particle stability decreases over the entire cross section and should result in widening of the alluvial channel. Secondary-circulation effects being primarily felt near the outer bank, the deviation angle λ thus relates to the widening process in alluvial channels.

The result of downward lateral forces is to destabilize the bed particles whereas upward lateral forces stabilize bed particles [Fig. 6.6(b)]. Under significant secondary flows, cross-sectional geometry becomes asymmetric, with the thalweg moving toward the outer bank, as sketched in Fig. 6.6. Stability can be maintained only through a heavier particle weight to counterbalance the higher erosive forces. Conversely, finer particles can deposit to form a point bar near the inner bank.

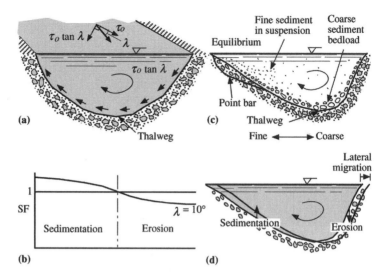

Figure 6.6. Stability and equilibrium in river bends.

To reach stability in graded bed material in which armoring is possible, coarser grains can be found near the outer bank and finer grains near the inner bank [Fig. 6.6(c)]. In the case of uniform erodible bed material, secondary flows will induce scour at the toe of the outer bank, leading to bank caving and lateral migration. Equilibrium is possible only when bank caving balances inside deposition on the point bar [Fig. 6.6(d)].

It is also important to note that the change in cross-sectional geometry depends on the magnitude of the streamline deviation angle λ. From Eq. (6.21) we obtain that the angle λ increases with flow depth h. At high flows, we thus expect to find the strongest secondary flows with the maximum scour potential near the outer bank. The asymmetry in the cross-sectional geometry should be observed during floods. At low flows, low values of the angle λ will tend to bring the thalweg position back closer to the channel centerline and the cross-sectional geometry should become more symmetrical.

6.5   Downstream hydraulic geometry

Flow discharge varies with time, and sediment transport is most active during floods. The morphology of alluvial channels should therefore depend on a high discharge value that contributes to the channel formation. This value should be high enough such that sediment transport is quite active, yet it should

No levees   Outer bank levees   Levees on both banks

Figure 6.7. Natural levees.

reoccur a sufficient number of times to maintain a channel free of vegetation. A bankfull discharge is determined by the discharge that a channel can convey when reaching the floodplain level. Natural levees result from the deposition of coarser fractions of the suspended load on the floodplain adjacent to the channel. As sketched in Fig. 6.7, levees can form near the outer bank of the channel or along both sides of alluvial channels. The reoccurrence frequency of bankfull flows is quite variable (Williams, 1978). Periods of return of 1.5 and 2 yrs are most often encountered in the literature, but mean annual floods, mean annual flows, and 5-yr floods have also been cited.

The dominant discharge is of sufficient magnitude and frequency to determine the dimension and geometry of an alluvial channel. For hydraulic-geometry relationships, it is taken to be the bankfull discharge, which has a period of return of approximately 1.5 yr in many natural channels. It is important to remember that the concept of dominant discharge cannot be precisely quantified. The dominant discharge of a river remains subjective and variable. The downstream hydraulic-geometry relationships developed in this section can provide only approximative values of width, depth, velocity, and slope.

The downstream hydraulic geometry of noncohesive alluvial channels can be analytically determined from the following four relationships. First, under steady-uniform bankfull flow conditions, the dominant discharge Q is

$$Q = WhV, \tag{6.22}$$

where the mean velocity vector V is taken normal to the cross-sectional area, h is the flow depth, and W is the bankfull width.

For channels with large width–depth ratios, the hydraulic radius R_h becomes equal to the flow depth h. Second, the power form of the resistance equation from Eq. (4.10) is

$$V = a\sqrt{g}\left(\frac{h}{d_s}\right)^m h^{1/2} S^{1/2}, \tag{6.23}$$

where g is the gravitational acceleration, d_s is the grain diameter, S is the slope, and the exponent $m = 1/\ln(12.2\,h/d_s)$. Note that the value of $m = 1/6$ corresponds to the Manning–Strickler resistance equation.

Third, the stability of noncohesive particles in straight alluvial channels is described by the relative magnitude of the downstream shear force and the weight of the particle. The ratio of these two forces defines the Shields number τ_* from Eq. 4.36:

$$\tau_* = \frac{hS}{(G-1)d_s},\tag{6.24}$$

where G is the specific gravity of sediment particles. The critical value of the Shields number, $\tau_* \approx 0.047$, identifies the beginning of motion of non-cohesive particles in turbulent flows over rough boundaries. For values of the Shields number below the critical value ($\tau_* \leq \tau_{*_c}$), the particles on the wetted perimeter of the alluvial channel are stable. Beyond this threshold ($\tau_* > \tau_{*_c}$), the particles enter motion and the rate of sediment transport increases with the Shields number. Two significant concepts are associated with the Shields number: (1) the threshold concept described by τ_{*_c} for the beginning of motion of noncohesive particles and (2) the concept that beyond the threshold value, the sediment-transport rate increases with the Shields number. Because the Shields number depends primarily on flow depth, it is thus associated with the vertical processes of aggradation and degradation in alluvial channels.

Fourth, for flow in bends, with the radius of curvature R being proportional to the channel width W, Eq. (6.21) shows that

$$\tan \lambda = b_r \left(\frac{h}{d_s}\right)^{2m} \frac{h}{W},\tag{6.21a}$$

where the value of $b_r = [(a^2 W)/(\Omega_R R)]$ is assumed to be constant.

Equations (6.21)–(6.24) contain 13 variables, namely W, h, V, S, Q, d_s, τ_*, $\tan \lambda$, g, a, b_r, G, and m. Four relationships enable the definition of the four dependent variables, W, h, V, and S, as functions of the others. Julien (1988) showed that Q, d_s, and τ_* were the primary independent variables. The variability in the other parameters is considered to be relatively small. Julien and Wargadalam (1995) determined approximative empirical values for the remaining parameters from a large data set, including data from 835 rivers and canals.

The downstream hydraulic geometry for noncohesive alluvial channels for hydraulically rough turbulent flows were derived for flow depth h in meters, surface width W in meters, average flow velocity V in meters per second, and friction slope S as

$$h = 0.133 \, Q^{\frac{1}{3m+2}} \, d_s^{\frac{6m-1}{6m+4}} \, \tau_*^{\frac{-1}{6m+4}},$$ (6.25a)

$$W = 0.512 \, Q^{\frac{2m+1}{3m+2}} \, d_s^{\frac{-4m-1}{6m+4}} \, \tau_*^{\frac{-2m-1}{6m+4}},$$ (6.25b)

$$V = 14.7 \, Q^{\frac{m}{3m+2}} \, d_s^{\frac{2-2m}{6m+4}} \, \tau_*^{\frac{2m+2}{6m+4}},$$ (6.25c)

$$S = 12.4 \, Q^{\frac{-1}{3m+2}} \, d_s^{\frac{5}{6m+4}} \, \tau_*^{\frac{6m+5}{6m+4}}$$ (6.25d)

from the equilibrium or dominant flow discharge Q in cubic meters per second, the median grain size $d_s = d_{50}$ in meters, and the Shields parameter $\tau_* = \gamma h S / (\gamma_s - \gamma) d_{50}$, given the resistance exponent m calculated from $m = 1/\ln(12.2\,h/d_{50})$. An example of the agreement among predicted and measured width, depth, velocity, and slope from Julien and Wargadalam (1995) is shown in Fig. 6.8.

The recommended calculation procedure for the downstream hydraulic geometry starts with the user selection of three independent variables. To include the effects of sediment transport, the user may want to calculate four dependent variables of average flow depth h in meters, surface width W in meters, average flow velocity V in meters per second, equilibrium slope S as a function of three known independent variables in discharge Q in cubic meters per second, median grain size d_s in meters, and dimensionless Shields number τ^*; Eqs. (6.25a)–(6.25d) are solved with the five-step procedure outlined below in Example 6.2.

When Manning's equation is applicable, $m = 1/6$, a simplified form of Eqs. (6.25) is obtained as

$$h \cong 0.133 \, Q^{0.4} \, \tau_*^{-0.2},$$ (6.26a)

$$W \cong 0.512 \, Q^{0.53} \, d_s^{-0.33} \, \tau_*^{-0.27},$$ (6.26b)

$$V \cong 14.7 \, Q^{0.07} \, d_s^{0.33} \, \tau_*^{0.47},$$ (6.26c)

$$S \cong 12.4 \, Q^{-0.4} \, d_s \, \tau_*^{1.2}.$$ (6.26d)

The hydraulic geometry of stable channels is obtained from Eqs. (6.26) when $\tau_* \cong 0.047$. Higher sediment transport implies higher velocity and slope and reduced width and depth.

This system of equations is sufficiently flexible to allow the user to use a different set of known independent variables. For instance, geomorphologists may prefer to calculate flow depth h, width W, mean velocity V, and Shields number τ_* as explicit functions of discharge Q in cubic meters per second,

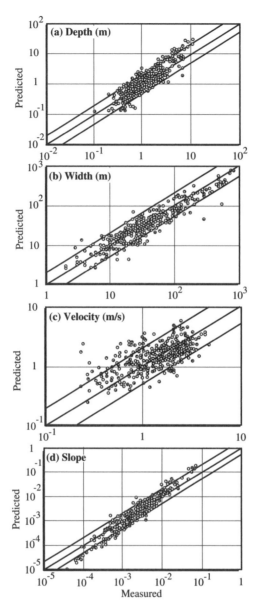

Figure 6.8. Downstream hydraulic-geometry method of Julien and Wargadalam (1995).

median grain size d_s in meters, and channel slope S. The empirically recalibrated equations of Julien and Wargadalam (1995) are

$$h = 0.2 \, Q^{\frac{2}{5+6m}} \, d_s^{\frac{6m}{5+6m}} \, S^{\frac{-1}{5+6m}}, \tag{6.27a}$$

$$W = 1.33 \, Q^{\frac{2+4m}{5+6m}} \, d_s^{\frac{-4m}{5+6m}} \, S^{\frac{-1-2m}{5+6m}}, \tag{6.27b}$$

$$V = 3.76 \, Q^{\frac{1+2m}{5+6m}} \, d_s^{\frac{-2m}{5+6m}} \, S^{\frac{2+2m}{5+6m}}, \tag{6.27c}$$

$$\tau_* = 0.121 \, Q^{\frac{2}{5+6m}} \, d_s^{\frac{-5}{5+6m}} \, S^{\frac{4+6m}{5+6m}}, \tag{6.27d}$$

where $m = 1/\ln(12.2 \, h/d_s)$. The equations are solved with the procedure given in Example 6.2.

For the particular case in which Manning's equation is acceptable, $m = 1/6$, the downstream hydraulic-geometry relationships simplify to

$$h = 0.2 Q^{0.33} \, d_s^{0.17} \, S^{-0.17}, \tag{6.28a} \, \blacklozenge\blacklozenge$$

$$W = 1.33 Q^{0.44} \, d_s^{-0.11} \, S^{-0.22}, \tag{6.28b} \, \blacklozenge\blacklozenge$$

$$V = 3.76 Q^{0.22} \, d_s^{-0.05} \, S^{0.39}, \tag{6.28c} \, \blacklozenge\blacklozenge$$

$$\tau_* = 0.121 Q^{0.33} \, d_s^{-0.83} \, S^{0.83}. \tag{6.28d} \, \blacklozenge\blacklozenge$$

It is observed that the channel width varies primarily with discharge whereas other parameters also affect flow depth and flow velocity. A plot of channel width vs. discharge is shown in Fig. 6.9.

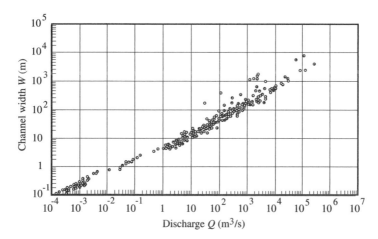

Figure 6.9. Channel width vs. discharge (after Kellerhals and Church, 1989).

Equation (6.28d) relates sediment transport by means of τ_* to the discharge, grain size, and slope. For instance, if we consider the following crude approximation for the unit bed-sediment discharge q_s in square meters per second as $q_s \cong 18\sqrt{g\, d_s^3}\, \tau_*^2$, the solution for the bed-material discharge $Q_s = W q_s$ is obtained from Eqs. (6.28b) and (6.28d):

$$Q_s\, d_s^{0.28} \simeq Q^{1.11}\, S^{1.44}, \tag{6.29} \blacklozenge$$

where Q_s is the bed-material load in cubic meters per second, d_s is the bed-material size in meters, Q is the bankfull discharge in cubic meters per second, and S is the bed slope. This approximation relates to downstream conditions and does not apply to at-a-station conditions.

This crude relationship shows the equilibrium between, on the right side, the product of hydraulic parameters (discharge and slope) and, on the left side, the product of sediment characteristics (sediment discharge and grain size). Lane (1955a) had proposed a similar form in which $QS \sim Q_s d_s$. Example 6.2 illustrates how the hydraulic geometry of a channel can be estimated from the discharge, grain size, and slope.

Example 6.2   Application to stable channel geometry. Calculate the downstream hydraulic geometry given $Q = 104\,\mathrm{m^3/s}$, $d_{50} = 0.056$ m, and $\tau_\theta^* = 0.047$ at the beginning of motion.

Step 1: Roughly estimate the flow depth, e.g., $h = 1$ m.

Step 2: From the flow depth and grain size calculate m from

$$m = \frac{1}{\ln\left(\dfrac{12.2\,h}{d_s}\right)} = 0.186.$$

Step 3: Calculate the exponents for flow depth from Eq. (6.25a), given $m = 0.186$:

$$h = a Q^b d_s^c \tau_\theta^{*d} = 0.133(104)^{0.39}\,(0.056)^{0.023}\,(0.047)^{-0.195} = 1.38\,\mathrm{m}.$$

Step 4: Repeat steps 2 and 3 with the calculated flow depth in step 3 until convergence:

$$m = 0.175 \text{ gives } h = 1.49\,\mathrm{m}, \quad \text{and} \quad m = 0.172 \text{ gives } h = 1.51\,\mathrm{m}.$$

Step 5: Calculate the channel width W, flow velocity V, and slope S by using the last value of m and the exponents of Q, d_s, and τ_θ^* in Eqs. (6.25), e.g., with

$m = 0.172$:

$$W = 0.512(104)^{0.534}(0.056)^{-0.335}(0.047)^{-0.267} = 36.4\,\text{m},$$

$$V = 14.7(104)^{0.068}(0.056)^{0.329}(0.047)^{0.466} = 1.87\,\text{m/s},$$

$$S = 12.4(104)^{-0.397}(0.056)^{0.994}(0.047)^{1.199} = 2.86 \times 10^{-3}.$$

6.6   Bars in alluvial rivers

Channel-bed configurations include bedforms and bars. Bedforms include ripples, dunes, and antidunes and remain submerged, except during droughts or in ephemeral streams. The reader is referred to Julien (1995) regarding bedform classification and formation. Bars refer to large bedform configurations that are often exposed during low flows. They are usually submerged at least once a year in order to prevent vegetation growth. Bars can be viewed as alluvial bed deposits that can be transported under high flow conditions. When bars do not get submerged approximately every year, vegetation grows and stabilizes the bars to form islands or to reduce the active channel width of a river.

Alternate bars form in straight channels with deposits alternating from the right bank to the left bank. As illustrated in Fig. 6.10(a), the wavelength of alternate bars Λ is proportional to the channel width W, and $\Lambda \simeq 2\pi W$. Alternate bars tend to form in channels where the Froude number is high and the Shields parameter is close to incipient motion. The height of alternate bars can reach the flow depth. The thalweg is said to wander or weave between both banks. Additional information regarding alternate bars can be found in Fujita and Muramoto (1982, 1985) and in Ikeda (1984).

Point bars are sketched in Fig. 6.10(b). Their formation process is akin to that of the secondary flows discussed in Section 6.4, and they usually are found near the inner bank of river bends. The increased particle stability near the inner bank induces sedimentation and fining of the bed deposits. The process of point-bar sedimentation is usually associated with the erosion of the outer bank and lateral migration of the river. During major floods, point-bar deposits can be remobilized to form chute cutoffs, as sketched in Fig. 6.10(c). At low flows, chute cutoffs may induce sedimentation in both river branches. The truncated point bar is also referred to as a middle bar.

Tributary bars form in the main channel near the confluence, with tributaries carrying a significant sediment load. As shown in Fig. 6.10(d) the tributary bar contributes to streamlining of the confluence of both streams. The

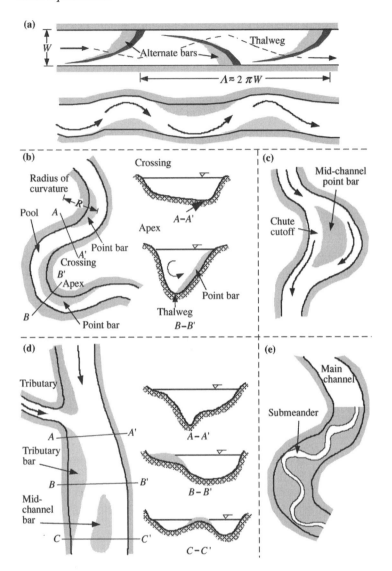

Figure 6.10. Bar formations in rivers.

tributary bars can be reworked, depending on the magnitude, sediment load, and timing of the floods in both channels. Figure 6.10(e) shows submeanders, which are defined as small meanders confined within the banks of a perennial stream channel. Submeanders are caused by very low flows compared with flood discharges.

6.7   River meandering

River meandering is characterized by a succession of alternating meander loops. A meander loop is the channel reach between two inflection points. A meander consists of a pair of loops in opposite directions. Numerous hypotheses have been suggested to explain the origin of meandering, including secondary flows, perturbation theory, and extremal hypotheses.

Meandering is basically a fluid mechanics problem in which vorticity plays a leading role. The motion of fluid in a curved channel is generally based on the equations of motion. Different approaches were proposed by Einstein (1926), Rozovskii (1957), Yen (1970), DeVriend (1977), Odgaard (1981), Nelson and Smith (1989), and others. Several attempts focus on the hydrodynamic stability of straight alluvial channels. A perturbation technique is used to determine whether small oscillations amplify or decay. Examples can be found in Callander (1969), Anderson (1967), Engelund and Skovgaard (1973), Parker (1976), and Ikeda et al. (1981).

Extremal hypotheses include the principle of minimum variance first proposed by Langbein and Leopold (1966). The minimization involves the adjustment of the planimetric geometry and the hydraulic factors of depth, velocity, and local slope. Yang (1976) stated that the time rate of energy expenditure explains the formation of meandering streams. Other studies by Maddock (1970) and Chang (1980) use the principle of minimum stream power. Chang (1979a) concluded that a meandering river is more stable than a straight one as it expends less stream power per unit channel length for the system. Julien (1985) treated meandering as a variational problem in which the energy integral corresponds to the functional of a variational problem, the solution of which is the sine-generated curve.

Consider the reach of a meandering alluvial river, as sketched in Fig. 6.11. Two systems of coordinates are defined: one rectilinear and one curvilinear. The down-valley axis X defines the rectilinear system along the centerline of the meandering pattern downstream of the valley slope. In the curvilinear system, the sinuous axis x follows the centerline of the meandering river path. The angle θ separates the directions x and X along the flow path. Measured from the river centerline, the radius of curvature R in the transversal direction y remains orthogonal to the downstream axis x. Both the magnitude and the direction of the radius of curvature R vary along the path of the channel width W and mean flow velocity V. The radius of curvature is minimum, R_{min}, at the apex and is maximum, $R = \infty$, at the crossing. A complete meander loop is denoted between O and M. At M, the river length is L and the meander length is Λ. The amplitude of the meander belt, or meander width, is W_m.

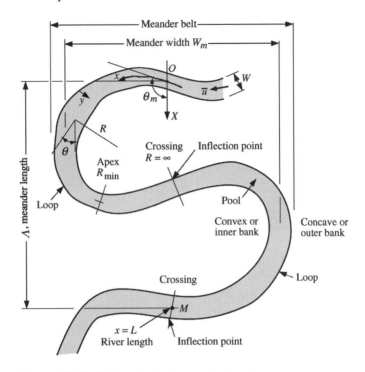

Figure 6.11. Definition sketch of a meandering river.

When plotting the orientation angle θ as a function of downstream distance x, Langbein and Leopold (1966) found that θ is a function of the maximum angle θ_m set at the origin, the downstream distance x and the river length L:

$$\theta = \theta_m \cos \frac{2\pi x}{L}. \tag{6.30}$$

This sine-generated curve is compared with an observed meandering pattern in Fig. 6.12.

The meander length Λ is computed from the following relationship:

$$\Lambda = \int_0^L \cos \theta \, \mathrm{d}x = \int_0^L \cos \left[\theta_m \cos \left(\frac{2\pi x}{L} \right) \right] \mathrm{d}x. \tag{6.31}$$

The sinuosity Ω, defined as $\Omega = L/\Lambda$, increases gradually with θ_m in radians,

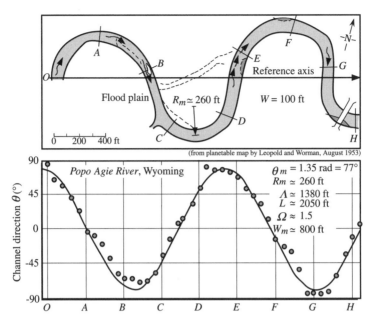

Figure 6.12. Meandering planform geometry (after Langbein and Leopold, 1966).

as illustrated in Fig. 6.13:

$$\Omega \equiv \frac{L}{\Lambda} = \frac{L}{\displaystyle\int_0^L \cos\theta \, dx} \cong 1 + \left(\frac{2\theta_m}{\pi}\right)^5 = 1 + 0.1\,\theta_m^5. \qquad (6.32)$$

The radius of curvature R can be obtained from $dx = R\,d\theta$; thus combining this with Eq. (6.30) gives

$$R = \frac{L}{2\pi\theta_m} \csc\left(\frac{2\pi x}{L}\right). \qquad (6.33)$$

The minimum radius of curvature R_m at the apex is obtained when csc $[(2\pi x)/L] = 1$ or

$$R_m = \frac{L}{2\pi\theta_m}. \qquad (6.34)$$

For a given meander length Λ, the minimum radius of curvature (which obviously corresponds to the maximum value of Λ/R_m) varies with θ_m as

$$\frac{\Lambda}{R_m} = \frac{2\pi\theta_m\Lambda}{L} = \frac{2\pi\theta_m}{\Omega}. \qquad (6.35)$$

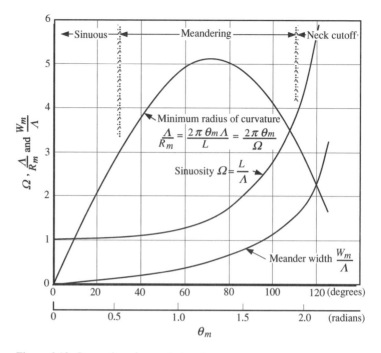

Figure 6.13. Properties of meandering rivers.

We easily obtain $R_m = [L/(2\pi\theta_m)] = [(\Omega \Lambda)/(2\pi\theta_m)]$. The ratio Λ/R_m varies with θ_m as shown in Fig. 6.13 and reveals that the minimum radius of curvature for a given meander wavelength corresponds to the maximum angle $\theta_m = 75° = 1.3$ rad. Consequently, the increase in radius of curvature beyond this point constitutes an extremely important feature because the radius of curvature controls the magnitude of the centrifugal force in bends.

Leopold et al. (1960) empirically observed that the meander length Λ is ~10 times the channel width W, as shown in Fig. 6.14(a). The properties of sine-generated curves are also supported by field evidence. For instance, the ratio of wavelength to minimum radius of curvature Λ/R_m for meandering streams in Fig. 6.13 varies between 3 and 5. Field measurements from Leopold and Wolman (1960) indicate an average ratio of 4.7, as shown in Fig. 6.14(b). Also, the mean radius of curvature $\bar{R}_m \simeq 2.3\,W$ is obtained from $\Lambda \simeq 10\,W \simeq 4.7\,\bar{R}_m$ in Figs. 6.14(a) and 6.14(b). From $\Lambda \simeq 10\,W$ and the above, we obtain

$$\frac{R_m}{W} \simeq \frac{5\,\Omega}{\pi\theta_m}. \tag{6.36}$$

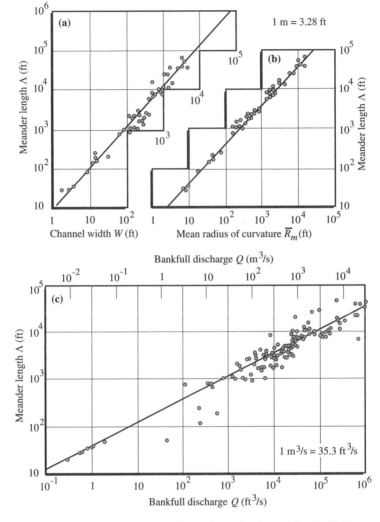

Figure 6.14. Empirical geometry of meanders (after Leopold et al., 1960 and Ackers and Charlton, 1970).

The meander width W_m as defined in Fig. 6.11 is evaluated analytically by the following integral: $W_m = 2 \int_0^{L/4} \sin \theta \, dx$. The ratio of meander width W_m to the wavelength Λ is a dimensionless measure of the amplitude:

$$\frac{W_m}{\Lambda} = \frac{2 \int_0^{L/4} \sin \left[\theta_m \cos \left(2\pi \frac{x}{L} \right) \right] dx}{\int_0^{L} \cos \left[\theta_m \cos \left(2\pi \frac{x}{L} \right) \right] dx}. \qquad (6.37)$$

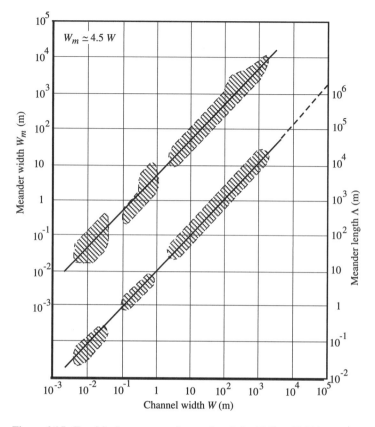

Figure 6.15. Empirical geometry of meanders (after Zeller, 1967a).

Equation (6.37) has been integrated numerically as a function of θ_m, and the ratio W_m/Λ is plotted in Fig. 6.13. The meander width increases rapidly when θ_m exceeds 90° and reaches the value 3.25 at the meander cutoff ($\theta_m = 125°$). Field results in Fig. 6.15 give empirical values $W_m = 4.5\,W \cong 0.45\,\Lambda$.

6.8   Lateral river migration

The effects of channel meandering on hydraulic geometry and sediment transport can be examined through the effect of sinuosity on channel slope. The energy gradient of the valley, S_{f0}, is given by the ratio of the energy loss ΔH over a meander wavelength Λ. The friction slope S_f of a meandering stream, however, corresponds to the energy loss ΔH over the sinuous stream length L:

$$S_f = \frac{\Delta H}{L} = \frac{\Delta H}{\Lambda} \frac{\Lambda}{L} = \frac{S_{f0}}{\Omega}. \tag{6.38}$$

The influence of the sinuosity of the channel width, depth, velocity, and Shields parameter can be examined through Eqs. (6.28). For instance, for the Manning–Strickler equation with $m = 1/6$, the flow depth $h \sim \Omega^{1/6}$, $W \sim \Omega^{0.22}$, $V \sim \Omega^{-0.39}$, and $\tau_* \sim \Omega^{-0.83}$. As a numerical example, when compared with a straight channel, a meandering channel with sinuosity $\Omega = 2$ would have a flow depth 12% larger and a channel width 16% larger. The flow velocity would decrease by 24%, and the shear stress would decrease by 44%. Sinuosity thus increases the cross-section area and decreases flow velocity and sediment transport.

Because sinuosity varies with θ_m [Eq. (6.32)], the ratio of the Shields parameter for a meandering channel τ_* is shown in Fig. 6.16 as a function of θ_m.

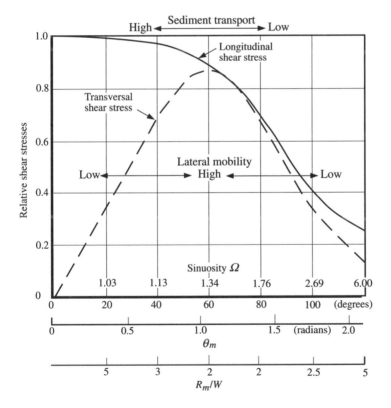

Figure 6.16. Relative longitudinal and transversal shear stresses.

When $\theta_m > 90°$, the Shields parameter of a meandering channel is less than half that of a straight channel. The decrease in the longitudinal Shields parameter as the sinuosity increases demonstrates that sinuous channels have a reduced ability to transport sediments compared with straight channels at any given valley slope.

The transversal shear stress has been shown to be proportional to tan λ and the square of the velocity. For instance, from Eq. (6.20), we obtain

$$\tau_{t0} \sim \frac{hV^2}{R},$$ (6.39)

where τ_{t0} is the bed shear stress in the transversal direction. The maximum transversal shear stress $\tau_{t\,max}$ corresponds to the minimum value of the radius of curvature $R_m = \Omega\Lambda/2\pi\theta_m$ from Eq. (6.35). Combining Eq. (6.39) with Eqs. (6.28a) and (6.28c) for a given discharge and grain size results in

$$\tau_{t\,max} \sim \frac{2\pi\theta_m\Omega^{-1/6}}{\Omega\Lambda\Omega^{0.77}} \sim \theta_m\Omega^{-1.61}.$$ (6.40)

The value of $\tau_{t\,max}$ from Eq. (6.40) is shown in Fig. 6.16 as a function of θ_m in radians, Eq. (6.40). High values of the transversal Shields parameter are found when $40° < \theta_m < 80°$. From Fig. 6.13, this corresponds to values of $\Lambda \sim 4\,R_m$. With $\Lambda \simeq 10\,W$ in Fig. 6.15, this leads us to the important conclusion that the lateral shear stress should be high when $R_m \sim 2.5\,W$.

Lateral migration in meandering rivers results from the erosion of the outer bank combined with equivalent sedimentation near the inner bank. This process is illustrated in Fig. 6.17(a). In natural rivers, erosion rates are variable and also depend on bank-material strength, cohesion, armoring, and vegetation. Typical patterns are

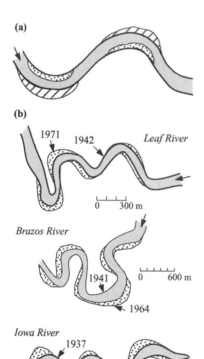

Figure 6.17. Examples of lateral migration.

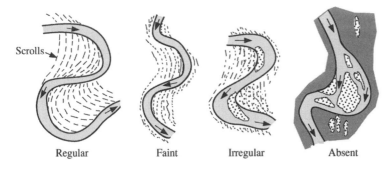

Figure 6.18. Types of meander scroll formation (after Richardson et al., 1990).

also sketched in Fig. 6.17(b). Widening is the result of erosion on one bank in excess of sedimentation near the opposite bank. Narrowing is the result of sedimentation near one bank in excess of the erosion rate of the opposite bank. The alluvium deposits near the inner bank sometimes form small ridges or terraces called scrolls caused by the reworking of point-bar deposits at low and high flows. Scrolls mark the successive positions of former meander loops, are often visible on aerial photographs, and are indicative of past inner-bank locations. Figure 6.18 illustrates several types of scrolls.

The radius of curvature to width ratio can be a useful tool for predicting erosion rates in river systems. Biedenharn et al. (1989) studied the effects of R_m/W and bank material on the erosion rates of 160 bends along the Red River in Louisiana and Arkansas. As indicated in Fig. 6.19, at R_m/W values greater than ~5, the erosion rates were generally low at 10–35 ft/yr (3–10 m/yr) with no discernible increasing or decreasing trend. The maximum erosion rates were observed in the R_m/W range of 2 to 4.

The relative migration rate, defined as the annual migration rate divided by the channel width, can be developed as a function of R_m/W. Nanson and Hickin (1986) studied the relative migration rate of 18 rivers in western Canada, including the Beatton River, British Columbia. As shown in Fig. 6.20, their study indicated that the migration rate of meanders is maximized at $2.0 < R_m/W < 4.0$. The relative migration rate can vary by 1 order of magnitude at a given value of R_m/W. The considerable scatter in the data suggests that the migration rate of meanders is extremely complex and may be a function of factors other than R_m/W.

The lateral stability of different stream reaches can be compared by means of a dimensionless erosion index. The erosion index is the product of its median bank erosion rate expressed in channel widths per year, multiplied by the percentage of reach along with erosion occurred, multiplied by 1,000.

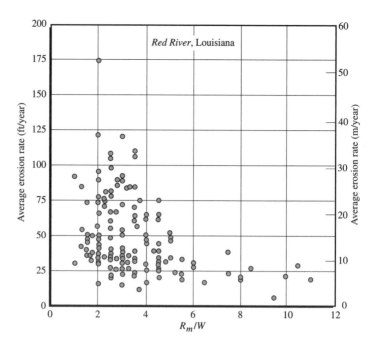

Figure 6.19. Lateral migration rates (after Biedenharn et al., 1989).

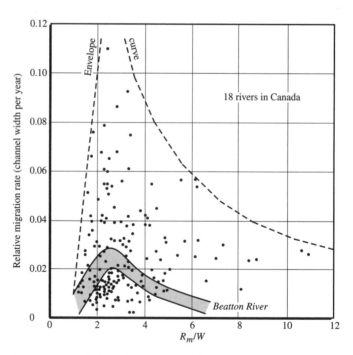

Figure 6.20. Relative migration rates (after Nanson and Hickin, 1986).

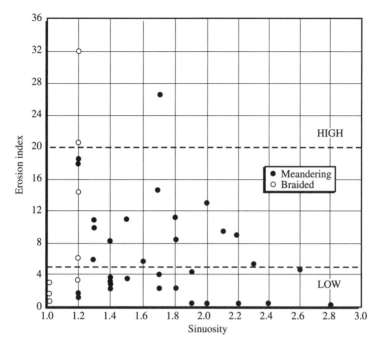

Figure 6.21. Erosion index (after Brice, 1984a).

Erosion indexes for 41 streams in the United States are plotted against sinuosity in Fig. 6.21. The length of most of these reaches is 25 to 100 times the channel width. The highest erosion-index values are for reaches with sinuosity of less than 2. It is clear from Fig. 6.16 that this corresponds to high lateral mobility. We can thus infer that lateral erosion of meandering channels is correlated to the value of the transversal shear stress. The erosion index value of 5 was suggested by Brice (1984a) as a boundary between stable and unstable reaches. Reaches having erosion-index values of less than 5 are unlikely to cause lateral erosion problems at bridges.

A neck cutoff is the natural result of lateral migration of a meandering pattern over a long period of time. Indeed, channel sinuosity increases as lateral migration of the outer bend progresses. When the channel sinuosity becomes very large, e.g., $\Omega > 3$, the downstream transport capacity, channel slope, and flow velocity are reduced. As a consequence, the risk of flooding is largely increased. Neck cutoffs form when the valley slope far exceeds the channel slope and the sediment-transport capacity is sufficient to cut through the neck of the long meander loops. Once the cutoff has occurred, silting at both ends of the

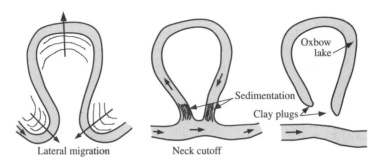

Figure 6.22. Oxbow-lake formation process.

meander loop results in clay plugs that isolate the oxbow lake. The oxbow-lake formation process through neck cutoffs is illustrated in Fig. 6.22. Neck cutoffs result in decreased sinuosity, increased slope, velocity, and sediment transport to be conveyed through or deposited in downstream bends. There may also be local incision and head cutting upstream of the neck cutoff. Over time, meander loops readjust their slope sinuosity and sediment transport to reach a renewed state of equilibrium.

Case Study 6.1   Sediment transport in the Fall River, Colorado, United States. Fall River is a meandering stream flowing through the Horseshoe Park area of Rocky Mountain National Park, Colorado (Fig. CS.6.1.1). During the late spring snowmelt, the daily discharge exceeds bankfull flow for several weeks through early summer. The bankfull flow of Fall River is ~ 7 m^3/s, and its winter lows are less than 0.5 m^3/s. The failure of the Lawn Lake Dam, located at the headwaters of Roaring River on July 15, 1982, is at the origin of a large alluvial fan, shown in Fig. CS.6.1.1. The alluvial fan supplies coarse sediment for bedload transport to the Fall River. The Fall River slope is gentle at 0.0013. Sediment loads as high as 0.6 kg/m s have been measured with sediment fractions ranging from 0.125 to 32 mm moving as bedload.

During field seasons 1986 and 1987, measurements were taken along two consecutive bends at 22 cross sections (Anthony, 1992) to determine the patterns of internal cross-sectional adjustments from high to low flows (Anthony and Harvey, 1991). Measurements included bed and free-surface topography, velocity profiles, and sediment transport by size fraction. Duplicated measurements at 1-m intervals across the channel included at least three vertical velocity

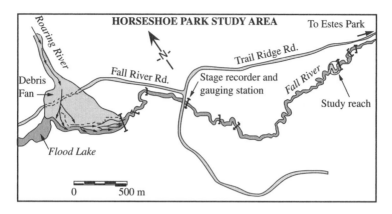

Figure CS.6.1.1. Location of the Fall River reach.

profiles (for both longitudinal and transverse flow) and six 1-min bedload measurements.

Topographic measurements included bed and water-surface elevations at 1-m intervals across the channel for each of the 22 cross sections identified in Fig. CS.6.1.2. At these high discharges, the cross-sectional geometry of this meandering channel alternates between a relatively rectangular shape at crossings to the most asymmetrical form near the bend apex. Near the apex, the cross-sectional geometry features (1) a deep thalweg along the outside of the bends, (2) a point-bar planform along the inside of the bends, and (3) a point-bar slope connecting the two.

At each cross-section, 2D velocity profiles were measured with a Marsh–McBirney current meter at 1-m intervals across the channel. Both transverse and longitudinal velocities were measured, starting from the bottom (with the current meter resting on the bottom) and then in approximately 10-cm intervals to the water surface. The data set of 1986 near bankfull discharge describes the flow pattern for the entire study reach. At each point in the profile, a 30-s sampling duration was allowed for measuring each velocity component. This sampling duration was extended to 1 min when turbulence was significant. The repeated sampling done in 1987 provided average-flow vectors for closely spaced points at each cross section. The streamline deviation angle λ was obtained from the 2D velocity measurements made near the channel bed. Both the velocity measurements close to the bed and those 10 cm above the bed were available for analysis. The measurements made 10 cm above the bed showed a more consistent pattern of flow direction and magnitude and were used to determine the streamline deviation angle λ.

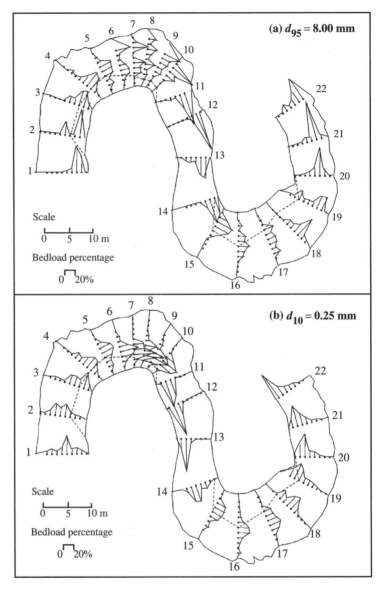

Figure CS.6.1.2. Bedload sediment transport in bends (after Julien and Anthony, in press.)

Sediment transport in the layer covering 3 in. (7.67 cm) above the bed was measured with a Helley–Smith sampler. Two 1-min bedload samples were collected at each vertical of each cross section, and the measurements were repeated three times. The bedload measured at each sampling location was later sieved to determine the particle-size distribution of moving material. Grain-size distributions from duplicate measurements were quite similar. Measured bed-load movement at each sampling location was divided into weights for each size fraction. At each cross section, the bedload movement for each size fraction was then summed over the entire cross section. For a given grain size, the percentage of the cross-section total material transported was then calculated for each sampling point. After this process was repeated at each cross section, a bedload percentage map over the entire reach was obtained for that particular size fraction. Bedload percentage maps were then produced for each size fraction, i.e., 16, 8, 4, 2, 1, 0.5, 0.25 and 0.125 mm. Typical bedload percentage maps are shown in Fig. CS.6.1.2 to represent coarse grains in transport ($d_{95} \cong$ 8 mm) and fine grains in transport ($d_{10} \cong 0.25$ mm).

The values of bedload percentages at each cross section were used to calculate the position of the center of mass of bedload transport for each size fraction. It is interesting to note in Fig. CS.6.1.2 that the location of the center of mass for fine grains is different from that of coarse grains. The lines linking the successive positions of the center of mass for different grain sizes are shown in Fig. CS.6.1.3. Near the crossing, the bedload center-of-mass curves for each

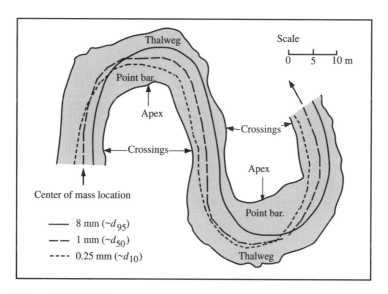

Figure CS.6.1.3. Mean bedload trajectory in Fall River (after Julien and Anthony, in press.)

size fraction are fairly parallel and oriented in the downstream direction. Near the apex, the bedload center-of-mass curves are shifting across the channel. The angle difference between coarse and fine grains can be calculated from the angle $\hat{\beta}$ of the particle-stability analysis in Example 6.1.

The particle-stability analysis determines the mean particle-direction angle β, given the particle grain size, sideslope angle, downstream slope angle, and the shear stress. The results of calculations with the method in Example 6.1 with different size fractions for the Fall River bend are shown in Fig. CS.6.1.4.

Finally, the cross-section geometry of the Fall River bends also depends on discharge. As shown in Fig. CS.6.1.5, increasing discharge causes erosion of the thalweg and sedimentation on the point bar. The larger flow depth increases the deviation angle λ according to Eq. (6.21). Conversely, as the

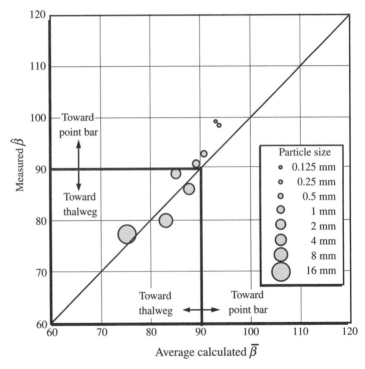

Figure CS.6.1.4. Particle-direction angle in Fall River (after Julien and Anthony, in press.)

(a) Increasing discharge

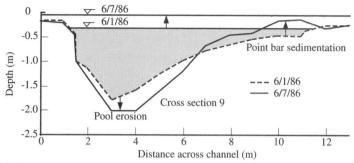

(b) Decreasing discharge

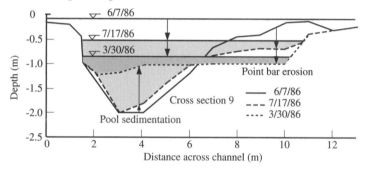

Figure CS.6.1.5. Changes in cross-section geometry: (a) increasing discharge and (b) decreasing discharge (after Anthony, 1992).

discharge decreases, the angle λ decreases and the cross section gradually becomes more rectangular through erosion of the point bar and sedimentation in the pools.

♦♦Exercise 6.1

With reference to Fig. 6.2, derive Eq. (6.9) from $\tau_c \sim F_s \tan \phi$ and $R/(F_s \cos \theta_1) = \tan \phi$.

♦Exercise 6.2

Derive Eq. (6.11) from Eqs. (6.9) and (6.10); then substitute Eq. (6.12) into Eq. (6.11) to demonstrate that it is indeed the solution.

♦Exercise 6.3

Derive the expression for the radius of curvature in Eq. (6.33) from $dx = Rd\theta$ and the sine-generated curve, Eq. (6.30).

♦Exercise 6.4

Examine the field observation in Figs. 6.9 and 6.14(c). Determine empirical ratios for the meander length as a function of channel width.

♦♦Problem 6.1

Calculate the particle-direction angle under an applied shear stress $\tau_0 = 10$ Pa, where the streamlines are deflected upward at $\lambda = -10°$. The downstream bed-slope angle is $\theta_0 = 0.05°$ and the sideslope angle is $\theta_1 = 10°$. Consider a particle of size 10 mm and compare the result with a 1-mm particle under identical flow conditions.

Problem 6.2

Define the ideal cross-section geometry for a 100-mm cobble-bed canal with all particles at beginning of motion. The slope of the channel is 0.01. Also estimate the flow discharge in this canal. (Assume that $\phi = 40°$, $n = 0.03$).

   Answers: $h_0 = 0.78$ m, $W = 2.9$ m, $A = 1.45$ m^2, $R_h = 0.42$ m, $V = 1.87$ m/s, $Q = 2.71$ m^3/s, $\tau_* = \frac{h\,S}{(G-1)\,d_s} = 0.047$, incipient motion

♦Problem 6.3

The Cache la Poudre River near Rustic, Colorado, has a bankfull discharge of 17.6 m^3/s, a width of 12.8 m, a depth of 0.65 m, and a slope of 0.0048 with a grain size of 150 mm. Compare the actual geometry with the regime equations and downstream hydraulic-geometry relationships.

♦♦Problem 6.4

Use the regime relationships to calculate the hydraulic geometry of an irrigation canal that conveys 5000 ft^3/s in a very fine gravel-bed channel. Compare with the hydraulic geometry for stable channels. (*Hint:* $\tau_* = 0.047$ for stable channels.)

   Answers: $V = 4.6$ ft/s, $R = 5.7$ ft, $P = 188$ ft, $S = 6.9 \times 10^{-4}$ from the regime equations; $V = 0.72$ m/s, $h = 1.78$ m, $W = 109$ m, $S = 1.3 \times 10^{-4}$ from Eqs. (6.26) with $d_s = 3$ mm.

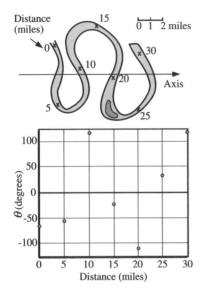

Figure P.6.5.1. Mississippi River near Greenville (after Langbein and Leopold, 1966).

♦♦**Problem 6.5**

From Fig. P.6.5.1, determine the orientation angle of the river at positions equally spaced by 1 mile and plot the values to complete the diagram. Fit a sine function and determine the values of θ_m and the origin of the downriver axis x. Determine L and Λ, measure the sinuosity, and compare with calculations based on θ_m and Eq. (6.32). Measure the minimum radius of curvature R_m and compare with values calculated from L and θ_m. Compare the measured meander width with the calculations, assuming the sine-generated curve.

♦♦**Problem 6.6**

From the information presented in Case Study 6.1, determine the following: downstream angle θ_0, sideslope angle θ_1 at high flow, bed shear-stress estimate τ_0, radius of curvature, and streamline deviation angle λ. Also calculate the downstream hydraulic geometry from the relationships of Section 6.5 and compare with field observations. Determine the river length L, the meander length Λ, the sinuosity, maximum planform deviation angle θ_m and meander width W_m, and compare with relationships in Section 6.7. Plot the field observations on Fig. 6.9, 6.13, 6.14 and 6.15, and determine whether lateral mobility should be high or low.

Answers: $\theta_0 = 0.0745$, $\theta_1 \simeq 15°$ from Fig. CS.6.1.5, $\tau_0 \cong \gamma\,h\,S \cong$ 12 Pa, $R_m \simeq 15\,\text{m}$, $\lambda \simeq \tan^{-1}$, $(11\,h/R) = 36°$ at high flow and $\lambda \simeq 10°$ at low flow. From $Q = 7\ \text{m}^3/\text{s}$, $d_s = 8$ mm and $S = 0.0013$, Eqs. (6.28) give $h \simeq$ 0.52 m, $W = 22$ m, $V = 0.55$ m/s, and $\tau_* = 0.05$. The actual cross-section geometry is deeper and narrower than calculated. Planform geometry, $L \simeq$ 103 m, $\Lambda \simeq 50$ m, $\Omega \simeq 2$, $\theta_m \simeq 95° = 1.66$ rad, and $W_m \simeq 45$ m. This gives $W_m \simeq 4.5\,W$, $\Lambda = 5\,W \simeq 3.5\,R_m$, $R_m \simeq 1.5\,W$, which is very low. Field measurements compare well with the figures. Based on Fig. 6.16, the lateral mobility should be fairly low.

♦Problem 6.7

Determine the sinuosity of the channel in Fig. P.6.7.1. Locate the inflection points and locate the clockwise and counterclockwise loops. Determine the number of loops and the number of meanders. Determine the average river wavelength, meander length, meander width, and length of the meander belt. Compare meander width with that of Fig. 6.13. Anticipate where neck cutoffs might occur and determine where the oxbow lakes would be. Discuss the impact of possible neck cutoffs on hydraulic geometry and sediment transport. Do you expect the lateral mobility of this river to be high or low?

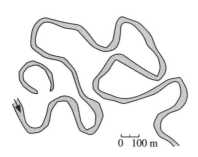

Figure P.6.7.1. Example of meandering channel.

7

River dynamics

Deviations from equilibrium conditions will trigger a dynamic response from the alluvial river system to restore the balance between inflowing and outflowing water and sediment discharges. Section 7.1 of this chapter deals with the dynamics of stream response to changes in water and sediment discharges. Sections 7.2 and 7.3 describe the dynamic response of alluvial systems to degradation and aggradation, respectively, particularly the effects on hydraulic geometry and channel morphology. Section 7.4 focuses on river confluences and branches. Finally, Section 7.5 provides guidelines on river databases, data sources, and field surveys.

7.1   River dynamics

Conceptually, the fluvial system of the watershed sketched in Fig. 7.1 can be divided into three main zones: (1) an erosional zone of runoff production and sediment source; (2) a transport zone of water and sediment conveyance; and (3) a depositional zone of runoff delivery and sedimentation. The second zone is characterized by near-equilibrium conditions between the inflow and the outflow of water and sediment. The bed elevation in this equilibrium zone is fairly constant and the hydraulic geometry is described in Chap. 6 and this section. The upper zone is characterized by net erosion of bed material and channel degradation. The dynamic response of degrading fluvial systems is discussed in Section 7.2. The lower zone is characterized by net sedimentation and channel aggradation. The dynamic response of aggrading fluvial systems is discussed in Section 7.3.

Conceptually, the hydraulic geometry of alluvial channels is related to stream characteristics that vary with time. As sketched in Fig. 7.2, the concept of dominant discharge used in Chap. 6 refers to bankfull discharge conditions or flood discharge with a period of return of $\sim$1.5 yr. The corresponding sediment discharge is divided into washload for size fractions of bed material finer than d_{10}. The bed-material discharge Q_{bv} corresponds to size fractions of

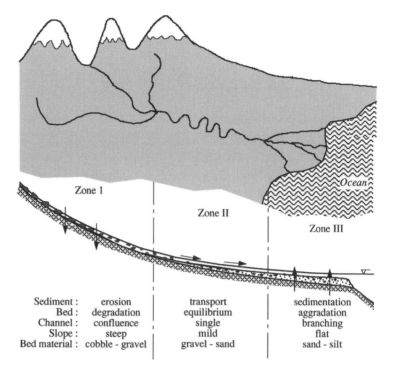

Sediment :	erosion	transport	sedimentation
Bed :	degradation	equilibrium	aggradation
Channel :	confluence	single	branching
Slope :	steep	mild	flat
Bed material :	cobble - gravel	gravel - sand	sand - silt

Figure 7.1. Erosion, transport, and sedimentation in a fluvial system.

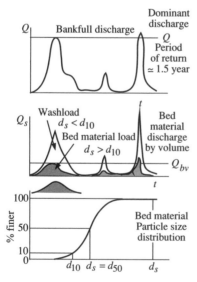

Figure 7.2. Time changes in discharge and sediment transport.

bed material coarser than d_{10}. The bed-material discharge is most important because it relates to changes in bed conditions and thus to changes in hydraulic geometry.

Section 6.4 dealt with equilibrium downstream hydraulic geometry. Accordingly, the relationship of Eqs. (6.26) among channel slope S, dominant discharge Q, grain size d_s, and Shields parameter τ_* can be rewritten as

$$SQ^{\frac{1}{2+3m}} = 1.24d_s^{\frac{5}{4+6m}}\tau_*^{\frac{4+6m}{5+6m}}. \qquad (7.1)$$

We remember that the unit of the bankfull discharge Q is cubic meters per second and that of the median grain diameter d_s is meters, whereas the slope S and the Shields parameter τ_* are dimensionless. After considering that the sediment discharge Q_s is proportional to the Shields number τ_*, we find that the product of slope and discharge on the left-hand side of Eq. (7.1) must be balanced by the product of grain size and sediment discharge on the right-hand side of Eq. (7.1), which is known as Lane's (1955) relationship, written as

$$QS \sim Q_s d_s. \tag{7.2}$$

This qualitative relationship states that equilibrium conditions exist between hydraulic conditions on the left-hand side of relation (7.2) and sediment conditions on the right-hand side of relation (7.2). Perturbations to one or several parameters in relation (7.2) will be balanced by a change in one or several of the remaining parameters.

A quantitative relationship between hydraulic and sediment variables is possible after defining a sediment transport relationship. It is important to use a bed-sediment discharge because riverbed changes are induced by erosion and/or sedimentation of bed material. The following empirical relationship for sand transport is used as a first approximation:

$$q_{bv} \simeq 18\sqrt{g}d_s^{3/2}\tau_*^2. \tag{7.3}$$

This relationship for the unit bedload discharge by volume q_{bv} in square meters per second is approximately valid for $0.1 < \tau_* < 1$, where the grain diameter d_s is in meters. Accordingly, the bedload discharge by volume is $Q_{bv} = q_{bv}W$, and from relations (6.27) and (7.3) we obtain

$$Q_{bv} \simeq 0.77Q^{\frac{6+4m}{5+6m}}d_s^{\frac{-2.5+5m}{5+6m}}S^{\frac{7+10m}{5+6m}}. \tag{7.4a}$$

Specifically for $m = 1/6$, this reduces to

$$Q_{bv}d_s^{0.28} \simeq 0.77Q^{1.11}S^{1.44}. \tag{7.4b} \blacklozenge\blacklozenge$$

We thus should apply this approximation only as a downstream hydraulic relationship among dominant discharge Q in cubic meters per second, the bed-material discharge by volume Q_{bv} in cubic meters per second, and the grain diameter d_s in meters. Relations (7.4) should not be used to define at-a-station properties such as sediment-rating curves.

The sediment concentration in milligrams per liter, $C_{\text{mg/l}} = 10^6 GC_v = 10^6 G(Q_{bv}/Q)$ can be obtained directly from relations (7.4) as

$$C_{\text{mg/l}} \simeq 2.65 \times 10^6 Q^{0.11}d_s^{-0.28}S^{1.44}. \tag{7.5}$$

This relationship deserves testing and should be accurate within an order of magnitude as long as the sediment-transport relationship and the approximation

$m = 1/6$ are applicable. Nevertheless, it clearly shows the dominant role played by the channel slope S in the conveyance of sediment in alluvial channels.

We can assess the effects of sediment discharge Q_{bv} on the downstream hydraulic geometry of sand-bed streams after substituting S into Eqs. (6.27) with the function of the bed-material discharge Q_{bv} in cubic meters per second, d_s in meters, and Q in cubic meters per second from relation (7.4b). We obtain the solution for the bankfull flow depth in meters, channel width W in meters, flow velocity V in meters per second, and Shields parameter τ_* after assuming that $m = 1/6$ and substituting S from relation (7.4b) into Eqs. (6.28):

$$h \simeq 0.19 Q^{0.46} d_s^{0.13} Q_{bv}^{-0.12}, \tag{7.6a}$$

$$W \simeq 1.3 Q^{0.62} d_s^{-0.15} Q_{bv}^{-0.15}, \tag{7.6b}$$

$$V \simeq 4 Q^{-0.08} d_s^{0.02} Q_{bv}^{0.27}, \tag{7.6c}$$

$$S \simeq 1.2 Q^{-0.77} d_s^{0.19} Q_{bv}^{0.69}, \tag{7.6d}$$

$$\tau_* \simeq 0.14 Q^{-0.31} d_s^{-0.67} Q_{bv}^{0.57}. \tag{7.6e}$$

Alternatively, the effects of bed-material sediment concentration $C_{mg/l}$ that correspond to the dominant discharge can be assessed after $Q_{bv} = 3.8 \times 10^{-7} C_{mg/l} Q$ is substituted into relations (7.6a)–(7.6e). We simply obtain downstream hydraulic-geometry relationships as functions of bankfull bed-material sediment concentration in milligrams per liter, d_s in meters, and Q in cubic meters per second:

$$h \simeq 1.1 Q^{0.34} d_s^{0.13} C_{mg/l}^{-0.12}, \tag{7.7a} \blacklozenge$$

$$W \simeq 12 Q^{0.47} d_s^{-0.15} C_{mg/l}^{-0.15}, \tag{7.7b} \blacklozenge\blacklozenge$$

$$V \simeq 0.075 Q^{0.19} d_s^{0.02} C_{mg/l}^{0.27}, \tag{7.7c} \blacklozenge$$

$$S \simeq 4.4 \times 10^{-5} Q^{-0.08} d_s^{0.19} C_{mg/l}^{0.69}, \tag{7.7d} \blacklozenge\blacklozenge$$

$$\tau_* \simeq 3 \times 10^{-5} Q^{0.26} d_s^{-0.67} C_{mg/l}^{0.57}. \tag{7.7e} \blacklozenge$$

At sediment concentrations of less than 1,000 ppm, sediment concentrations in milligrams per liter C_{ppm} can be used instead of $C_{mg/l}$. The difference between $C_{mg/l}$ and C_{ppm} is less than 10% at concentrations of less than 145,000 ppm.

Problem 7.2 illustrates how relations (7.7) can be used to estimate the characteristics of an alluvial channel, based on discharge, grain size, and bed-material concentration.

We can thus infer from relations (7.6) that an increase in dominant discharge Q^+ is expected to cause a significant increase in bankfull width W^+ and in depth h^+, a significant decrease in slope S^-, and a less-pronounced decrease in Shields parameter τ_*^-. An increase in dominant sediment discharge Q_{bv}^+ corresponds to a significant increase in slope S^+ and Shields parameter τ_*^+, a less-pronounced increase in velocity V^+, and slight decreases in channel width W^- and flow depth h^-. The effects of increases in grain size d_s^+ are comparatively less significant, except for a decrease in Shields parameter τ_*^-. In summary, we can expect the following dynamic responses of alluvial systems to perturbations in water and sediment discharges:

$$Q^+ \rightarrow W^+ h^+ S^- \tau_*^-, \qquad (7.8a)$$

$$Q_{bv}^+ \rightarrow S^+ \tau_*^+ V^+, \qquad (7.8b)$$

$$d_s^+ \rightarrow \tau_*^-. \qquad (7.8c)$$

We thus note that the downstream hydraulic geometry in width and depth primarily depends on discharge. Increases in water and sediment discharges exert counterbalancing effects on the channel slope and the Shields parameter. Grain-size effects are comparatively small. The opposite effects are of course obtained for decreasing values of the parameters.

As sketched in Fig. 7.3, changes in hydraulic geometry take place through reworking of the alluvium in which rivers flow. Flow depth increases as the bed degrades, and flow depth decreases through aggradation. Channel widening will occur through bank erosion. Channel narrowing will probably take place through shoaling and the formation of bars and islands, followed by incision of the main channel. In general, shoals are submerged sandbars at low flows. Sandbars and gravel bars are submerged at high flow, and islands are covered with vegetation. Changes in flow velocity are quite naturally linked to bedforms and the ability of a stream to transport sediment. Changes in bed slope, on the other hand, are quite problematic in the sense that they require adjustment of bankfull conditions and changes in the floodplain.

As an example, consider an alluvial fine sand-bed channel that has a bankfull width $W_1 = 300$ ft, a flow depth $h_1 = 10$ ft, slope $S_1 = 8$ ft/mile, and a flow velocity $V_1 = 10$ ft/s. Determine the expected change in downstream hydraulic geometry if the dominant flow discharge is decreased by 50%. The bed-material size and the sediment concentration are expected to remain the

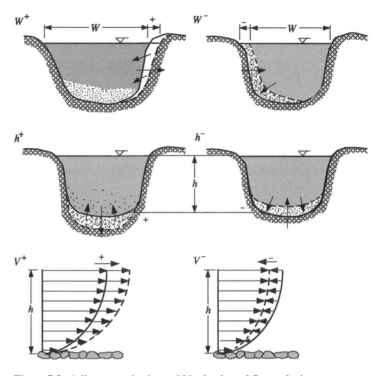

Figure 7.3. Adjustments in river width, depth, and flow velocity.

same. The changes in hydraulic-geometry characteristics with $Q_2/Q_1 = 0.5$ are calculated from relations (7.7): $h_2/h_1 \cong (0.5)^{0.34} = 0.80$, $W_2/W_1 \cong (0.5)^{0.47} = 0.72$, $V_2/V_1 \cong (0.5)^{0.19} = 0.87$, and $S_2/S_1 \cong (0.5)^{-0.08} = 1.05$.

Accordingly, the flow depth is expected to decrease to ~8 ft, the bankfull width should decrease to ~210 ft, the flow velocity should decrease to ~9 ft/s, and the bed slope should slightly increase to ~8.5 ft/mile for the new equilibrium condition. These expected changes only can be used as a first approximation.

7.2   Riverbed degradation

Channel degradation refers to the general lowering of the bed elevation that is due to erosion. In some cases, the bed material is fine and degradation will result in channel incision, which is covered in Subsection 7.2.1. In other cases, the material is sufficiently coarse to form an armor layer that prevents further degradation, as discussed in Subsection 7.2.2.

7.2.1   Incised rivers

Slope adjustments refer to streams that would require either a steeper or a milder slope for reaching equilibrium between incoming and outgoing water and sediment discharges. Stated in simple terms, when the outgoing exceeds the inflowing sediment load, alluvial streams will scour bed material and degrade. Degradation results in channel incision and milder slopes, as sketched in Fig. 7.4.

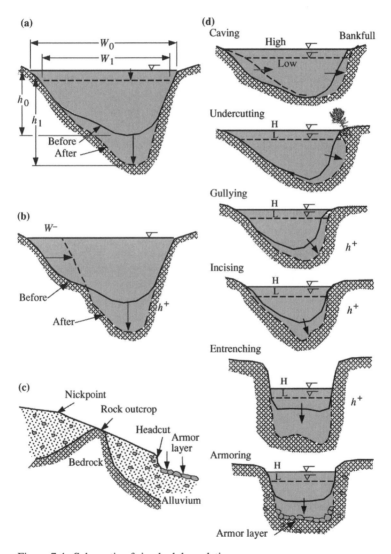

Figure 7.4. Schematic of riverbed degradation.

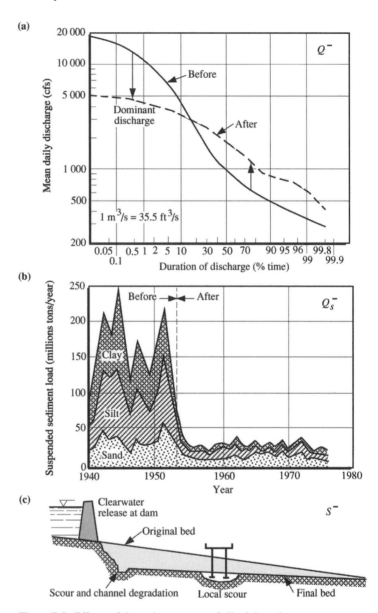

Figure 7.5. Effects of dams downstream of alluvial reaches.

Incised channels tend to be narrow and deep compared with equilibrium conditions. Channel incision will take place until equilibrium condition is reached. Incised channels are typical of upland areas whereby the sediment-transport capacity increases in the downstream direction. Rills are small-scale channels found in upland areas. Gullies are larger-scale features also found in upland areas. Conventionally, rills can be crossed by farm machinery whereas gullies cannot.

In rivers, channel incision is found in arroyos and canyons. Arroyos are ephemeral channels in arid areas with flashy hydrographs that carry large sediment loads during short periods of time. Many arroyos dry out in the downstream direction as a result of infiltration and evaporation. The sediment load eventually deposits on the channel bed downstream of arroyos to form wide-shallow streams. Canyons are usually deeply entrenched in vertical bedrock walls. Incised channels typically are narrower and deeper then equilibrium channels and are characterized by a shortage of sediment. Channel degradation also causes the banks to become unstable and subject to failure. Gully-like incised channels become very unstable, and bank erosion may become a significant source of sediment to the channel. Incised channels can often be found where the stream slope increases in the downstream direction. Knickpoints indicate points with a sudden change in bed slope. Headcuts usually refer to sudden drops in bed elevation. Headcuts usually start downstream, and their upstream migration is a characteristic feature of incised channels.

As sketched in Fig. 7.5, artificial structures such as dams alter the equilibrium between the flow of water and sediment in alluvial channels. Reservoirs tend to decrease the magnitude of flood flows and increase low flows. The clearwater release from the dam also causes the reach below the dam to degrade in the form of a wedge starting below the dam. The magnitude and the extent of the degradation below dams depend on the reservoir size and operation and on the size and availability of alluvium below the dam. Incision tends to be deepest in sandy materials and is subject to armoring in gravels.

Degradation of the main river stem at river confluences causes headcutting and degradation in the tributaries. As sketched in Fig. 7.6, the headcut propagates upstream from the confluence and can cause severe stability problems in structures on shallow foundations such as bridges and some grade-control structures. The ensuing gullying in a tributary can cause significant bank instabilities and channel widening.

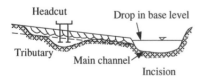

Figure 7.6. Schematic of headcut migration.

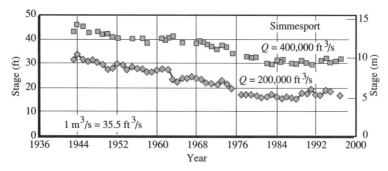

Figure 7.7. Specific-gauge record of the Atchafalaya River at Simmesport (after U.S. Army Corps of Engineers, 1999).

Specific-gauge records are often used to determine whether a stream tends to aggrade or degrade over time. A specific-gauge record is the water-surface elevation that corresponds to a given discharge. When gauge records are available for a long period of time, plotting the gauge elevation at given discharges can detect long-term river trends. For instance, the specific-gauge record of the Atchafalaya River is shown in Fig. 7.7. Both discharges show a gradual lowering of the water-surface elevation with time. Over a period of 50 yrs, the water-surface elevation is ~15 feet lower than in the mid-1950s. When the procedure is repeated at several stream gauges along a river, longitudinal water-surface profiles at a specific discharge can be obtained. The example in Fig. 4.7(a) illustrates the water-surface lowering in the Atchafalaya River from Simmesport to Chicot Pass and the rising in water-surface elevation between Chicot Pass and Morgan City. It should be noted that specific-gauge records do not provide information on bed elevation, and, per se, aggradation or degradation changes cannot be determined from specific-gauge records. Changes in bedforms, bed material, resistance to flow, loop-rating effects, and channel widening can alter the trends observed with specific gauges.

It is possible to observe opposite trends at different discharges; for instance, specific gauges may indicate a decreasing trend at a low discharge and an increasing trend at a high discharge. It is thus recommended to compare cross sections over time in order to confirm any trend detected with specific gauges.

7.2.2   Riverbed armoring

Armoring of the bed layer refers to coarsening of the bed-material size as a result of degradation of well-graded sediment mixtures. The selective erosion of finer particles of the bed material leaves the coarser fractions of the mixture on the bed to induce coarsening of the bed material. When the applied bed shear stress is sufficiently large to mobilize the larger bed particles, degradation continues; when the applied bed shear stress cannot mobilize the coarse bed particles, an armor layer forms on the bed surface. The armor layer becomes coarser and thicker as the bed degrades until it is sufficiently thick to prevent any further degradation. The armor layer is representative of stable bed conditions and can be mobilized only during large floods. A riverbed is sometimes said to be paved when the armor layer can be mobilized only during exceptional floods. Three conditions need to be satisfied to form armor layers: (1) the stream must be degrading, (2) the bed material must be sufficiently coarse, and (3) there must be a sufficient quantity of coarse bed material. Relative to the first condition, the sediment-transport capacity must exceed the sediment supply such that the stream attempts to scour the bed.

The second condition can be quantified as follows from the Shields diagram. The incipient condition of motion with $\tau_{*_c} \simeq 0.05$ can be rewritten in terms of minimum grain size at the beginning of motion.

$$d_{sc} \simeq 10\,hS, \tag{7.9a} \blacklozenge$$

where d_{sc} is the minimum grain diameter, h is the flow depth during floods, and S is the channel slope. The units of grain size are the same as those of the flow depth. Alternatively, we can estimate the flow depth corresponding to the beginning of motion by

$$h \simeq d_{sc}/10\,S. \tag{7.9b}$$

The third condition refers to the fraction of material Δp_c coarser than d_{sc} available in the bed material. When this percentage is large, the armor layer will form rapidly and the extent of degradation will be minimal. When this percentage is low, a large volume of bed material will be scoured before the armor layer can form. The effect of the armor layer in this case will be limited. Quantitatively, we can consider that an armor layer of approximately twice the grain size will stabilize the bed. The scour depth Δz that will form an armor layer equal to $2d_{sc}$ can be estimated from

$$\Delta z = 2d_{sc} \left(\frac{1}{\Delta p_c} - 1 \right). \tag{7.10}$$

The scour depth becomes very large when Δp_c is small, and it is therefore important to have a particle-size distribution that is representative of the sublayer, including clay, sand, and gravel layers. Example 7.1 illustrates some characteristics of river degradation and armoring.

Once an armor layer has formed, it plays a very important role in channel stability and morphology. Indeed, the riverbed is stable except under large floods, and the armor layer protects the bed against further degradation. Its removal through natural and/or artificial activities can cause significant stream-instability problems. For instance, gravel mining operations remove the coarse armor layer from bed streams at low flows. The damage can often be seen during subsequent floods in which large discharges cause severe degradation in the channel. Headcut development and upstream migration can also cause the failure of upstream structures such as bridges. Case study 7.1 illustrates the complex effects of dams, land-use changes, and gravel mining on river-channel morphology.

Example 7.1   Application to riverbed degradation. The Meuse River in The Netherlands is 250 km long. The mean annual discharge is 230 m^3/s near Maastricht and exceeds 3,000 m^3/s during major floods. The Meuse is more or less controlled by weirs to enable river navigation, and the lower stretches of the Meuse have dikes. From Maastricht to Maasbracht, the Meuse meanders over shallow gravel banks, and barge traffic is possible only in a parallel canal. The river is easily navigable between Maasbracht and Lith. Near Maasbracht, the river shows a sharp transition from a gravel-bed to a sand-bed river with d_{50} decreasing from ~16 to 3 mm and the bed slope decreasing from 48 to 10 cm/km. Figure E.7.1.1 from Murrillo-Muñoz (1998) shows several characteristics of the bed material of the Meuse River.

At a flow depth of 3 m, the critical grain size at the beginning of motion is calculated from the slope and relations (7.9). In this case, $d_{sc} \simeq 12 \times 3\,\mathrm{m} \times 48 \times 10^{-5} = 17$ mm for the upper reach and $d_{sc} \simeq 12 \times 3\,\mathrm{m} \times 10 \times 10^{-5} = 3$ mm for the power reach. It is interesting to note that these size fractions approximately correspond to d_{50} of the bed material. To examine whether armoring is possible during floods, the critical grain sizes at a flood flow depth $h = 6$ m are 25 and 7 mm. Considering the particle-size distribution of the lower reach, particle sizes exceeding 35 mm cannot be found in large quantities and degradation can be expected during floods. Considering that only ~3% of material, $\Delta p_c \simeq 0.03$, is coarser than 35 mm in the lower reach, the extent of degradation calculated from Eq. (7.10) is $\Delta z \simeq 2.3$ m, which is comparable with the degradation measurements shown in Fig. E.7.1.1. Note that Eq. (7.10) is very sensitive to low values of Δp_c and an infinite degradation depth is obtained when $\Delta p_c \to 0$. It is therefore very important to carefully determine the particle-size distribution of coarse bed fractions. When several meters of degradation are expected, borings are required for examining the substrate and looking for possible gravel and cobble layers that can limit the extent of

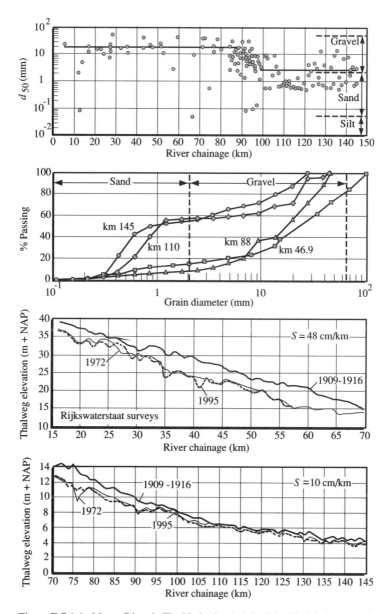

Figure E.7.1.1. Meuse River in The Netherlands (after Murillo-Muñoz, 1998).

degradation. The presence of bedrock outcrops along the river profile can be most useful when severe degradation is expected.

In the field, quick estimates of the flow depth required for beginning of motion of the bed material are obtained by

$$h_m \simeq 10\, d_s / S_{cm/km},\tag{E.7.1.1} \blacklozenge$$

where the flow depth h is in meters, the median grain diameter is in millimeters, and the slope is in centimeters per kilometers.

In this example, the flow depth for the upper reach with $d_s \simeq 16$ mm and $S = 48$ cm/km is approximately 3 m. The flow depth for the lower reach with $d_s \simeq 3$ mm and $S = 10$ cm/km is also approximately 3 m. Downstream fining in this river seems to follow the empirical rule that d_s/S is constant.

Case Study 7.1   Dynamic changes of Dry Creek, United States. Dry Creek is a major tributary to the Russian River just south of Healdsburg, California (Fig. CS.7.1.1). The Russian River Basin drains approximately 235 km^2 and Dry Creek has a drainage area of 34 km^2.

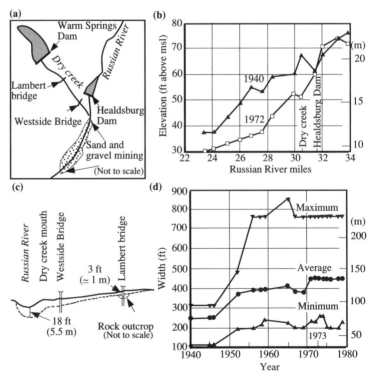

Figure CS.7.1.1. Russian River in California (after Simons and Julien, 1983).

Healdsburg Dam serves as a control on the Russian River, and Dry Creek will be even more significantly altered when the Warm Springs Dam is completed and put in operation.

Significant land-use changes are also evident from the aerial photographs. The Dry Creek valley is agricultural, with citrus fruits as the major crop. The upland areas are rugged, consisting primarily of hilly and mountainous terrain. Recently this condition has been aggravated by record storms and wildfire, both contributing to abnormal runoff volumes. Other land-use changes include some urbanization and the conversion of forest lands to agricultural and grazing lands.

Dry Creek is located in a valley of high relief (>300 m) on a well-defined wide floodplain averaging 1.2 km in width (Fig. CS.7.1.1). The low-flow channel is incised within a wider-flow channel bordered by low scarps. The low-flow channel has a sinuosity of approximately 1.20, whereas the wider flood flow channel is straight (sinuosity approximately 1.05). Aerial photographs indicate that Dry Creek is extremely active. This corroborates the findings in Fig. 6.16. Although the overall channel sinuosity remains approximately the same from year to year, a comparison of the location of meander bends and crossovers reveals that their positions change with time. Channel boundaries are alluvial, being composed of sand and gravel. The low-flow channel is locally braided and locally anabranched. Both the low-flow and the flood-flow channels are equiwidth with the development of wide point bars prevalent along the low-flow channel. Tree cover is generally less than 50% of the bankline, and cut banks are evident. Bank material is generally noncohesive silt, sand, gravel, and cobbles.

Dry Creek is a braided stream that is overloaded with sands and gravel. The deposition of sediments from the overloaded condition creates gravel-bar formation and aggradation of the channel bed. The presence of the gravel bars enhances lateral migration of the channel. These conditions create severe bank-erosion problems because of the highly erodible bank. Bank-erosion problems were detected more than 30 km upstream of the confluence with the Russian River and have been documented from records and photographs as early as 1940. In addition, the similarity of degradation and lateral migration tendencies between the areas upstream and downstream of the Lambert Bridge (control) indicates the significance of natural channel instability.

Gravel mining in the Russian River and other tributaries such as Dry Creek has been an important industry in Sonoma County since the early 1900s. In recent years, such activities have continued in Dry Creek and in the Russian River near the mouth of Dry Creek. There is evidence of gravel mining near the confluence of Dry Creek and the Russian River, resulting in a general deepening of the Russian River. This is due to the closing of the

Healdsburg Dam in 1952 and the Coyote Dam in 1958. The lowering of the base level in the Russian River has induced a general lowering of the base of Dry Creek.

The channel of Dry Creek downgraded significantly in the 1950s and 1960s because of the drop in the base level of the Russian River. This drop was the result of in-stream gravel mining and the construction and operation of the Healdsburg and Coyote Dams. A total streambed elevation drop of 7 m has been recorded in a 0.6-km length near the Healdsburg Dam since 1940. The current Healdsburg Dam structure, completed in 1952, has maintained the past upstream bed elevation and is acting as a control that prevents upstream headcuts in the Russian River. However, the base-level drop in the Russian River initiated a headcut in Dry Creek that has propagated a total distance of 13 km upstream from the mouth. The headcut was controlled when rock just upstream of the Lambert Bridge was exposed in 1972–1977. Calculations indicate that the headcut traveled at an estimated rate of 0.6 km/yr, requiring 23 yrs to travel the entire 13 km. With the deeper channel system and with controlled flooding, higher banks became exposed to attack by the flowing water. Abnormal flooding and fire sequence produced record runoff and sediment that has caused the deeply incised lower ends of Dry Creek to begin to widen again from extensive bank erosion. Analysis of aerial photographs indicates a significant increase in channel width.

Qualitative analysis reveals that the erosion problems along Dry Creek are complicated. The accelerated channel degradation has been primarily caused by (1) lowering of the base level that is due to the construction of dams on the Russian River; (2) increased runoff resulting from record storms, complicated by land-use changes and wildfires; and (3) gravel mining activity on the Russian River and Dry Creek.

7.3   Riverbed aggradation

Channel aggradation refers to a gradual bed-elevation increase that is due to bedload sedimentation. This section considers braiding in Subsection 7.3.1 and alluvial fans and deltas in Subsection 7.3.2.

7.3.1   Braided rivers

When the inflowing sediment discharge exceeds the outgoing sediment capacity, alluvial channels tend to deposit their sediment load throughout the reach. Streams carrying mostly washload will not change their morphology because the sediment overload will be carried downstream to settle in lakes, reservoirs, or estuaries. Streams carrying most of their sediment load in suspension change their morphology gradually as the excess sediment load settles in the downstream direction. The riverbed material size becomes gradually finer in the downstream direction. From Lane's relationship, downstream fining is usually accompanied by a downstream decrease in bed slope. On the other hand, streams that carry predominantly bedload material will respond quite rapidly to a change in sediment-transport capacity. A decrease in transport capacity induces direct settling on the bed of alluvial channels.

As sketched in Fig. 7.8, the settling of bedload forces aggrading channels out of the bankfull conditions. The flow spreads on the floodplain with accumulation of the bed-sediment load to form natural levees on a wide floodplain. There is a tendency for the stream to widen and become very shallow with bars subjected to rapid changes in morphology. At high flows, braided streams have a low sinuosity and often appear to be straight. At low flows, numerous small channels weave through the exposed bars. These streams are known to braid as the bed slope increases through aggradation. The flow velocity of braided streams is high, and the bed material can be easily mobilized. Braided streams are rather unstable in that they are prone to severe lateral migration, frequent shifts, and changes in cross-section geometry. The bars of braided streams are generally submerged once a year and are devoid of vegetation.

Islands are different from bars in that they are stabilized by vegetation and rivers with multiple islands are anastomosed. Anastomosed rivers are usually more stable than braided channels because vegetation straightens the banks and stable islands control the flow between the branches. During floods, vegetated islands trap sediment and aggrade.

Because braided channels require large bedload transport, most braided rivers are steep, and therefore, at a given discharge, braided rivers should be steeper than meandering rivers. Several criteria based on bankfull discharge and slope are shown in Fig. 7.9. The range of slope variability, however, is quite extensive, and it remains difficult to separate braiding from meandering channels solely on the basis of bankfull discharge and slope. An alternative approach is based on the width–depth ratio. In general, braided channels have a width–depth ratio in excess of 100. Case Study CS.7.3 illustrates several features of a large braided river.

A single channel with given dominant discharge is thus thought to meander on mild slopes and braid on steep slopes. This concept has been expanded by Lane (1957) who proposed a slope-discharge relation for sand-bed channels. Empirically, braided channels were observed when $SQ^{1/4} > 0.01$ and channels

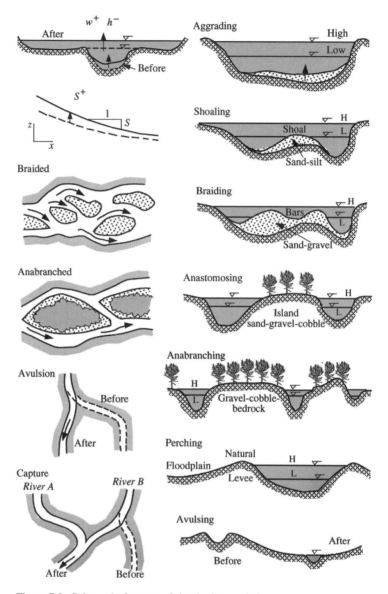

Figure 7.8. Schematic features of riverbed aggradation.

were meandering when $SQ^{1/4} < 0.0017$, given the slope S and the dominant discharge Q in cubic feet per second. Unfortunately, this criterion is not always valid. Lane's diagram is considered only as an index describing the morphological pattern.

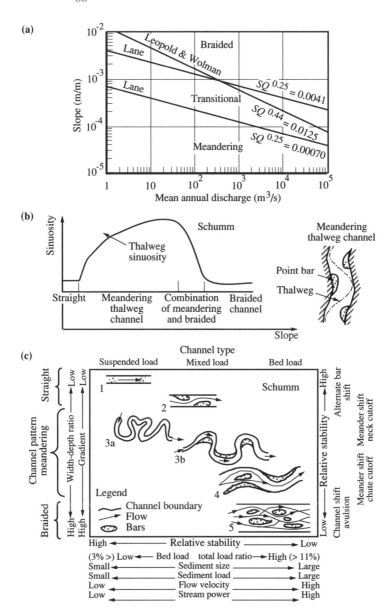

Figure 7.9. Braiding and meandering: (a) predictors of Lane (1957) and Leopold and Wolman (1960), (b) sinuosity diagram of Schumm (1977), and (c) relative stability diagram of Schumm (1977).

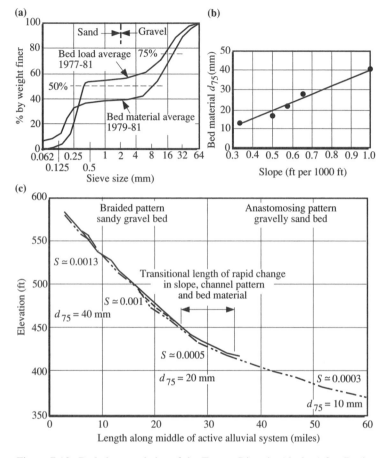

Figure 7.10. Bed characteristics of the Tanana River in Alaska (after Buska et al., 1984).

Direct measurements of bedload transport and comparisons with transport capacity are too uncertain to provide any good indication of a stream tendency to braid. One useful indicator is that braiding often occurs near a sudden decrease in slope, provided that the stream carries excess bedload material. An example of a sudden decrease in bed slope with downstream fining and braiding is shown in Fig. 7.10 for the Tanana River in Alaska. In this case, from Buska et al. (1984), channel aggradation takes place with selective deposition of coarser material upstream. Note that the bed slope is proportional to the grain size. The magnitude of slope and grain size is quite comparable with those of the degrading Meuse River in Case Study 7.1.

7.3.2   Alluvial fans and deltas

Alluvial fans are found where steep mountain channels reach valley floors, as sketched in Fig. 7.11. The sudden break in bed slope causes the bed material

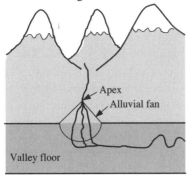

Figure 7.11. Sketch of an alluvial fan.

transported by the river to deposit. The accumulation of debris usually takes a conical shape. The volume of material in the alluvial fan is indicative of the sediment-transport capacity of the stream through geologic times. The aggradation takes place on the riverbed and on natural levees between the apex of the alluvial fan and the valley floor. An example of an alluvial fan is discussed in Case Study 7.2.

Through aggradation and natural levee formation, a river raises its own bed elevation above the surrounding floodplain to form a perched river. Perched rivers are stable as long as they cannot breach their levees. Perched rivers are prone to avulsion in which rivers select a new flow path that can be located up to hundreds of kilometers away from their original river courses. Old channels of perched rivers rapidly dry out, and the process of aggradation and natural levee formation starts at the new river location.

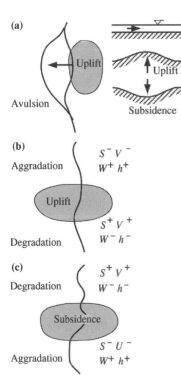

Figure 7.12. Sketch of possible tectonic effects.

To some extent, tectonic activities exert a similar influence on alluvial river morphology as aggradation and degradation (sketched in Fig. 7.12). Uplift and subsidence on the side of a river may result in river perching and may result in river avulsions. Uplift along the water course should cause aggradation and possible braiding upstream and degradation and possible incision downstream. Subsidence along a river course should cause the opposite effects, with degradation upstream and aggradation downstream.

Deltas are observed when rivers reach large lakes, reservoirs and oceans. The sediment deposits extend in a deltaic form into the water, and the aggradation in the upstream river causes the deposits to spread laterally in the lower reaches of the river. The features of deltas and alluvial fans are quite similar where the valley floor is replaced with the water level and deltas are much flatter than alluvial fans. The delta deposits are usually fine (finer than sand) as opposed to alluvial fans (coarser than sand). Rivers transporting large quantities of washload may remain sinuous and meander to the river mouth. Rivers transporting large quantities of bed material tend to braid.

Case Study 7.2   Lawn Lake Dam failure, Colorado, United States. Jarrett and Costa (1986) documented the failure of Lawn Lake Dam in Rocky Mountain National Park, Colorado, on the morning of July 15, 1982. The dam released 674 acre-ft ($8.3 \times 10^{-5} \, m^3$) of water at an estimated peak discharge of 18,000 ft^3/s (500 m^3/s) down the Roaring River valley. Three people were killed and damages totaled $31 million. The probable cause of failure was deterioration of the lead caulking used for the connection between the outlet pipe and the gate valve. The resulting leak eroded the earthfill, and progressive piping led to failure of the embankment. Floodwaters from Lawn Lake Dam overtopped a second dam, Cascade Lake Dam, located 6.7 miles downstream, as shown in Fig. CS.7.2.1. Cascade Lake Dam, a 17-ft-high, concrete gravity, 12.1-acre-ft capacity dam, failed by toppling with 4.2 ft of water flowing over its crest. The flood continued down the Fall River and caused extensive damage from overbank flow to the city of Estes Park.

Peak discharges were determined with a variety of indirect methods including a dam-break model. Peak discharges for the flood were estimated at (1) 18,000 ft^3/s from Lawn Lake Dam; (2) 12,000 ft^3/s at Horseshoe Falls where Roaring River joins the Fall River; (3) 7,210 ft^3/s into Cascade Lake Dam at the east end of Horseshoe Park; (4) 16,000 ft^3/s after the failure of Cascade Lake Dam; (5) 13,100 ft^3/s ~1 mile downstream from Cascade Lake Dam; and (6) 8,520 ft^3/s just upstream from Estes Park. Maximum depths ranged from 6.4 to 23.8 ft, maximum widths ranged from 97 to 1,112 ft, and mean velocities ranged from 3.3 to 12.6 ft/s. Travel times of the flood were determined from eyewitness accounts. The leading floodwave took 3.28 h to travel 12.5 miles (average 3.8 miles/h). Flood peaks were 2.1 to 20 times the 500-yr flood for selected locations along the flood path. It probably was the largest flood that these basins have experienced since the retreat of the glaciers several thousands years ago.

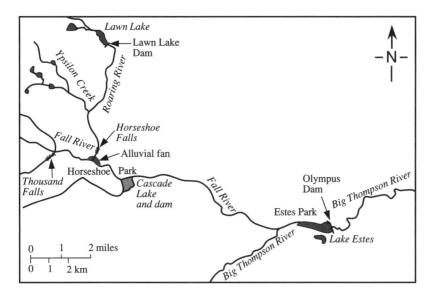

Figure CS.7.2.1. Lawn Lake Dam failure.

In the Roaring River valley, alternate river reaches were either scoured or filled, depending on valley slope. Channels were widened tens of feet and scoured from 5 to 50 ft locally. Generally, reaches steeper than 7% were scoured and reaches less than 7% were filled. In the Roaring River, 56% of the channel reach was scoured, some by as much as 50 ft, and 44% was filled with coarse sediments, 2 to 8 ft thick.

An alluvial fan extending 42.3 acres contained 364,600 yd^3 of material deposited at the mouth of the Roaring River. The fan had a maximum thickness of 44 ft and an average thickness of 5.3 ft. A 452-ton boulder measuring 14 × 17.5 × 21 ft moved with the flood and was deposited on the alluvial van. Down the flow axis, average particle sizes changed from 7.5-ft boulders at the fan apex to fine sand and silt within a distance of 1,900 ft. The alluvial fan dammed the Fall River, forming a 17-acre lake upstream from the fan.

7.4   River confluences and branches

The effects of channel narrowing and widening are quite important in alluvial rivers. Indeed, a widening stream tends to aggrade and a narrowing tends to degrade. The combination of these two mechanisms stabilizes the river width, and this is the primary reason why rivers tend to have fairly constant widths over long river reaches.

River confluences in equilibrium have to convey the water and sediment discharges in the downstream direction. When the sediment concentration and particle size are the same, relations (7.7) can be used to determine the expected changes in hydraulic geometry from an increase in discharge while keeping d_s and C_{ppm} constant. We thus obtain that river confluences will cause a significant increase in bankfull width, an increase in flow depth and shear stress, and a slight increase in flow velocity. The channel slope should also slightly decrease. The example of the confluence of the Ganges and Jamuna Rivers flowing into the Padma River is shown in Case Study 7.3. In particular, Table CS.7.3.1 confirms the expected changes in width, depth, velocity, and slope.

River branches under equilibrium condition have to convey water and sediment discharges in the downstream direction. From relations (7.7), with constant grain size and sediment concentration, the opposite changes are expected from river branching: (1) significant decrease in river width, (2) moderate decrease in flow depth and shear stress, (3) slight decrease in flow velocity, and (4) slight increase in riverbed slope.

Deviations from equilibrium conditions are possible from river captures that result from a new confluence as a result of avulsion of a tributary. At a new river capture, we can expect degradation and channel widening downstream of the new confluence. Likewise, flow diversions out of alluvial channels will trigger sedimentation below the diversion. Channel aggradation, narrowing, reduced flow velocity, and gradual steepening of the river are the expected changes in channel branches.

As an example, consider the sediment diversion of the Mississippi River into the Atchafalaya River. Approximately 22% of the flow discharge of the Mississippi River is diverted into the Atchafalaya River. The sand concentration in the Mississippi River varies largely with discharge, as shown in Fig. 7.13(a). Sand concentrations remain less than approximately 200 ppm at all discharges less than 1,000,000 ft^3/s. Sand concentrations increase approximately linearly with discharge. The effect of a flow diversion should be to decrease the sand concentration downstream of the diversion.

An analysis of sediment transport by size fraction and a sediment budget are shown in Fig. 7.13(b). The very fine sand and fine sand fractions are in reasonable equilibrium, and the inflowing sediment load equals the outflowing sediment load. For the medium sand and coarse sand fractions, the inflowing sand load of $\sim$19 million yd^3/yr far exceeds the outflowing of 2.25 million yd^3/yr. We thus expect the sedimentation of $\sim$17 million yd^3 of medium and coarse sand, 0.25 mm $< d_s <$ 1 mm. In a river reach that is 262 miles long and a river

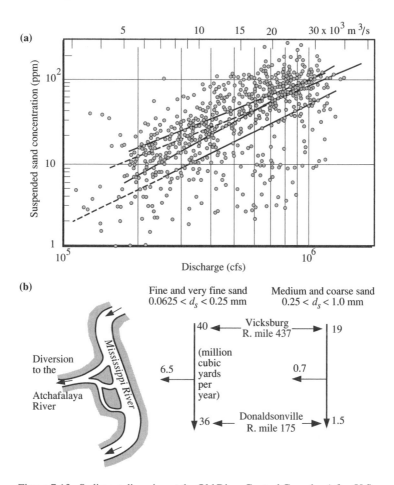

Figure 7.13. Sediment diversion at the Old River Control Complex (after U.S. Army Corps of Engineers, 1999). (cfs, cubic feet per second.)

that is ~2,000 ft wide, the accumulation of sediment represents an average accumulation of sand of 0.16 ft/yr. The tendency toward braiding is possible, but the Mississippi River has a low width–depth ratio (~2,000 ft wide and 50 ft deep). At this rate, it would take 100 yr to raise the bed by 16 ft.

Case Study 7.3 Alluvial changes of the Jamuna River, Bangladesh. The Jamuna River is the lowest reach of the Brahmaputra River in Bangladesh. It drains an area of 550,000 km², and the mean annual discharge is 20,000 m³/s. It is a large braided sand-bed river, and the number of braids at low flows

Table CS.7.3.1. *Characteristics of the Jamuna, Ganges, and Padma Rivers*

River	Drainage area (km^2)	Mean annual discharge (m^3/s)	Bankfull discharge (m^3/s)	Slope (cm/km)	d_s (mm)	Width (m)	Depth (m)	Velocity (m/s)
Jamuna	550,000	20,000	48,000	7.5	0.20	4,200	6.6	1.70
Ganges	1,000,000	11,000	43,000	5	0.14	3,700	6.5	1.78
Padma	1,550,000	28,000	75,000	4.5	0.10	5,200	7.5	1.93

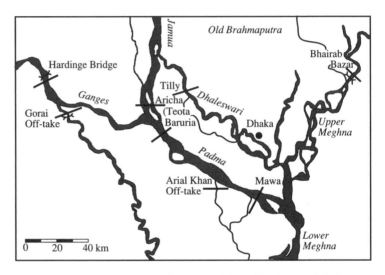

Figure CS.7.3.1. Large river confluences and branching in Bangladesh.

varies between 2 and 3, as shown in Fig. CS.7.3.1. The total width of the braided channel pattern varies between 5 and 17 km. At the confluence with the Ganges, the average annual flood is ~60,000 m^3/s and low-flow discharges vary between 4,000 and 12,000 m^3/s. The maximum discharge recorded in 1988 reached 100,000 m^3/s. The water-surface slope decreases from 10 to 6 cm/km and the bed material is quite uniform with $d_{50} \cong 0.25$ mm near the Indian boarder and 0.16 mm at the confluence with the Ganges River (See Table CS.7.3.1).

Klaassen et al. (1993) report that the Jamuna River is quite active, with frequent channel shifts and lateral migration rates E frequently exceeding 500 m/yr. The shifting rate of the first-order channel is 75 to 150 m. Bank-erosion rates of second-order channels of 250 to 300 m are common. Lateral migration rates exceeding the channel width W in 1 yr have been measured, and, in general, E/W decreases with R/W, where R is the channel radius of curvature. An example of lateral shift is shown in Fig. CS.7.3.2. Significant changes in cross-section geometry can take place within a few years. Examples of braid confluences and branching are shown in Fig. CS.7.3.3.

The confluence shown in Fig. CS.7.3.3 (a) did not move significantly upstream or downstream. In general, slackwater zones are behind wide bars and prone to quick deposition within a year or so. Confluences thus tend to be streamlined. The branching shown in the same figure migrates upstream at a rate of ∼900 m/year. In general, symmetrical branches move upstream.

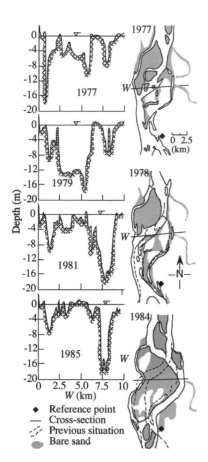

Asymmetrical bifurcations are characterized by one dominant channel in the main downstream direction. The small channel tends to bifurcate at a right angle from the main flow direction. These smaller channels tend to get smaller in size and eventually disappear.

The characteristics of the Ganges and the Padma Rivers are also shown in Table CS.7.3.1 for comparison. The reader will note that the sediment size is finer in the Padma River, and the slope is accordingly reduced.

The width, depth, flow velocity, and coarse sediment transport of the Jamuna River are shown in Fig. CS.7.3.4. The figures indicate a significant variability in measurements that are typical of measurements in alluvial rivers. As much as there is debate as to what the dominant discharge of a stream really is, the reader will note that, at a given discharge near 50,000 m^3/s, the variability in width and depth is approximately a factor of 2. This illustrates the fact that the dominant discharge and downstream hydraulic-geometry relationships are not exact features of alluvial rivers.

Figure CS.7.3.2. Lateral migration of the Jamuna River (after Klaassen et al., 1993).

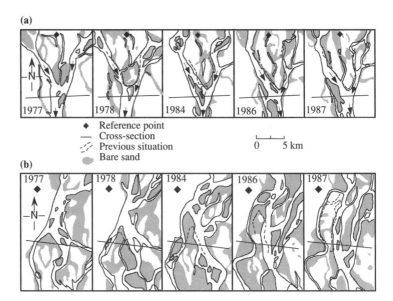

Figure CS.7.3.3. Confluences and branching of the Jamuna River (after Klaassen et al., 1993).

7.5   River databases

River databases for the analysis of channel stability include (1) historical developments, (2) maps and photos, (3) streamflow data, (4) sediment data, and (5) field surveys. Historical information affecting channel morphology and stability should be reviewed. Upstream basin information includes land-use changes, flow diversions, and artificial structures for flood control, irrigation, and navigation.

Topographic maps at various scales, depending on the stream size, indicate the nature of the drainage area and the fluvial system. The stream planform geometry, longitudinal profile, and estimates of channel slope can be obtained from topographic maps with contours. Aerial photographs are most useful in examining sediment deposits. The comparison of several sets of aerial photographs enable the evaluation of the lateral migration rates of alluvial channels. Geographic information systems (GIS) are useful in examining topography, soil types, and land-use data. The analysis of lateral migration in very large rivers, like the Jamuna River, is sometimes possible with GIS data. Satellite imagery is sometimes useful in the analysis of watershed data and some turbidity and flow patterns in lakes and estuaries.

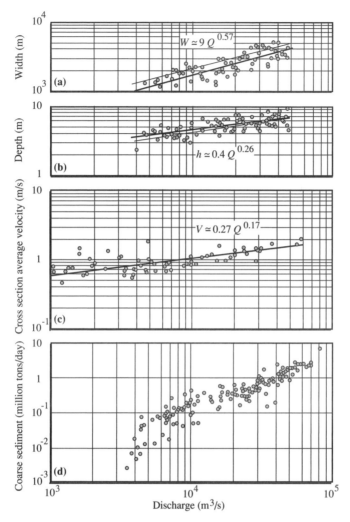

Figure CS.7.3.4. Hydraulic geometry of the Jamuna River (after FAP24, 1996).

Streamflow data include discharge data on a daily basis for the entire period of record. The entire flow-discharge record can be used to determine the flow-duration curves and for flood-frequency analysis. Hydrometric gauges are useful in determining the stage–discharge relationship. Loop-rating effects that are due to aggradation/degradation, bedform changes, and/or hydrodynamic effects can be examined from stage–discharge relationships. It is also useful to examine specific-gauge records to detect aggradation/degradation trends over long periods. The bankfull discharge should normally fit within the range of

1 to 5 yr in the flood-frequency analysis. Extrapolations to periods of return far exceeding the length of record can be misleading.

Sediment data include bed material and sediment transport; the particle-size distribution of bed material should be determined as accurately as possible. The variability in bed material in alluvial rivers can be high. Several samples at different locations are often desirable. In degrading channels, careful attention should be paid to the coarse fractions of the surface material and underlying deposits. Sieve analyses are best suited to fine-grained streams, and gravel-bed and cobble-bed streams require the examination of large volumes of bed material to determine the median grain size.

Suspended sediment records should indicate the flux-average sediment concentration and the sediment load. Sediment load by size fractions is most valuable to separate washload from bed-material load. Sediment concentration profiles enable evaluation of the Rouse parameter. Sediment budgets by size fractions are sometimes most useful in determining the different patterns of sediment transport for washload and bed-material load.

Field surveys are most effective after a review of maps and photos. It is very important to gather additional information on the cross-sectional geometry of the river, including bankfull conditions and floodplain elevation, land use, and vegetation. Field notes should include indications of actual upstream basin conditions, and recent changes in vegetation, land use, and sediment sources should be reviewed. Riverbed, banks, profile, and planform should be examined for particle-size distribution of bed material, sediment deposits, aggradation/degradation, headcutting, and bedrock control. Bank stability and mode of failure, stratigraphy and seepage, lateral migration, and vegetation should be considered. River hydraulics includes velocities, high-water marks on bridge piers, structures, and the floodplain; river choking, debris, ice cover and ice jams, and flow controls provide important information for water-surface calculations with hydraulic models. Flights over the river are quite informative on the overall planform stability of rivers during floods. Case Study 7.4 illustrates how river databases can be used to examine the dynamics of river systems by means of the analysis of mass curves.

Case Study 7.4   Bernardo reach of the Rio Grande, New Mexico, United States. The 10-mile-long Bernardo reach of the Middle Rio Grande in New Mexico is included in the habitat designation for two federally listed endangered species: the silvery minnow and the southwestern willow flycatcher. A complete database of flow and sediment transport in the Rio Grande below Cochiti Dam has been assembled at Colorado State University by G. Richard,

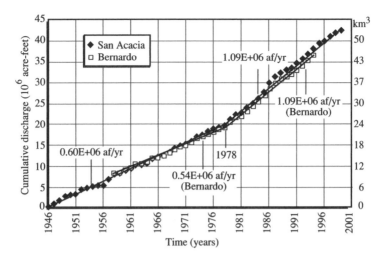

Figure CS.7.4.1. Mass curve for the discharge of the Rio Grande (after Richard et al., 2000). (af/yr, acre-ft per year.)

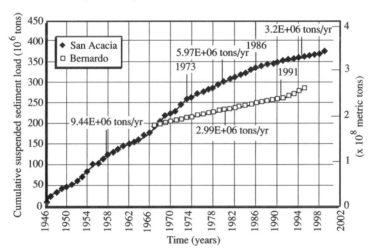

Figure CS.7.4.2. Mass curve for the suspended sediment of the Rio Grande (after Richard et al., 2000).

C. Leon, and T. Bauer in collaboration with D. Baird at the U.S. Bureau of Reclamation. Besides the complete data reports (Leon et al., 1999, and Bauer et al., 2000), the geomorphic analysis of Richard et al. (2000) shows a 500-ft-wide river with a fine sand bed at $d_{50} \simeq 0.3$ mm. The channel width varies from 150 to 1,200 ft within a 10-mile-long reach. A discharge mass curve at Bernardo and San Acacia in Fig. CS.7.4.1 represents the cumulative runoff

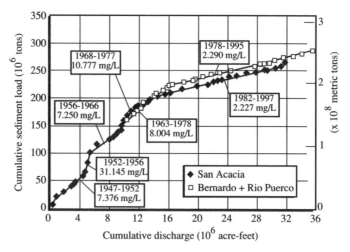

Figure CS.7.4.3. Double mass curve of the Rio Grande (after Richard et al., 2000).

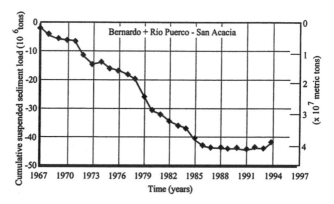

Figure CS.7.4.4. Sediment accumulation in the Bernardo reach of the Rio Grande (after Richard et al., 2000).

volume as a function of time. During the same period, the sediment mass curves in Fig. CS.7.4.2 describe the cumulative sediment yield as a function of time. The influence of Cochiti Dam is clearly identified after 1975.

Double mass curves represent the cumulative sediment load as a function of the cumulative runoff water volume. The case of Bernardo reach of the Rio Grande is illustrated in Fig. CS.7.4.3. The slope of double mass curves defines average sediment concentrations in suspension. In the case of the Rio Grande, the slope reduction reflects the effects of Cochiti Dam in terms of reduced sediment concentration after 1975.

Mass difference curves show the difference between sediment inflow and sediment outflow to a given reach. It can be determined from the sum and the differences between the sediment curves of all tributaries and branches of a given river reach. Mass difference curves provide information on the net sediment balance on a river reach over time and indicate whether a river is aggrading $(+)$ or degrading $(-)$. The case of the Bernardo reach of the Rio Grande is illustrated in Fig. CS.7.4.4, and the negative values indicates degradation of the river reach, which has been confirmed with field measurements of the cross-sectional geometry. It is important to include the contribution of all tributaries in the analysis of mass difference curves.

◆Exercise 7.1

Derive relations (7.4a) and (7.4b) from Relations (6.27) and (7.3).

Exercise 7.2

Derive hydraulic-geometry relations (7.6a)–(7.6e) from combining Relations (6.28) and (7.4).

◆Exercise 7.3

Derive hydraulic-geometry relations (7.7a)–(7.7e) from Relations (7.6) and the definition of C_{ppm}.

◆Exercise 7.4

Calculate the annual rate of bed aggradation in the Mississippi River from the data provided in the river branching section, Section 7.4.

Problem 7.1

Determine the combined effects of a 50% decrease in water discharge Q^- and a 200% increase in sediment discharge Q_{bv}^+ on channel width, flow depth, flow velocity, slope, and Shields parameter.

Answers: New flow depth $\sim$67% of initial flow depth $h \simeq 0.67\,h_0$, $W \simeq 58\%W_0$, $V \simeq 1.27V_0$, $S \simeq 2.75S_0$, and $\tau_* \simeq 1.84\tau_{*_0}$.

◆Problem 7.2

Estimate the hydraulic geometry of an alluvial stream at a bankfull discharge of 4,500 ft^3/s with $d_{50} = 0.5$ mm and a bed-material concentration of 150 ppm.

Answers: With $Q = 127$ m^3/s, $d_s = 0.0005$ m, and $C_{mg/l}$ 150, we obtain from relations (7.7a)–(7.7e) $h \simeq 1.2$ m, $W \simeq 170$ m, $V = 0.63$ m/s, $S \simeq 2.2 \times 10^{-4}$, and $\tau_* \simeq 0.3$.

Problem 7.3

Varmint Creek drains a 320-square-mile watershed (see also Problem 7.2). The mean annual rainfall is 45 in. The largest known peak discharge in a 45-yr record was 26,000 ft^3/s in 1929. The 10-yr suspended sediment record indicates a mean annual yield of 48,000 tons, mostly a washload of silt and clay. The channel slope is ~2.5 ft/mile. The 2-yr flood peak is 4,500 ft^3/s. The corresponding bed-sediment concentration is estimated at 150 ppm. Compare the measured peak discharge and sediment yield with those of comparable watersheds of the same drainage area.

◆Problem 7.4

The Jamuna River is a large braided river with a median grain size of 0.2 mm. The river conveys ~48,000 m^3/s at bankfull conditions and the corresponding bed-material discharge is approximately 2.6 million tons per day. Estimate the downstream hydraulic geometry of the river.

Answers: Calculate by using Relations (7.6a)–(7.6e) with $Q = 48,000$ m^3/s, $d_s = 0.0002$ m, and $Q_{bv} = 11.6$ m^3/s to give $h \simeq 6.7$ m, $W \simeq 2,500$ m, $V \simeq 2.8$ m/s, $S \simeq 3.2 \times 10^{-4}$ and $\tau_* \simeq 6$. Field measurements in Table CS.7.3.1 indicate $h \simeq 6.6$ m, $W \simeq 4,200$ m, $V \simeq 1.7$ m/s, $S \simeq 7.5 \times 10^{-5}$, and $\tau_* \simeq 15$. The calculated slope far exceeds the measured slope, and the stream may be aggrading and braiding.

◆◆Problem 7.5

From the information in Case Study 7.1 and Fig. CS.7.1.1, estimate the minimum grain size at the beginning of motion during a flood of the Meuse River at 100,000 ft^3/s.

Answers: The flow depth is ~25 ft during floods, or $h \simeq 7.6$ m, and the slope is ~4 ft/mile, or $S \simeq 7.6 \times 10^{-4}$. We obtain from Relations (7.9) $d_{sc} \simeq 12\,hS = 0.069$ m or 69 mm. This indicates that all sand and gravel sizes are in motion during floods. Only cobbles are stable and can armor the riverbed if available in sufficiently large quantity.

◆Problem 7.6

With reference to the Lawn Lake Dam failure in Case Study 7.2, determine the following: (1) the flow depth required for mobilizing a 2-m boulder in Roaring River and (2) the volumetric sediment concentration of the Roaring River flow.

Answers: (1) From Relation (7.9b) with $S = 0.07$ and $d_s = 2$ m, we obtain a flow depth of $\sim$2.8 m; (2) a volumetric sediment concentration of $\sim$33% is obtained from the alluvial fan volume and the Lawn Lake water release. It can be classified as hyperconcentrated flow.

Problem 7.7

With reference to the Jamuna River in Case Study 7.3, determine the planform geometry from the planform predictors based on discharge and slope. Also use the width–depth ratio as an indicator.

◆Problem 7.8

Anticipate the effects of a 30% flow reduction and 30% sediment diversion on the hydraulic geometry (W, h, V, S, τ_*) of a large meandering river.

◆◆Problem 7.9

With reference to the Jamuna, Ganges, and Padma Rivers' confluences in Case Study 7.3, apply the hydraulic-geometry relationships and compare with field measurements of W, h, V, and S. Compare the slopes upstream and downstream of the confluence. Would a river confluence in equilibrium require an increase or decrease in slope in the downstream direction?

◆◆Problem 7.10

The Bernardo reach of the Rio Grande in New Mexico features a channel width ranging from 150 to 1,200 ft with an average of $\sim$500 ft. The average reach slope is 80 cm/km, a median grain size of 0.3 mm, and a sandload up to 10,000 tons per day at discharges of $\sim$5,000 ft^3/s. Compare with the range of channel width and slope calculated at a sediment concentration of sand varying from 500 to 2,000 mg/l.

Answers: From Relations (7.7) and $C = 500$ mg/l, $W = 163$ m and $S = 4.6 \times 10^{-4}$.

8

River stabilization

River-stabilization structures are designed to protect the riverbanks and prevent lateral migration of alluvial channels through bank erosion. River-stabilization methods can be classified according to two different approaches: (1) strengthening the banks and (2) reducing hydrodynamic forces. This chapter first examines the bank stability of alluvial streams in Section 8.1. Bank-protection methods through strengthening the banks with riprap are discussed in Section 8.2, and other bank-strengthening methods are covered in Section 8.3. Flow-control structures offer an alternative approach by reducing the hydrodynamic forces applied against the riverbanks. The flow-control structures covered in Section 8.4 aim at gaining control over the flow depth and the direction and magnitude of flow velocity near the river banks. Some engineering considerations are discussed in Section 8.5.

8.1   Riverbank stability

Bank stability is examined in this section. First, the processes are reviewed in Subsection 8.1.1, followed by conceptual solutions for slope reduction in Subsection 8.1.2 and subsurface drainage in Subsection 8.1.3.

8.1.1   Bank-erosion processes

Processes of bank erosion are directly linked to the lateral migration of alluvial channels. Bank erosion is the result of flowing water that applies active forces met by the passive forces of the bank material to resist motion. As discussed in Chap. 6, the hydrodynamic forces in river bends induce secondary flow where the free-surface streamlines are deflected toward the outer bank and the near-bed streamlines are deflected toward the inner bank. Along a cross section, the streamlines are deflected downward near the outer bank and deflected upward on the point bar. The resulting effect is to decrease the

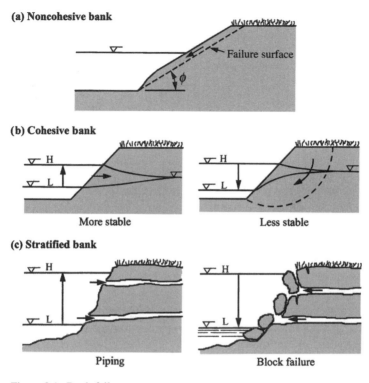

Figure 8.1. Bank-failure types.

stability of sediment particles and cause degradation near the outer bank. On the other hand, the particle-stability increases and aggradation is expected near the point bar.

The scour at the toe of the outer bank shifts the thalweg to the outside of river bends and causes steepening of the outer bank. Increased steepening of the outer-bank material causes bank failure. Three modes of failure are typical of alluvial rivers, as sketched in Fig. 8.1. With noncohesive granular material, grain removal at the toe of the outer-bank induces sliding of the granular material as soon as the bank angle exceeds the angle of repose of the material [Fig. 8.1(a)]. In the case of cohesive bank material, rotational failure is typical and the presence of tension cracks may accelerate the bank-erosion process [Fig. 8.1(b)]. In alluvial streams flowing in stratified deposits, the underlying noncohesive material is mobilized, thus leaving the overlying cohesive material unsupported and subject to tension cracks and cantilever failure [Fig. 8.1(c)]. In general, the most erosive banks are sandy and silty, whereas the least erosive are clayey and gravelly.

Factors affecting streambank failure include hydraulic parameters that control the active forces such as discharge magnitude and duration, velocity, and applied shear-stress magnitude and orientation. Additional active forces that are due to seepage, piping, surface waves, and ice can also contribute to bank erosion. Passive forces relate to bank-material size, gradation, and cohesion. Biological factors such as vegetation can play a significant stabilizing role. Artificially-induced activities such as urbanization, drainage, floodplain farming and development, boating and commercial navigation, and water-level fluctuations from hydropower generation can have detrimental effects on bank stability.

Changes in channel geometry through bank erosion are particularly significant during floods. The sediment-transport capacity is often several orders of magnitude greater during floods than at intermediate or low flows. Most cases of riverbank instability in alluvial rivers take place during the small percentage of time when the dominant discharge is exceeded. The analysis of flow-duration curves and sediment-duration curves in Section 5.6 is useful in the analysis of bank stability.

Landslides refer to the downslope movement of earth and organic materials [Fig. 8.1(b)]. Active forces are involved in mass wasting. These forces are associated with the downslope gravity component of the slope mass. Resisting these downslope forces are the shear strength of the Earth's materials and any additional contributions from vegetation by means of root strength or human slope-reinforcement activities. When a slope is acted on by a stream or river, an additional set of forces is added. These forces are associated with removal of material from the toe of the slope, fluctuations in groundwater levels, and vibration of the slope. A slope may fail if stable material is removed from the toe. When the toe of a slope is removed, the slope loses more resistance by buttressing than it does by downslope gravitational forces. The slope materials may then tend to move downward into the void in order to establish a new balance of forces or equilibrium.

The presence of water in riverbanks and its movement toward or away from the river affects bank stability and bank erosion in various ways. The outflow of water from the river into the adjacent banks stabilizes the riverbanks. Rivers that continuously seep water into the banks tend to have smaller widths and larger depths for a particular discharge. The converse is true of rivers that continuously gain water by an inflow through their banks. The inflow destabilizes riverbanks.

Piping is another phenomenon common to the alluvial banks of rivers. With stratified banks [e.g., Fig. 8.1(c)], flow is induced in more permeable layers by changes in river stage and by wind- and boat-generated waves. If the flow through the permeable lenses is capable of dislodging and transporting fine particles from the permeable lenses, the material is slowly removed, undermining portions of the bank. Without this foundation material to support the overlying layers, a block of bank material drops down and results in the development of tension cracks, as sketched in Fig. 8.1(c). These cracks allow surface flows to enter, further reducing the stability of the affected block of bank material.

Mass wasting is an alternative form of bank erosion. If the bank becomes saturated and possibly undercut by flowing water, blocks of the bank may slump or slide into the channel. Mass wasting may be further aggravated by construction of homes on riverbanks, operation of equipment on the floodplain adjacent to the banks, added gravitational force resulting from tree weight, saturation of banks, and increased infiltration into the floodplain.

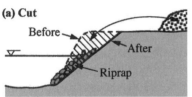

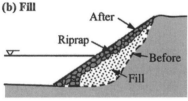

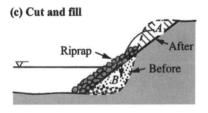

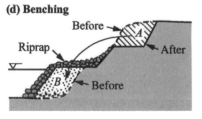

Figure 8.2. Slope-reduction methods.

8.1.2 Slope reduction and benching

The weight of soil at the toe of the slope counterbalances the weight of soil in the upper part of the slope and aids the shear strength of the soil in resisting against failure.

Direct slope reduction is an excavation method whereby soil is removed from the slope to flatten the slope angle, as shown in Fig. 8.2(a). This method is usually preferred when adequate space is available on the floodplain. Where there is insufficient space, the slope may be flattened by use of fill material, as shown in Fig. 8.2(b), or by cut-and-fill operations, as shown in Fig. 8.2(c). Slope, or buttress, benching is also an indirect method of slope reduction. It differs from the method described above only in the final shape of the slope section. The benching method produces a benched or a series of stepped sections, as shown by Fig. 8.2(d). The end effect of this method is a gross reduction in the angle of the slope.

8.1.3   Subsurface drainage

The control of groundwater within a slope is often a feasible means of stabilizing the slope. The control of groundwater may be achieved by two methods: (1) the prevention of infiltration of surface water into soils, and (2) the provision for subsurface drainage to remove the water from the soil mass. The first method is generally accomplished by merely providing adequate surface drainage. The second method uses various subsurface-drainage techniques.

Subsurface drains are most effective when only a small quantity of water is required to be removed to affect stabilization. A thin confined aquifer can be intercepted to reduce the artesian pressure. Horizontal drains consist of a slotted pipe to remove excessive water from cut slopes experiencing stability problems. Figure 8.3(a) shows an idealized concept for the application of horizontal drains to streambank stability.

Stabilization of highway cut slopes frequently involves the use of a combination of horizontal and vertical drainage systems. Vertical drains can be utilized in riverbank stabilization, as sketched in Fig. 8.3(b). The main function of a drain-well installation is to intercept groundwater moving toward the bank and thus to relieve pore pressures that would otherwise develop in a saturated fine-grained bank material. Usually this high-cost concept is applicable only in cases in which high groundwater levels exist above maximum river elevations.

A drainage trench is shown conceptually in Fig. 8.3(c). Its function is similar to that of the vertical drain to intercept any groundwater before it reaches the bank slope. A slotted or perforated pipe is placed at the bottom of the trench to drain the water intercepted by the trench. As with vertical drains, this method would be effective for bank stabilization only in cases in which high groundwater levels occur above maximum river elevations.

8.2   Riverbank riprap revetment

Several engineering methods can be used to strengthen riverbanks against erosion. Methods commonly encountered include the use of riprap or large stones that are not easily removed from the banks. Besides riprap in this section, the other methods considered in Section 8.3 are vegetation, gabions, blocks and

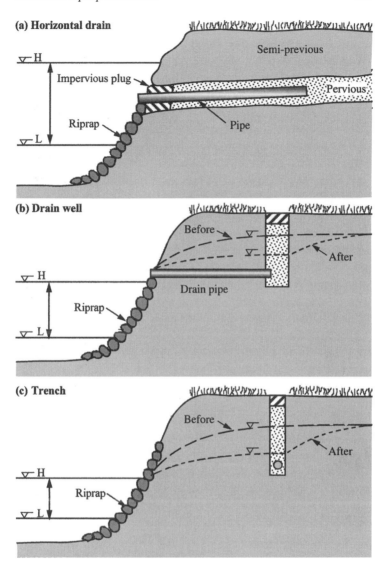

Figure 8.3. Riverbank drainage methods.

rocks, rock-fill trenches, windrow revetment, soil cement, fences, bulkheads, and mattresses.

When economically available in sufficient size and quantity, rock riprap is usually the most widely used material for bank protection. A rock riprap blanket is flexible and is neither impaired nor weakened by slight movement of the bank resulting from settlement or other minor adjustments. Local damage or loss is easily repaired by the placement of more rock. Construction is not complicated, and special equipment and construction practice are not necessary. Riprap is usually durable and recoverable and may be stockpiled for future use. Locally available riprap usually provides a cost-effective alternative to many other types of bank protection. The appearance of rock riprap is natural, and after a period of time vegetation will grow between the rocks. Finally, wave runup on rock slopes is usually less than on other types of structures. The important factors to be considered in designing rock riprap blanket protection are: (1) the velocity (both magnitude and direction) of the flow or shear stress in the vicinity of the rock; (2) the sideslope of the bankline being protected; (3) the density of the rock; (4) the angle of repose for the rock, which depends on stone shape and angularity; (5) the durability of the rock; (6) the riprap blanket thickness; (7) the filter needed between the bank and the blanket to allow seepage but to prevent erosion of bank soil through the blankets; (8) the blanket stabilization at the toe of the bank; and (9) the blanket must be tied into the bank at its upstream and downstream ends.

Two methods to determine the size of rock riprap required for streambank stabilization are presented: (1) shear-stress method in Subsection 8.2.1 and (2) flow-velocity method in Subsection 8.2.2. Specifications are given for riprap gradation in Subsection 8.2.3 and for filters in Subsection 8.2.4. Protection against riprap failure is then considered in Subsection 8.2.5.

8.2.1   Shear-stress method

Riprap stability on a sideslope is a function of (1) the magnitude and the direction of the flow velocity or shear stress in the vicinity of the streambank, (2) the sideslope angle, and (3) the properties of the rock including size, density, and angularity. The functional relationship that determines the stability factor of a particular stone under given hydrodynamic forces has been analyzed previously in Chap. 6. The stability factor SF can be determined from Eqs. (6.3), (6.5), (6.6), and (6.8), as detailed in Example 6.1.

On the outer bank of natural meandering channels, the streamlines are deflected downward at an angle λ that can be estimated from $\tan \lambda \cong 11\,h/R$, where h is the mean flow depth and R is the bend radius of curvature. A simplified approach to the calculation of the rock size that corresponds to the beginning of motion on a sideslope θ_1 is possible when the following approximations are acceptable: (1) flow in the downstream direction, i.e., in the absence of secondary flows, or $\lambda \cong 0$; (2) the downstream bed slope is negligible, $\theta_0 \cong 0$; (3) the specific gravity of the rock is close to 2.65; and (4) the viscous drag is small compared with the lift force.

The effective rock size d_m required for stabilizing a riverbank under applied shear stress τ_0 is estimated from Lane's relationship as

$$d_m = \frac{\tau_0}{\tau_{*c}\,(\gamma_s - \gamma)\left[\cos\theta_1\sqrt{1 - \dfrac{\tan^2\theta_1}{\tan^2\phi}}\right]} \qquad (8.1a) \blacklozenge$$

or

$$d_m = \frac{\tau_0}{\tau_{*c}\gamma\,(G - 1)\left[\sqrt{1 - \dfrac{\sin^2\theta_1}{\sin^2\phi}}\right]}, \qquad (8.1b) \blacklozenge$$

where τ_0 is the applied shear stress, γ_s and γ are the specific weight for the rock and water, respectively, θ_1 is the sideslope angle, ϕ is the angle of repose of the rock riprap, and τ_{*c} is the critical value of the Shields number. Examples 8.1 and 8.2 illustrate how to determine riprap size and particle stability. It must be considered that the term in brackets of Eqs. (8.1) becomes very small when the sideslope angle becomes approximately equal to the angle of repose of the material. This tends to require extremely large material on steep banks.

Example 8.1 Calculation of riprap size in straight channels. Determine the rock riprap size required for stabilizing the banks of a straight river given the river width $W = 300$ m, a flow depth of $h = 7$ m, and a channel slope of $S = 60$ cm/km. The bank slope is $\theta_1 = 30°$, the rock density G is 2.7, and the angle of repose is $\phi = 40°$.

Step 1: The shear stress applied on the particle is $\tau_0 = \gamma h S = 9{,}810$ N/m^3 × 7 m × 60 × 10^{-5} = 41 Pa.

Step 2: The term in brackets of Eqs. (8.1) is calculated as

$$\eta = \cos\theta_1\sqrt{1 - \left(\frac{\tan\theta_1}{\tan\phi}\right)^2} = \cos 30\left[1 - \left(\frac{\tan 30}{\tan 40}\right)^2\right]^{0.5} = 0.63.$$

Note that very large riprap sizes may be obtained when the term η becomes very small.

Step 3: The effective riprap size is obtained from Eqs. (8.1) assuming $\tau_{*c} = 0.047$ as

$$d_m \cong \frac{\tau_0}{\eta\tau_{*c}\,(\gamma_s - \gamma)} = \frac{41\,\text{Pa}}{0.63 \times 0.047\,(2.65 - 1)\,9{,}810\,\text{N}}\,\text{m}^3 = 0.085\,\text{m},$$

or ~3.3 in.

Example 8.2 Application to riprap stability in a bend. The mean flow velocity in a river bend reaches 18 ft/s (5.5 m/s) during floods at a Manning coefficient of $n = 0.03$, the river is 300 ft wide (91 m) and 10 ft deep (3.0 m) with a radius of curvature of 700 ft (215 m). Determine the stability factor of 6-in. crushed rock riprap at a sideslope of $1V:2.25H$, or $\theta_1 = 24°$. The angle of repose of the material is $\phi = 40°$ and the specific gravity $G = 2.65$.

Step 1: The slope is calculated from Manning's formula

$$S = \tan \theta_0 = \left(\frac{nV}{1.49 \, h^{2/3}} \right)^2 = \left(\frac{0.03 \times 18}{1.49 \times 10^{2/3}} \right)^2 = 6.1 \times 10^{-3}$$

or 610 cm/km, or $\theta_0 = 0.35°$.

Step 2: The shear stress is obtained from $\tau_0 = \gamma h S = 62.4 \times 10 \times 6.1 \times 10^{-3} =$ 3.8 lb/ft^2 or 182 Pa.

Step 3: From the method in Chap. 6, the angle $\theta = \tan^{-1} (\sin \theta_0 / \sin \theta_1) = 0.86°$.

Step 4: The factor $a_\theta = \sqrt{\cos^2 \theta_1 - \sin^2 \theta_0} = \sqrt{\cos^2 24 - \sin^2 0.35} = 0.91$.

Step 5: The streamline deviation angle is $\lambda \simeq \tan^{-1} (11h/R) = \tan^{-1} [(11 \times 10)/700] = 8.9°$.

Step 6:

$$\eta_0 = \frac{21\tau_0}{(G-1)\gamma d_s} = \frac{21 \times 3.8 \text{ lb-ft}^3}{\text{ft}^2 \, (1.65) \, 62.4 \text{ lb} \times 0.5 \text{ ft}} = 1.55.$$

Step 7: The angle β, assuming $M = N$, is

$$\beta = \tan^{-1} \left\{ \frac{\cos(\lambda + \theta)}{\left[\dfrac{2\sqrt{1 - a_\theta^2}}{\eta_0 \tan \phi} + \sin(\lambda + \theta) \right]} \right\}$$

$$= \tan^{-1} \left\{ \frac{\cos(8.9 + 0.86)}{\left[\dfrac{2\sqrt{1 - 0.91^2}}{1.55 \tan 40} + \sin(8.9 + 0.86) \right]} \right\} = 50.7°.$$

Step 8:

$$\eta_1 = \eta_0 \left[\frac{1 + \sin(\lambda + \beta + \theta)}{2} \right]$$

$$= 1.55 \left[\frac{1 + \sin(8.9 + 50.7 + 0.86)}{2} \right] = 1.45.$$

Step 9: The stability factor is

$$SF = \frac{a_\theta \tan\phi}{\eta_1 \tan\phi + \sqrt{1 - a_\theta^2} \cos\beta}$$

$$= \frac{0.91 \tan 40°}{1.45 \tan 40° + \sqrt{1 - 0.91^2} \cos 50.7°} = 0.52$$

The stone is unstable because SF < 1. As a first approximation, assuming $\lambda = 0$ and $\theta_0 = 0$, the stone size at incipient motion is estimated from Eqs. (8.1) as

$$d_m \cong \frac{\tau_0}{\sqrt{1 - \frac{\sin^2\theta_1}{\sin^2\phi}} 0.047(\gamma_s - \gamma)}$$

$$= \frac{3.8 \text{ lb-ft}^3}{\text{ft}^2 \sqrt{1 - \frac{\sin^2 24°}{\sin^2 40°}} 0.047 \times 1.65 \times 62.4 \text{ lb}} \simeq 1.0 \text{ ft or } 30 \text{ cm.}$$

Calculation steps 6–9 account for secondary flow and can be repeated for different stone sizes until SF = 1.

Based on flow velocity, the first estimate of riprap size can be obtained from Eqs. (8.4) in Subsection 8.2.4 solved for d_s as

$$d_s = \frac{V_c^2}{K_c^2} \frac{\sin\phi}{2(G-1)g(\sin^2\phi - \sin^2\theta_1)^{1/2}}$$

$$= \frac{(18)^2 \text{ ft}^2\text{s}^2 \sin 40}{\text{s}^2(1.2)^2 2 \times 1.65 \times 32.2 \text{ ft}\sqrt{\sin^2 40 - \sin^2 24}} = 2.7 \text{ ft.}$$

In this case, this value can also be obtained from Fig. 8.4 in Subsection 8.2.2.

8.2.2 Velocity method

The stone size needed to protect a streambank from erosion by a current that is moving parallel to the embankment can also be determined as a function of flow velocity. The diameter d_{50} is that of a spherical stone that would have the same weight as the 50% size of the stone. For stone riprap, the velocity at the top of the stone, called velocity against the stone, v_s, is related to the shear velocity u_* as $v_s = 5.75\, u_*$.

From the incipient motion condition of material with angle of repose ϕ, for hydraulically rough flow conditions,

$$\tau_{*_c} = \frac{u_{*_c}^2}{(G-1)gd_s} = 0.06 \tan \phi,$$

we obtain the critical velocity against the stone v_{sc} as

$$v_{sc} = 5.75\, u_{*_c} = \sqrt{2(G-1)gd_s \tan \phi} \,. \tag{8.2}$$

The velocity against the stone v_s also relates to the mean flow velocity V as a function of the flow depth h and the stone diameter d_s as $V = v_s \log (4\,h/d_s)$. After combining this with Eq. (8.2), we obtain the critical mean flow velocity V_c as

$$V_c = K_c \sqrt{2(G-1)gd_s} \,, \tag{8.3a}$$

where

$$K_c = \log \left(\frac{4h}{d_s} \right) \sqrt{\tan \phi} \,. \tag{8.3b}$$

The critical mean flow velocity thus depends on relative submergence h/d_s and angle of repose ϕ. It is noticed that, for riprap, $\sqrt{\tan \phi} \simeq 1$ and $K_c \simeq 1.2$, when $h/d_s \simeq 5$.

The critical mean flow velocity for representative riprap design conditions ($h \simeq 5\, d_s$ or $K_c = 1.2$) is shown in Fig. 8.4. This graphic should be used when $h < 10\, d_s$. At larger flow depths ($h > 10\, d_s$), the shear-stress method is preferable. The reader is referred to Maynord (1992) and Abt (Colorado State University, 2001, pers. comm.) for recent developments on riprap design procedures based on flow velocity.

On a sideslope without secondary flows, the critical mean flow velocity V_c can be approximated according to Lane's approach by

$$V_c \simeq K_c \sqrt{2(G-1)gd_s} \left[1 - \frac{\sin^2 \theta_1}{\sin^2 \phi} \right]^{1/4}. \tag{8.4} \blacklozenge$$

Typical curves at an angle of repose $\phi = 40°$ are shown in Fig. 8.4 for sideslope angles up to $33°$. It is observed that sideslope effects become very significant when $\theta_1 > 20°$. Gradations calculated with the U.S. Army Corps of Engineers (1981) method can be obtained from Table 8.1.

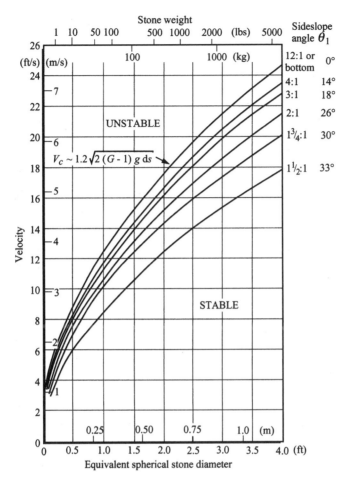

Figure 8.4. Particle-stability diagram.

8.2.3 Riprap gradation

The concept of a representative grain size for riprap is fairly simple. A uniformly graded riprap with a median size d_{50} scours to a greater depth than a well-graded mixture with the same median size. The uniformly distributed riprap scours to a depth at which the velocity is less than that required for the transportation of d_{50} size rock. The well-graded riprap, on the other hand, develops an armor plate; that is, some of the finer materials, including sizes up to d_{50} and larger, are transported by the high velocities, leaving a layer of large rock sizes that can not be transported under the given flow conditions. Thus the size of rock representative of the stability of the riprap is determined by the

Table 8.1. *Riprap weight from the U.S. Army Corps of Engineers (1981)*

d_{100} max (in.)	Stone weight (lb) for percent lighter by weight[a]						d_{30} max (ft)	d_{90} max (ft)
	100		50		15			
	max	min	max	min	max	min		
12	86	35	26	17	13	5	0.48	0.70
15	169	67	50	34	25	11	0.61	0.88
18	292	117	86	58	43	18	0.73	1.06
21	463	185	137	93	69	29	0.85	1.23
24	691	276	205	138	102	43	0.97	1.40

[a] Assuming $G = 2.65$, or $\gamma_s = 165$ lb/ft^3 = 26 kN/m^3, 1 ft = 12 in. = 30.5 cm. The relationship between diameter and weight is based on a spherical shape.

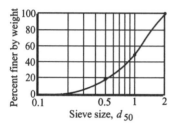

Figure 8.5. Suggested riprap gradation curve.

larger sizes of rock. The representative grain size d_m for riprap is larger than the median rock size d_{50}. The effective size d_m approximately corresponds to d_{65}, and we can use $d_m \cong 1.25\ d_{50}$. The computations of the representative grain size d_m for the recommended gradation are illustrated in Fig. 8.5 in terms of d_{50}. Recommended gradations from the U.S. Army Corps of Engineers are presented in Tables 8.1 and 8.2.

Riprap consisting of angular stones is more suitable than that consisting of rounded stones. Control of the gradation of the riprap is almost always made by visual inspection. When necessary, poor gradations of rock can be used as

Table 8.2. *Suggested riprap size gradation*

Percent finer by weight	Sieve diameter ($\times d_{50}$)	Stone diameter ($\times d_{50}$)
0	0.25	—
10	0.35	0.28
20	0.50	0.43
30	0.65	0.57
40	0.80	0.72
50	1.00	0.90
60	1.20	1.10
70	1.60	1.50
90	1.80	1.70
100	2.00	1.90

riprap, provided that a filter is placed between the riprap and the bank or bed material.

Considering the practical problems of quarry production, a gradation band is usually specified by the U.S. Army Corps of Engineers (1981) rather than a single gradation curve, and any stone gradation within the limits is acceptable. The Corps criteria for establishing gradation limits for riprap are as follows: (1) the lower limit of d_{50} stone should not be less than the size of stone required for withstanding the design shear forces; (2) the upper limit of a d_{50} stone should not exceed five times the lower limit of a d_{50} stone, the size that can be obtained economically from the quarry or the size that satisfies layer-thickness requirements; (3) the lower limit of d_{100} stone should not be less than two times the lower limit of d_{50} stone; (4) the upper limit of d_{100} stone should not exceed five times the lower limit of d_{50} stone, the size that can be obtained economically from the quarry or the size that satisfies layer-thickness requirements; (5) the lower limit of d_{15} stone should not be less than $1/16$ the upper limit of d_{100} stone; and (6) the bulk volume of stone lighter than the d_{15} stone should not exceed the volume of voids in the structure without this lighter stone.

The riprap thickness should not be less than (1) 12 in. (30 cm) for practical placement, (2) less than the diameter of the upper limit of d_{100} stone, or (3) less than 1.5 times the diameter of the upper limit d_{50} stone, whichever is greater. If riprap is placed under water, the thickness should be increased by 50%, and if it is subject to attack by large floating debris or wave action it should be increased 6–12 in. (15–30 cm).

Riprap placement is usually accomplished by dumping directly from trucks. If riprap is placed during construction of the embankment, rocks can be dumped directly from trucks from the top of the embankment. Rock should never be placed by dropping down the slope in a chute or pushed downhill with a bulldozer. These methods result in segregation of sizes. With dumped riprap there is a minimum of expensive handwork. Draglines with orange peel buckets, backhoes, and other power equipment can also be used advantageously to place riprap.

8.2.4   Filters

Filters are used under the riprap revetment to allow water to drain easily from the bank without carrying out soil particles. Filters are required when the d_{15} of the riprap gradation exceeds five times the d_{85} of the bank material. Filter blankets must meet two basic requirements: stability and permeability. The filter material must be fine enough to prevent the base material from escaping through the filter, but it must be more permeable than the base material. There is no standard filter that can be used in all cases. Two types of filters are commonly used: gravel filters and synthetic filter cloths.

Gravel filters consist of a layer, or blanket, of well-graded gravel placed over the embankment or riverbank before riprap placement. Sizes of gravel in the filter blanket should be from 3/16 in. (5 mm) to an upper limit, depending on the gradation of the riprap, with maximum sizes of approximately 3–3¹/₂ in. (90 mm). The filter thickness should not be less than 6–9 in. (20 cm). Filters that are one-half the thickness of the riprap are quite satisfactory. Suggested specifications for gradation are as follows:

$$\frac{d_{50}\,(\text{filter})}{d_{50}\,(\text{base})} < 40, \qquad\qquad (8.5a) \; \blacklozenge$$

$$5 < \frac{d_{15}\,(\text{filter})}{d_{15}\,(\text{base})} < 40, \qquad\qquad (8.5b) \; \blacklozenge$$

$$\frac{d_{15}\,(\text{filter})}{d_{85}\,(\text{base})} < 5. \qquad\qquad (8.5c) \; \blacklozenge$$

If the base material is a fine-grained cohesive soil, such as fat or lean clay, these requirements are not applicable, and the stability criterion is that the d_{15} size of the filter cannot exceed 0.4 mm. Multiple filters may be used when the base material is very fine. In such a case, each layer must satisfy the stability and permeability requirements relative to the underlying layer. A detailed filter design calculation example is presented in Example 8.3.

Synthetic filter cloths (plastic cloth and woven plastic materials) are also used as filters, replacing a component of a graded filter. Numerous plastic filter fabrics exist with a wide variation in size of number of openings and in strength and durability of material. Opening areas of 25%–30% appear desirable to minimize the possibility of clogging and to reduce head loss. It is often desirable to place a protective blanket of sand or gravel on the filter or to take care in placing the rock to that the filter fabric is not punctured. The sides and the toe of the filter fabric must be sealed or trenched so that the base material does not leach out around the filter cloth. Care is also required in joining adjacent section soft filter fabric together.

Example 8.3   Application to filter design. The following filter design example involves the properties of the base material and the riprap given in Table E.8.3.1.

The riprap does not contain sufficient fines to act as the filter because

$$\frac{d_{15}\,(\text{riprap})}{d_{85}\,(\text{base})} = \frac{100}{1.5} = 67,$$

Table E.8.3.1. *Sizes of materials*

Base material	Riprap
Sand	Rock
$d_{85} = 1.5$ mm	$d_{85} = 400$ mm
$d_{50} = 0.5$ mm	$d_{50} = 200$ mm
$d_{15} = 0.17$ mm	$d_{15} = 100$ mm

which is much greater than 5, the recommended upper limit [requirement (8.5c)]. Also

$$\frac{d_{15} \text{ (riprap)}}{d_{15} \text{ (base)}} = \frac{100}{0.17} = 600,$$

which is much greater than 40, the recommended upper limit [requirement (8.5b)]. The properties of the filter to be placed adjacent to the base, from requirement (8.5c), are as follows:

(a) $\dfrac{d_{50} \text{ (filter)}}{d_{50} \text{ (base)}} < 40$ so d_{50} (filter) $< (40)(0.5) = 20$ mm.

(b) $\dfrac{d_{15} \text{ (filter)}}{d_{15} \text{ (base)}} > 5$ so d_{15} (filter) $> (5)(0.17) = 0.85$ mm.

(c) $\dfrac{d_{15} \text{ (filter)}}{d_{15} \text{ (base)}} < 40$ so d_{15} (filter) $< (40)(0.17) = 6.8$ mm.

(d) $\dfrac{d_{15} \text{ (filter)}}{d_{85} \text{ (base)}} < 5$ so d_{15} (filter) $< (5)(1.5) = 7.5$ mm.

Thus, with respect to the base,

$$0.85 \text{ mm} < d_{15} \text{ (filter)} < 6.8 \text{ mm},$$
$$d_{50} \text{ (filter)} < 20 \text{ mm}.$$

The properties of the filter to be placed adjacent to the riprap are as follows:

(a) $\dfrac{d_{50} \text{ (riprap)}}{d_{50} \text{ (filter)}} < 40$ so d_{50} (filter) $> 200/40 = 5$ mm.

(b) $\dfrac{d_{15} \text{ (riprap)}}{d_{15} \text{ (filter)}} > 5$ so d_{15} (filter) $< 100/5 = 20$ mm.

(c) $\dfrac{d_{15} \text{ (riprap)}}{d_{15} \text{ (filter)}} < 40$ so d_{15} (filter) $> 100/40 = 2.5$ mm.

(d) $\dfrac{d_{15} \text{ (riprap)}}{d_{85} \text{ (filter)}} < 5$ so d_{85} (filter) $> 100/5 = 20$ mm.

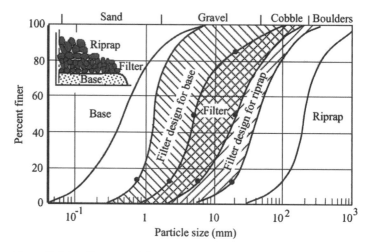

Figure E.8.3.1. Example of filter design.

Therefore, with respect to the riprap, the filter must satisfy these requirements:

$2.5\,\text{mm} < d_{15}\,\text{(filter)} < 20\,\text{mm}$,

$d_{50}\,\text{(filter)} > 5\,\text{mm}$,

$d_{85}\,\text{(filter)} > 20\,\text{mm}$.

These riprap filter requirements along with those for the base material are shown in Fig. E.8.3.1. Any filter having sizes represented by the double cross-hatched area is satisfactory. For example, a good filter could have these sizes:

$d_{85} = 40$ or 20 mm,

$d_{50} = 10$ mm,

$d_{15} = 4$ mm.

8.2.5   *Preventing riprap failure*

A major shortcoming in riprap design techniques is the assumption that riprap failure is due only to particle erosion. Riprap failure modes are identified as follows: (1) particle erosion; (2) translational slide; (3) slump; and (4) sideslope failure, as illustrated in Fig. 8.6.

Particle erosion by flowing water is the most commonly considered erosion mechanism [Fig. 8.6(a)]. Particle erosion can be initiated by abrasion,

(a) Particle erosion

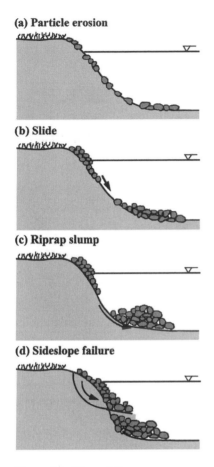

(b) Slide

(c) Riprap slump

(d) Sideslope failure

Figure 8.6. Riprap-failure types.

impingement of flowing water, eddy action/reverse flow, local flow acceleration, freeze/thaw action, ice, or toe erosion. Probable causes of particle erosion include: (1) stone size that is not large enough or is reduced by abrasion; (2) individual stones that are removed by impact; (3) a bank sideslope that is so steep that the angle of repose of the riprap material is easily exceeded; and (4) riprap gradation that is too uniform. The solution requires coarser riprap material and careful consideration of riprap gradation and angularity of the material.

A translational slide is a riprap failure caused by the downslope mass movement of stones. The initial phases of a translational slide are indicated by cracks in the upper part of the riprap blanket that extend parallel to the channel. Slides are usually initiated by channel-bed degradation that undermines the toe of the riprap blanket. It has been suggested that the presence of a filter blanket may provide a potential failure plane for translational slides. The causes of translational slides include: (1) a bank sideslope that is too steep; (2) the presence of excess hydrostatic pore pressure; and (3) loss of material at the toe of the riprap blanket. The solution requires strengthening the toe of the riprap blanket and possible use of larger stones near the channel bed.

Modified slump failure of riprap [Fig. 8.6(c)] is the mass movement of material within only the riprap blanket. Probable causes of modified slump are: (1) a bank sideslope that is too steep; and (2) lack of toe support. The solution requires adding coarse material at the toe of the embankment and reducing the sideslope angle in the upper part of the embankment.

A sideslope failure is a rotation–gravitational movement of material along a surface of rupture that has a concave upward curve [Fig. 8.6(d)]. The cause

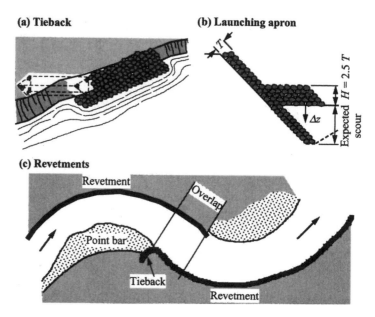

Figure 8.7. Riprap revetments.

of slump failures is related to shear failure of the underlying base material that supports the riprap. Probable causes of slump failures are: (1) excess pore pressure in the base material; and (2) sideslopes that are too steep. The solution requires reduction of the embankment slope or possibly draining the base material. Riprap should not be used at slopes steeper than $1V{:}1.5H$.

The upstream and the downstream ends of the blanket should be tied into the bank to prevent stream currents from unraveling the blanket. The most common method to tie into the bank is to dig a trench at both ends of the blanket (Fig. 8.7). The depth of a trench should be twice the blanket thickness, and the bottom width of the trench should be three times the thickness.

The most effective method to prevent undermining is the launching apron sketched in Fig. 8.7(b). A flexible "launching apron" is laid horizontally on the bed at the foot of the revetment, so that when scour occurs the materials will settle and cover the side of the scour hole on a natural slope. This method is recommended for cohesionless channel beds in which deep scour is expected. In cohesive channel beds, bank revetment should be continued down to the expected worst scour level and the excavation then refilled.

Alternatives to the launching apron include: (1) excavating and continuing the revetment down to a nonerodible material or to below the expected scour level; and (2) driving a "cutoff wall" of sheet piling from the toe of the revetment down to a nonerodible material or to below the expected scour level.

8.3   Riverbank protection

This section discusses bank-protection measures other than riprap: vegetation (Subsection 8.3.1), windrows and trenches (Subsection 8.3.2), sacks and blocks (Subsection 8.3.3), gabions and mattresses (Subsection 8.3.4), articulated concrete mattresses (Subsection 8.3.5), soil cement (Subsection 8.3.6), and retaining walls (Subsection 8.3.7).

8.3.1   Vegetation

Vegetation is probably the most natural method for protecting streambanks. It is less expensive than most structural methods and it improves environmental conditions for wildlife. The presence of vegetation below the water surface can effectively protect a bank in two ways. First, the root system helps to hold the soil together and increases overall bank stability by forming a binding network. Vegetation takes water from the soil, providing additional capacity for infiltration, and may improve bank stability by water withdrawal. Second, the exposed stalks, stems, branches, and foliage provide resistance to the streamflow, causing the flow to lose energy by deforming the plants rather than by removing soil particles. Dense vegetation reduces flow velocities and induces deposition. Above the water surface, vegetation prevents surface erosion by absorbing the impact of falling raindrops.

Vegetation is generally divided into two broad categories: grasses and woody plants (trees and shrubs). The grasses are less costly to plant on an eroding bank and require a shorter period of time to become established. Woody plants offer greater protection against erosion because of their more extensive root systems; however, under some conditions the weight of the plant will offset the advantage of the root system. On very high banks, tree roots do not always penetrate to the toe of the bank. If the toe becomes eroded, the weight of the tree and its root mass may cause a bank failure.

Using planted vegetation for streambank erosion control also has its limitations. These may include the following: (1) their failure to grow; (2) they are subject to undermining; (3) they may not withstand alternate periods of wetting and drying for varied durations; (4) they may be uprooted by freezing and thawing of ice; and (5) they may suffer wildlife or livestock damage.

Native plants should normally be used because they have become adapted to the climate, soils, and other ecological characteristics of the area. Exotic plants, in contrast, are often met with local opposition. Plants chosen should have some tolerance to flooding. A mixture of grasses, herbs, shrubs, and trees should be used to provide a diversity of wildlife habitats. Some nitrogen-fixing plants may be required in poor soils and difficult climates.

Streambank zones depend on the ability of certain plants to tolerate various durations of flooding and their attributes of dissipating wave and current energies. The splash zone located between normal high water and normal low water is the zone of highest stress. The splash zone is exposed frequently to wave wash, erosive river currents, ice and debris movement, wet–dry cycles, and freezing-thawing cycles. This section of the bank would be inundated throughout most of the year (at least 6 months/yr).

The bank zone lies above the normal high-water level; yet this site is exposed periodically to wave wash, erosive river currents, ice and debris movement, and traffic by animals or humans. The site is inundated for at least a 60-day duration once every 2–3 yrs. The water table in this zone frequently is close to the soil surface because of its closeness to the normal river level.

The terrace zone, inland from the bank zone, is usually not subjected to erosive action of the river except during occasional flooding.

For the splash zone, only herbaceous semiaquatic plants, such as reeds, rushes, and sedges, are suggested for planting. These types of plants can tolerate considerable flooding and are more likely to live in this zone. Reeds also protect streambanks in various ways. With their roots, rhizomes, and shoots, they bind the soil under the water, sometimes even above the water. In the reed zone along the riverbank, they form a permeable underwater obstacle that slows down the current and waves by friction, thereby reducing their impact on the soil.

In the bank zone, both herbaceous and woody plants are used. These should still be quite flood tolerant and able to withstand partial to complete submergence for up to several weeks. Various willows can be used in this zone. In periods of high water, the upper branches of such shrubs reduce the speed of the current and increase friction and thereby decrease the erosive force of the water. The branches of such shrubs have a great resilience, springing back after currents subside.

The terrace zone is less significant for bank protection because it is less often flooded, and thus less easily eroded. The terrace zone contains native grasses, herbs, shrubs, and trees that are slightly less flood tolerant than those in the bank zone. The tree species also become taller and more massive. Trees are noted for their value in stabilizing banks of streams and rivers. The banks of some rivers have not been eroded for durations of 100–200 yrs because heavy tree roots bind the alluvium of floodplains. A combination of trees, shrubs, and grasses in this zone will not only serve as an integrated plant community for erosion control, but will improve wildlife habitat diversity and aesthetic appeal.

Grasses can be planted by hand seeding, sodding (transplanting clumps of grass or herbaceous plants, sprigging (planting plant stems or rhizomes), or by mechanical spreading mulches consisting of seed, fertilizer, and other organic mixtures. Several commercial manufacturers now market erosion-control matting that will hold the seed and soil in place until new vegetation can become established. The matting is generally installed by hand and secured to the bank where plantings have been made to prevent erosion. A fence should be placed along the top bank if livestock requires access to the stream.

We construct reed rolls by combining sections of sod, rhizomes, and shoots, and enclosing them within a wire net, and placing all components in a trench. Usually the sod must be held in place with wire netting or stakes. Shrublike willow, dogwood, and alder transplants or 1-yr-old rooted cuttings are effectively used in this zone and can augment the sodding practice.

Hydroseeding can be a useful and effective means of direct seeding, particularly on steep slopes. Often barges with hydroseeders mounted on them can be floated on the stream and used adjacent to the site. Seeds should be blown on first in a water slurry and then mulches applied following seeding to reduce soil moisture loss. The mulch also will tend to tie down and cover the seeds and reduce immediate surface soil erosion by wind and water. Case Study 8.1 illustrates the effectiveness of vegetation protection on the neck of Thompson Bend, Mississippi River.

Case Study 8.1   Thompson Bend on the Mississippi River, United States. Thompson Bend is located on the right descending bank of the Mississippi River between river miles 30 and 45, above the confluence of the Mississippi and Ohio Rivers; see Fig. CS.8.1.1. The river flows in a gooseneck encompassing approximately 10,000 acres (40 km^2) of valuable agricultural land. At the throat

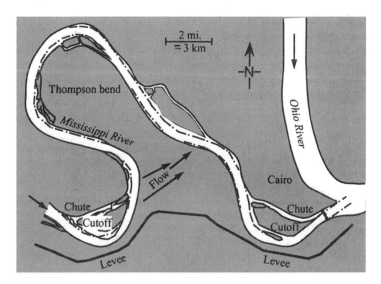

Figure CS.8.1.1. Thompson Bend of the Mississippi River.

of the bend, the overland distance is approximately 2 miles. The river distance along the thalweg is approximately 14 miles (22.5 km). During large floods the river naturally tries to flow straight across Thompson Bend. The water surface drop along the thalweg is 7 ft (2.1 m) for a slope 0.5 ft/mile (9.5 cm/km) along the river and 3.5 ft/mile (66 cm/km) across the neck.

In the early 1980s, severe erosion of the upper bankline began along the right descending bank in the upper reaches of the bendway. In addition, localized surface erosion reached an estimated rate of 40,000 (tons/acre)/event [16,000 (tons/hectare)/event]. Continued erosion could have allowed for development of a chute cutoff across the bend. This would have impaired navigation to the steep slopes and resulting high velocities in the new channel across the bend. It would also have destroyed thousands of acres of valuable farmland and changed the river regime for miles upstream and downstream.

Thompson Bend clearly illustrates the vital importance that vegetation exerts on controlling overbank scour. The revegetation process began in 1985 and early 1986. The results were immediately evident during the fall flood of 1986. Very little erosion was observed. The area was tested again in the flood of 1990, when very little erosion occurred. The Thompson Bend also suffered very little visible damage during the Great Flood of 1993, when record high stages occurred, and the duration of the overland flows reached an unprecedented 130 days. However, the flood took its toll on the vegetation. Numerous trees that were inundated for most of the 130 days died. The flood of 1994 did not allow any significant revegetation to occur. Subsequently, the flood of 1995, the second highest ever at Cape Girardeau, produced some visible scour for the first time since the revegetation was initiated. However, the erosion was very minor compared with the massive amounts of scour that occurred in the early 1980s.

8.3.2   Windrows and trenches

A windrow revetment consists of piling a sufficient supply of erosion-resistant material on the existing land surface along the bank. Trenches are similar except that the material is buried as sketched in Fig. 8.8(b). Windrows and trenches permit the area between the natural riverbank and the windrow to erode through natural processes until the erosion reaches and undercuts the supply of riprap. As the rock supply is undercut, it falls onto the eroding area, thus giving protection against further undercutting and eventually halting further landward movement.

High banks tend to produce a nonuniform revetment alignment and have a tendency for large segments of the bank to break loose and rotate slightly. Comparatively, low banks simply slough into the stream.

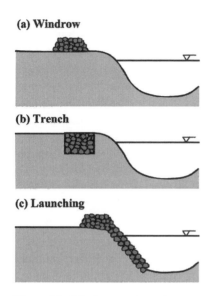

(a) Windrow

(b) Trench

(c) Launching

Figure 8.8. Windrows and trenches.

The velocity and the stream characteristics dictate the size of stone forming a windrow revetment. The size of stone must be large enough to resist being transported by the stream. An important design parameter is the ratio of the relative thickness of the final revetment to the stone diameter. Large stone sizes will require more material than smaller stone sizes to produce the same relative thickness. A well-graded stone is important to ensure that the revetment does not fail from leaching of the underlying bank material. The stream velocity was found to have a strong influence on the ultimate sideslope of the revetment. In general, the greater the velocity, the steeper the sideslope of the final revetment.

8.3.3   Sacks and blocks

Burlap sacks filled with soil or sand–cement mixtures have long been used for emergency work along levees and streambanks during floods (Fig. 8.9). The sacks protect streambanks for which riprap of suitable size and quality is either not available or must be transported over long distances. Although most types of sacks are easily damaged and will eventually deteriorate, sacks filled with sand–cement mixtures can provide long-term protection if the mixture has set up properly.

Precast cellular blocks can be manufactured with locally available sand, cement, and aggregates or can be obtained from commercial sources. Blocks are durable, exceeding riprap in freeze/thaw resistance and are less likely to be encapsulated and lifted off the bank by ice. Also, most designs provide easy pedestrian access to the water's edge, and may be more aesthetic than competing materials. Channel boundary roughness is less than with many other techniques, and wave runup is less than that for smooth concrete surfaces. Cellular blocks are cast with openings to provide for drainage and to allow vegetation to grow through the blocks, thus permitting the root structure to strengthen the bank. Fabric or a gravel blanket can be used as a filter under the blocks if there is any danger that the bank soil will be eroded through the block

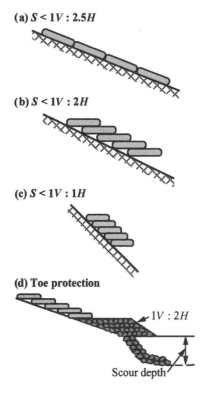

Figure 8.9. Sacks and blocks.

openings by streamflow or seepage. Hand placement is frequent instead of the use of specialized equipment. After the blocks have been placed, the revetment offers sufficient flexibility to conform to minor changes in bank shape. Weepholes should be included in the revetment design to allow drainage of groundwater and prevent pressure buildup that could cause revetment failure.

Sacks do have certain advantages over stone riprap, as follows: (1) they allow possible placement on steep slopes; (2) they use locally available materials; (3) they result in a smooth boundary, if channel conveyance is a major consideration; and (4) the cobblestone effect may be considered to be more aesthetic.

Sack revetments also have disadvantages compared with stone riprap: (1) they are highly labor intensive, and thus are generally more costly than stone; (2) they have a tendency to act monolithically and therefore do not have as much flexibility as riprap and are more susceptible to excess hydrostatic pressure; (3) uniformly sized sacks are not as effective against erosion and leaching as well-graded stones, and therefore a sack revetment is more likely to require a filter material; and (4) synthetic bags may be vulnerable to environmental hazards such as fire, ice, vandalism, and livestock traffic.

The preferred placement technique is sketched in Fig. 8.9. A rule of thumb is to consider flat placement only if the bank slope is flatter than $1V:2.5H$. On slopes of $1V:2H$, the bags should be overlapped by being placed with the long dimension pointing toward the bank, whereas on slopes steeper than $1V:2H$, the bags should be overlapped with the short dimension pointed toward the bank. The maximum slope should be $1V:1H$.

8.3.4   Gabions and mattresses

Gabions are patented rectangular wire boxes (or baskets) filled with relatively small-sized stone, usually less than 8 in. (20 cm) in diameter. Where flow velocities are such that small stones would not be stable if used in a riprap blanket, the wire boxes provide an effective restraint. Limiting recommended maximum velocity for use of gabions ranges from 8 to 15 ft/s (2 to 5 m/s), depending on the manufacturer.

The baskets are commercially available in a range of standard sizes and are made of heavy galvanized steel wire with a polyvinyl chloride (PVC) coating when used in a corrosive environment. They are supplied at a job site folded flat and are assembled manually, by use of noncorrosive wire. The baskets are normally 0.5 m deep × 1 m × 2 m and are set on a graded bank for revetments. A filter blanket or filter cloth is used where required to prevent leaching of base material and undermining of the baskets.

Box gabions are normally stacked on relatively steep slopes to form a massive structure capable of resisting the forces of both river flows and also unstable bankline materials. The flexibility of their mesh and filler stone allows them to maintain their structural integrity even after some degree of displacement, undercutting, or settlement. Box gabion structures generally are aligned either along the streambank toe to form a retaining wall for the bank materials or out from the bank to form dikes for diverting flows away from the bank. Examples are shown in Fig. 8.10.

Mattress gabions are shaped into shallow, broad baskets and are tied together side by side to form a continuous blanket of protection. They are normally placed on a smoothly graded riverbank slope.

Gabions and mattresses are among the more expensive methods of stream-bank erosion protection. However, their record of satisfactory performance is making them more and more popular. Unit costs of bankline length protected vary widely, depending on the complexity of the protection design.

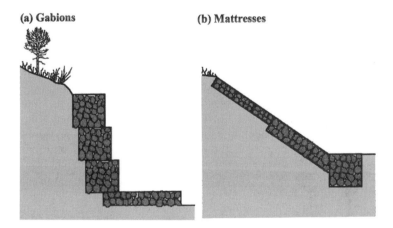

Figure 8.10. Gabions and mattresses.

Gabion structures need to be periodically inspected, and proper maintenance ensures reliable performance. Wire mesh is subject to damage from floating debris, water pollutants, corrosion, wear from high-velocity sediments, and vandalism.

Any cracks or breaks in the PVC coating will expose the wire to corrosive elements, thereby providing negligible advantage over standard galvanized gabions. Freezing temperatures have caused the coating to lose its ductility. Fabricating, filling, and lacing gabions under these conditions have caused an unsatisfactory number of coating cracks and splits.

8.3.5   *Articulated concrete mattresses*

In large rivers, precast concrete blocks held together by steel rods or cables can be used to form flexible articulated mats, sketched in Fig. 8.11. Block sizes may vary to suit the contour of the bank. It is particularly difficult to make a continuous mattress of uniformly sized blocks to fit sharp curves. The open spacing between blocks permits removal of bank material unless a filter blanket of gravel or plastic filter cloth is placed underneath. For embankments that are subjected to only occasional flood flows, the spaces between blocks may be filled with earth and vegetation can be established.

The use of articulated concrete mattresses (ACMs) has been limited primarily to the Mississippi River. This is due to the large cost of the plant required for the placement of the mattress beneath the water surface. The present ACM originated from experiments started in 1915 to develop a flexible and permanent underwater willow mattress. After many failures and discouragements, the concrete mattress sketched in Fig. 8.11 was developed. The basic unit of this mattress is 4 ft wide (1.3 m) × 25 ft long (7.6 m) and 3 in. (7.5 cm) thick.

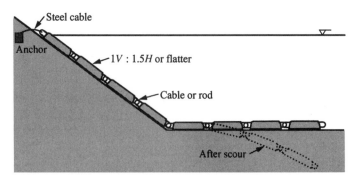

Figure 8.11. Articulated concrete mattresses.

The articulated concrete is flexible, strong, and durable, and ensures complete coverage of the riverbank when properly placed. However, ~8% of its surface area is open, which permits fines to pass through. The open areas are undesirable but necessary to facilitate placement of the mattress in swift deep water, to relieve hydrostatic pressure, and to provide the required flexibility. Yearly maintenance runs approximately 2% of inplace work. This type of revetment has an excellent service record and is considered the standard for the Lower Mississippi River.

8.3.6   Soil–cement

The use of concrete, soil–cement, or roller-compacted concrete generally comes into play when the design and/or site conditions preclude more conventional flexible techniques. Although concrete generally implies permanency, the material itself can be manufactured and placed to respond and withstand changing field conditions. The use of concrete-related materials offers a cost-effective alternative that has proven to be effective under a wide variety of field conditions. The key to success, however, is proper recognition of the rigid nature of the material in a flexible boundary channel.

In areas where riprap is scarce, the combination of on-site soil with cement provides a practical alternative. Figure 8.12 sketches a typical soil–cement construction for bank protection. For use in soil–cement, soils should be easily pulverized and contain at least 5%, but not more than 35%, silt and clay (material passing through the No. 200 sieve). Finer soils usually are difficult to pulverize and require more cement as do 100% granular soils that have no material passing through the No. 200 sieve. Soil–cement can be placed and compacted on slopes as steep as $2H:1V$. Best results have been achieved on slopes no steeper than $1V:3:H$.

A stairstep construction is recommended on channel embankments with relatively steep slopes. Placement of 6–9-ft (2–3 ms)-wide horizontal layers of soil–cement (6-in.- or 15-cm-thick layers) can progress more rapidly than a

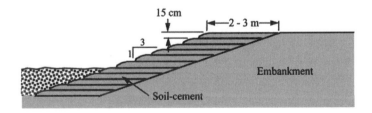

Figure 8.12. Soil–cement.

large quantity of fill material. Special care should be exercised to prevent raw soil seams between successive layers of soil–cement. If uncompleted embankments are left at the end of the day, a sheepsfoot roller should be used on the last layer to provide an interlock for the next layer. The completed soil–cement installation must be protected from drying out for a 7-day hydration period. After completion, the material has sufficient strength to serve as a roadway along the embankment.

When velocities exceed 6–8 ft/s and the flow carries sufficient bedload to be abrasive, the aggregates should contain at least 30% gravel particles retained on a No. 4 (4.75-mm) sieve. It should be emphasized that soil–cement provides rigid bank protection. The depth of the bank protection should be sufficient to protect the installation from the anticipated total scour.

A soil–cement blanket with 8%–15% cement may be an economical and effective streambank protection method for use in areas where vegetation is difficult to establish in sandy bank material. However, soil–cement has three major disadvantages; impermeability, low strength, and susceptibility to temperature variations. If the bank behind the blanket becomes saturated and cannot drain, failure may occur. Also, because a soil–cement blanket is relatively brittle, very little, if any, traffic (vehicular, pedestrian, or livestock) can be sustained without cracking the thin protection veneer. In northern climates the blanket can easily break up during freeze–thaw cycles.

8.3.7   Retaining walls

Retaining walls are near-vertical structures designed to prevent streambank erosion or failure. Vertical retaining walls provide a substantial increase in waterfront land area and often improve the access to water.

Although some soils may be relatively stable in a vertical embankment under dry conditions, wet soils are unstable in a vertical embankment. Retaining walls, have been classified into three distinct types, discussed below: (1) gravity walls; (2) cantilever walls; and (3) sheet-piling walls.

Gravity walls are massive walls that rely on their mass to restrain the movement of soil. The walls are constructed to such proportions that any developed soil or hydrostatic pressures that would tend to cause movement or failure of the wall are resisted by the weight of the wall and resultant shear forces that develop at the base of the wall. Examples of gravity walls are shown in Fig. 8.13.

The concrete walls in Fig. 8.13(a) are usually designed such that no tension is developed within the concrete and no reinforcing steel is required for their construction. Variations of this type of wall may consist of stone masonry, as

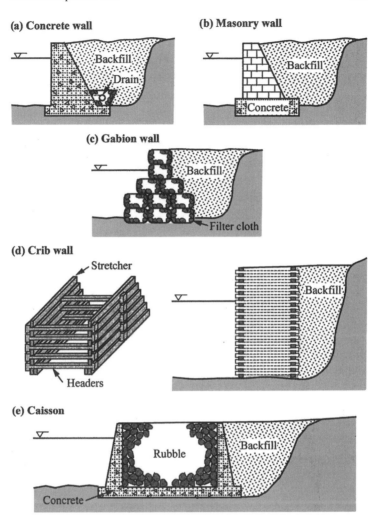

Figure 8.13. Gravity walls.

shown in Fig. 8.13(b). This type of wall has been successfully adapted to the construction of very large walls.

Retaining walls constructed of stone-filled gabions in Fig. 8.13(c) are another type of gravity wall. As discussed in Subsection 8.3.4, gabions are baskets made of metal mesh or geotextiles that are filled with stone and have wide usage in hydraulic structures. To form a wall, the gabions are simply stacked atop one another in such a number as to provide sufficient mass to retain the soil within a streambank.

The timber crib walls in Fig. 8.13(d) are similar to common fence construction except that all wood should be treated with a preservative to minimize deterioration that is due to repetitive wetting and drying or insect activity and the toe of the crib should always be protected with riprap. The most common cause of failure is scour around the pilings, followed by the structure tipping over because of the pressure of the bank behind the bulkhead. The fill material should drain freely. Filter fabric or gravel can be placed as a filter behind openings in the fence to prevent fine soils from leaching through. The bulkhead should be tied into the bank at the upstream and the downstream ends of the structure to prevent flanking from flow behind the bulkhead. The crib walls in Fig. 8.13(d) are constructed of interlocking structural members of treated wood, precast reinforced concrete, or metal, usually galvanized steel or aluminum. The structural components of a crib wall interlock to form "boxes" that are filled with stone, blocks, sand, or rubble to provide the mass necessary for stability.

Caissons, or large "boxes" of reinforced concrete in Fig. 8.13(e) are constructed on land and then floated to the site and flooded and sunk to form a continuous wall. The caisson is then filled with concrete or compacted sand to provide the mass required for stability.

A cantilever wall refers to a reinforced concrete base with a stem wall cantilevered upward from the base. The stem is designed to resist the lateral earth and hydrostatic forces. The soil above the base provides mass to resist movement. Reinforced concrete is required for strengthening cantilevered walls. The wall stem may be supported or stiffened by buttresses fronting the wall or by counterforts behind the wall, as shown in Fig. 8.14.

The simplest form of cantilever wall would be wooden posts driven into the streambed with wooden lagging or timbers placed behind the posts or nailed or bolted to the posts in order to retain the soil within the banks. This type of construction is restricted to small walls, although larger walls have been constructed in a similar manner by use of steel H piling rather than posts and with timber lagging or reinforced concrete panels placed between the flanges of the piling in order to retain the soil.

Sheet-piling walls are sometimes referred to as flexible walls or flexible bulkheads. As shown in Fig. 8.15, the walls are normally constructed by driving of the sheet piling and then excavation of the earth fronting the wall. Although wooden and reinforced concrete sheet piling have been used, steel sheet piling is most commonly used. Because of the limited stiffness of the sheet piling and the resulting large deflections, the height of these walls is somewhat restricted.

Anchors increase the allowable height of flexible cantilevered walls. Earth anchors are constructed to provide a horizontal restraining force to the wall. Walls of this sort are often referred to as "anchored" or "tied-back" walls, as

(a)

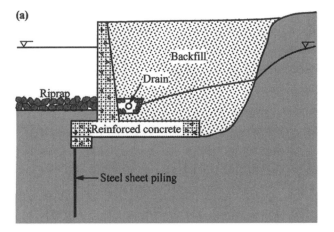

(b)

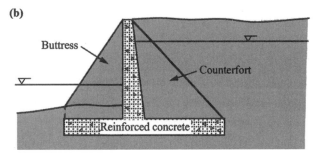

Figure 8.14. Reinforced cantilevered walls.

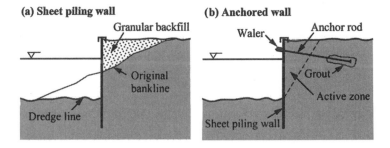

Figure 8.15. Sheet-piling walls.

shown in Fig. 8.15(b). Anchors may be constructed by the drilling of horizontal or angled holes through the wall and into the retained soil a distance beyond any zone of potential active movement of the soil. Threaded steel rods are then pressure grouted into the holes and bolts are torqued against the wall to tension the rods and to apply a restraining force on the wall. Anchors may also

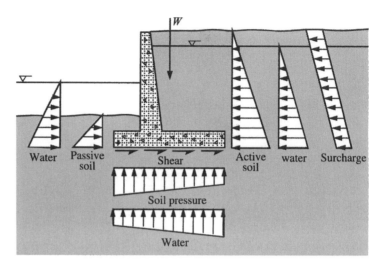

Figure 8.16. Diagrams of forces that act on a retaining wall.

be provided when a structural member, or "deadman," is buried beyond the zone of active movement within the soil and a rod connected to that member is tensioned against the wall. In some cases, when the walls serve as components of a hydraulic structure or a cofferdam, it may be possible to brace against the face of the wall and restrain movement.

Retaining-wall stability. The external forces that act on a retaining wall are sketched in Fig. 8.16. The lateral earth pressures that act behind the structure are generally referred to as active. The earth pressures that resist movement in front of the wall are referred to as passive. Hydrostatic pressures result from groundwater behind the wall and water within the channel. For the development of active earth pressure, the assumption is made that the wall moves or deflects enough to allow for the development of shearing resistance under the structure. The pressure exerted under the wall is necessary to maintain equilibrium of the soil mass. The active and the passive pressures represent the limits of a broad spectrum of stress conditions. When a wall is restrained and not free to move, "at-rest" earth-pressure conditions are assumed to exist. These conditions lie between the active and passive limits. For design purposes, the limiting conditions of active and passive earth pressures are normally selected for free-standing and cantilevered walls. For braced or restrained walls, an at-rest condition may be assumed, often in concert with the limiting conditions.

The active and the passive earth pressures are usually computed based on methods originally proposed by Rankine and modified by others. The at-rest earth pressures are usually computed by use of factors developed from physical tests or empirical means. Lateral earth pressures resulting from granular soils are usually more accurately predicted than those resulting from cohesive soils. For this reason, and for purposes of drainage, granular material is usually selected as backfill for retaining walls. With the rise and fall of stream stages, a difference in water elevation across the retaining wall may develop. The differences in water elevation will then result in an imbalance in hydrostatic forces that act on the wall. Further, hydrostatic forces acting upward on the base of the wall will lessen the effective weight of the structure, thereby lessening its ability to retain the soil. The stability of free-standing walls is a relatively simple problem of statics, once the forces acting on a wall have been defined.

Groundwater control is required for preventing imbalances in water elevations across a retaining wall. Whenever conditions allow, the backfill for a wall should be a clean, free-draining granular material. Drains or weepholes through the wall should be provided to ensure a prompt lowering of the groundwater behind the wall with the lowering of the stage within the stream. Erosion control is critical to the stability of all retaining walls. Under submerged conditions, there is always the potential for the loss of material at the toe of the retaining wall.

8.4   River flow-control structures

Flow-control structures are designed to reduce hydrodynamic forces against streambanks by controlling the direction, velocity, or depth of flowing water. Among the most important properties of a flow-control structure is its degree of permeability. As used here, the term "permeable" means that a structure has definite openings through which water is intended to pass. An impermeable structure may deflect a current entirely, whereas a permeable structure may serve mainly to reduce the flow velocity. Structures made of riprap, or filled with riprap, have some degree of permeability, but these are classed as impermeable because they act essentially as impermeable barriers to a rapidly moving current of water. Types of flow-control structures include hardpoints (Subsection 8.4.1), spurs (Subsection 8.4.2), guidebanks (Subsection 8.4.3), retards (Subsection 8.4.4), dikes (Subsection 8.4.5), jetties (Subsection 8.4.6), fences (Subsection 8.4.7), vanes (Subsection 8.4.8), bendway weirs (Subsection 8.4.9), and drop structures (Subsection 8.4.10).

8.4.1   Hardpoints

Hardpoints consist of stone fills spaced along an eroding bank line (Fig. 8.17). The structures protrude only short distances into the river channel and are supplemented with a root section extending landward into the bank to preclude

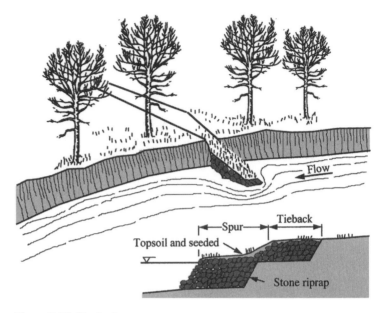

Figure 8.17. Hardpoint.

flanking, should excessive erosion persist. The majority of the structure cannot be seen as the lower part consists of rock placed underwater, and the upper part is covered with topsoil and seeded with native vegetation. The structures are especially adaptable in long, straight reaches not subject to direct attack.

8.4.2 Spurs or groynes

A spur, also called a groyne, is a structure or embankment projected a fair distance from the bank into a stream to deflect flowing water away from the bank. Spurs prevent erosion of the bank and establish a more desirable width and channel alignment. By deflecting the current away from the bank and causing sediment deposits, a spur or a series of spurs may protect the streambank more effectively and at a lower cost than riprap. Also, if the location of any scour is moved away from the bank, failure of the riprap on the spur can often be repaired before damage is done to structures along and across the rivers. Very short spurs are similar to hardpoints. Long spurs or groynes may also be called spur dikes, and very long spurs can be referred to as dikes and jetties (Subsections 8.4.4 and 8.4.5). Spurs are also used to channelize a wide, poorly defined stream into a well-defined channel that neither aggrades nor degrades, thus maintaining its location from year to year. Spurs on streams with suspended sediment discharge induce sedimentation to establish and maintain the new alignment.

(a) Spurs or groynes

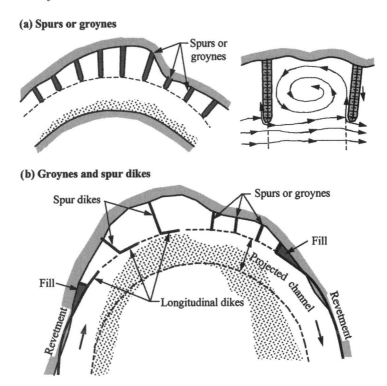

(b) Groynes and spur dikes

Figure 8.18. Spurs of groynes.

As the spur length increases, the scour depth at the spur tip increases, the severity of flow deflection increases, and the length of channel bank protection increases. The projected length of impermeable spurs should be held to less than 15% of the channel width at bankfull stage. The projected length of permeable spurs should be held to less than 25% of the channel width.

The spacing of spurs in a bank-protection scheme is a function of the spur's length, angle, and permeability, as well as the channel bend's degree of curvature (see Fig. 8.18). As a rule of thumb, the spacing should be 3–5 times the projected length. Reducing the spacing between individual spurs results in a reduction of the local scour at the spur tip and causes the flow thalweg to stabilize farther away from the concave bank toward the center of the channel. Spurs angled downstream produce a less severe constriction of flows than those angled upstream or normal to the flow. Retardant spurs should be designed perpendicular to the primary flow direction.

The spur height should be sufficient to protect the regions of the channel bank affected by the erosion processes active at the particular site. If the design flow stage is lower than the channel bank height, spurs should be designed to a height no more than 3 ft (1 m) lower than the design flow stage. If the design flow stage is higher than the channel bank height, spurs should be designed to bank height. Permeable spurs should be designed to a height that will permit passage of heavy debris over the spur crest and not cause structural damage. When possible, impermeable spurs should be designed to be submerged by approximately 3 ft (1 m) under their worst design flow condition, thus minimizing the impacts of local scour and flow concentration at the spur tip and the magnitude of flow deflection.

Permeable spurs should be designed with level crests unless bank height or other special conditions dictate the use of a sloping crest design. Impermeable spurs should be designed with a slight fall toward the spur head to allow different amounts of flow constriction with stage (particularly important in narrow channels). A simple straight spur head form is recommended. The spur head or tip should be as smooth and rounded as possible. Smooth, well-rounded spur tips help minimize local scour, flow concentration, and flow deflection.

8.4.3   Guidebanks

Guidebanks are placed at bridge crossings near the ends of approach embankments to guide the stream through the bridge opening, as shown in Fig. 8.19. Constructed properly, flow disturbances, such as eddies and cross flow, will be eliminated to make a more efficient waterway under the bridge. They are also used to protect the highway embankment and reduce or eliminate local scour at the embankment and adjacent piers. The recommended shape of a guidebank is

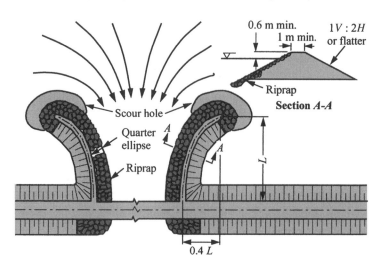

Figure 8.19. Guidebank.

a quarter ellipse with a major to minor axis ratio of 2.5. The major axis should be approximately parallel to the main flow direction.

For some flow conditions, a short curved guidebank on one side and a long straight bank on the other may be the best solution. The crest elevation should be 1 ft (30 cm) higher than the elevation of the design flood, with the effect of the contraction of the flow taken into consideration; this is because the design flow should not overtop the guidebank. Besides improved erosion protection, guidebanks provide a more efficient water flow (less head loss) through a bridge opening. They also decrease scour depth and move the scour hole away from the abutments.

8.4.4   Retards

A retard is a low permeable structure located near the toe of the bank slope parallel to the streamflow, as shown in Fig. 8.20. The function of a retard is to decrease velocity behind the structure and eliminate erosive secondary currents, thereby inducing deposition and growth of vegetation. Retards are most successful on streams that carry a large bed material load.

Occasional tieback connections to the bank are desirable and are mandatory where the retard is located any distance away from the bank. A satisfactory structure height is usually 1/3 to 2/3 of the streambank height. A top elevation

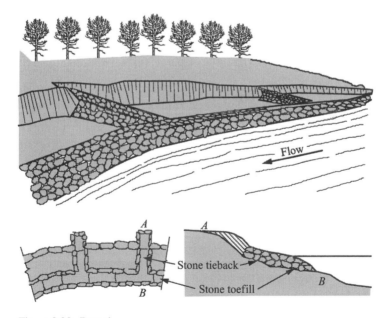

Figure 8.20. Retard.

that is less than the bank height allows drift during high flows to pass over and lodge against the bank rather than against the structure.

Retards have several advantages over the basic technique of sloping the bank and paving with stone, as follows: (1) they can be adapted to a wide range of conditions; (2) the channel alignment can be improved; (3) they are usually less costly; and (4) little if any bank grading is required, which simplifies rights-of-way acquisition and material disposal problems.

Retards are less desirable than stone paving in the following respects: (1) they offer no direct immediate improvement in bank-slope stability; (2) they offer no direct immediate prevention of erosion by overbank drainage and return flows; (3) they are subject to damage by ice, drift, fire, vandalism, and deterioration from the elements; (4) they may reduce channel capacity, particularly after vegetation is established; (5) they interfere with local access to the stream channel; and (6) they may not be aesthetically pleasing.

Pile retards can be made of concrete, steel, or timber. They may be used in combination with bank-protection works such as riprap. The retard then serves to reduce the velocities sufficiently so that either smaller riprap can be used, or riprap can be eliminated. The design of timber pile retards is essentially the same as timber pile dikes (Subsection 8.4.5).

8.4.5   Dikes

There are two principal types of dikes: (1) stone fill dikes; and (2) timber pile dikes. Stone fill dikes are very long spur dikes or groynes. Except for their extended length, their characteristics are described in Subsections 8.4.2 and 8.4.3.

The design approach for dikes and retards requires a comparison of recent and old aerial photographs to identify bank-caving rates and to become familiar with the overall stream-meander pattern. A preliminary determination of upstream and downstream termination points and general structural alignment can then be made. A site inspection then confirms the general stream characteristics, survey of cross sections, bed and bank material, soil boring if piles are to be driven, and drift carried by the stream. The usual alignment problem is to stabilize the outside of a bend. It is not mandatory to have a constant radius of curvature, but the alignment should be smooth without abrupt changes in radius, as sketched in Fig. 8.21(a).

Unless a retard is located in a zone of very mild attack, toe protection is highly desirable [Fig. 8.21(b)]. Toe deepening can cause structural failure and/or loss of filler material from the retard. Toe protection is more effective than excess penetration of piles or posts. Toe protection should be heavier at and downstream of the point of maximum attack determined from aerial photographs and site inspection and survey. Extending the heavier toe protection downstream of this

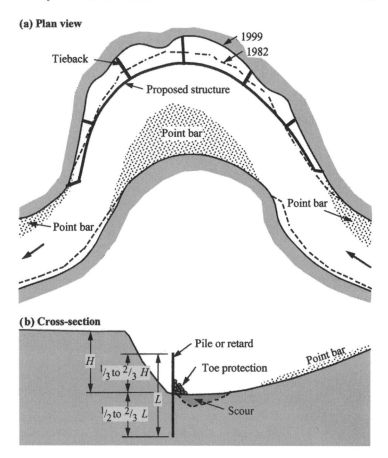

Figure 8.21. Bank-stabilization sketch.

point is desired because impingement points usually tend to migrate downstream with time. The simplest method is to windrow aggraded stone along the channel side of the retard at a uniform rate per linear foot, varying the rate along the bend if appropriate. A rule of thumb is that 1 ton of stone per foot will accommodate 3 ft (1 m) of scour.

Mistakes in selecting upstream and downstream termination points are a common reason for retard failure because points of attack migrate down valley with time, which can be detected from comparative aerial photographs. Therefore, the downstream termination should be downstream of the existing knickpoint of bank stability. Conversely, the upstream terminating point should correspond to the downstream end of the adjacent point-bar deposition zone. The usual termination method is to turn the structure azimuth back into the bank at an angle. The top elevation can be kept constant.

Dikes and retards allow the channel alignment to be modified if that is a project requirement. They are well suited to the incremental construction approach. Because dikes and retards extend into the channel, they are subject to severe attack. Therefore inspection and maintenance are essential. Channel capacity at high flow is decreased initially. The channel will usually adjust by forming a deeper, though narrower, cross section, and the ultimate effect may even be an increase in capacity. Because conservative assumptions on future deposition and vegetative growth would be necessary, dikes or retards must be approached with caution on flood-sensitive projects.

Timber pile dikes may consist of closely spaced single, double, or multiple rows of timber piles, as shown in Fig. 8.22. Wire fence may be used in conjunction with pile dikes to collect debris and thereby cause effective reduction of velocity. Double rows of timber piles can be placed together to form timber

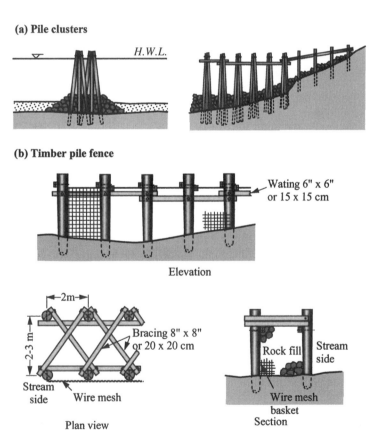

Figure 8.22. Timber pile dikes.

cribs, and rocks may be used to fill the space between the piles. Timber pile dikes are vulnerable to failure through scour. This can be overcome if the piles can be driven to a large depth to achieve safety from scour, or the base of the piles can be protected from scour with dumped rock in sufficient quantities.

The arrangement of timber piles depends on the velocity of flow, quantity of suspended sediment transport, and depth and width of the river. If the velocity of flow is large, timber pile dikes are not likely to be very effective. On the other hand, they are quite effective in moderate flow velocities with high concentrations of suspended sediments. The deposition of suspended sediments in the pile dike field is a necessary consequence of reduced velocities. If there is not sufficient material in suspension, or if the velocities in the dike fields are too large for deposition, the permeable timber pile dikes will be only partially effective in training the river and protecting the bends. The length of each dike depends on channel width, position relative to other dikes, flow depth, and available pile lengths. Generally, pile dikes are not used in large and deep rivers. On the other hand, banks of wide shallow rivers can be successfully protected with dikes. The spacing between dikes varies from 3 to 20 times the length of the upstream dike, with closer spacing favored for best results.

8.4.6 Jetties

Jetty fields add roughness to a channel or overbank area to train the mainstream along a selected path. The added roughness along the bank reduces the velocity and protects the bank from erosion. Jetty fields are usually made up of steel jacks tied together with cables. Both lateral and longitudinal rows of jacks are used to make up the jetty field, as shown in Fig. 8.23. The lateral rows are usually angled approximately 45° to 70° downstream from the bank. The spacing varies, depending on the debris and sediment content in the stream, and may be 50 to 250 ft (15 to 80 m) apart. Jetty fields are effective only if there is a significant amount of debris carried by the stream and the suspended sediment concentration is high. When jetty fields are used to stabilize meandering rivers, it may be necessary to use jetty fields on both sides of the river channel because in flood stage the river may otherwise develop a chute channel across the point bar.

Steel jacks are devices with basic triangular frames tied together to form a stable unit. The resulting framework is called a tetrahedron. The tetrahedrons are placed parallel to the embankment and cabled together with the ends of the cables anchored to the bank. Wire fencing may be placed along the row of tetrahedrons. The aesthetical value of steel jacks is minimal, particularly in rivers used for boating, fishing, rafting, and kayaking.

(a) Jetty field

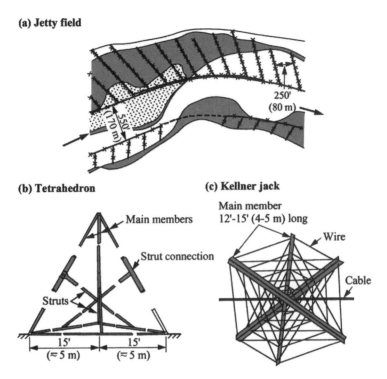

(b) Tetrahedron

(c) Kellner jack

Figure 8.23. Jetties.

8.4.7   Fences

Fencing can be used as a low-cost bank-protection technique on small to medium size streams that are usually wide and shallow (see Fig. 8.24). Special structural design considerations are required in

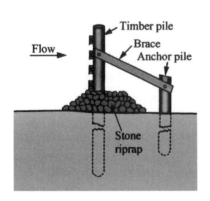

Figure 8.24. Fences.

areas subject to ice and floating debris. Both longitudinal (parallel to stream) fence retards and transverse (perpendicular to stream) fences have been used in the prototype with varying degrees of success: (1) the channel gradient must be stable and not be steep (subcritical flow); (2) toe scour protection can be provided by extending the support posts well below the maximum scour expected or by placing loose rock at the base of the fence to launch

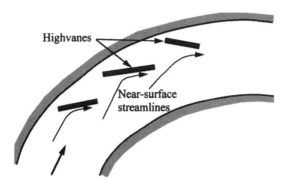

Figure 8.25. Vane dikes.

downward if scour occurs at the toe; (3) tiebacks to the bank are important to prevent flanking of the fence and to promote deposition behind the fence; (4) fence retards generally reduce attack on the bank so that vegetation can become established; and (5) metal or concrete fences are preferred because of ice damage and fire loss of wooden fences.

8.4.8   Vanes

Vane dikes are structures designed to guide the flow away from an eroding bank line (see Fig. 8.25). The structures can be constructed of rock or other erosion-resistant material, the tops of which are constructed below the design water-surface elevation and would not connect to the high bank. Water is free to pass over or around the structure, with the main thread of flow near the surface directed away from the eroding bank. The findings from model investigations of these structures include the effects of various vane dike orientation, vane dike length, and gap length. The ends of the dikes are subjected to local scour, and appropriate allowance should be made for loss of dike material into the scour hole. The rock size to be used for the dike depends on availability of material. Large rocks are generally used to cover the internal section constructed with smaller rocks or earthfill. Side slopes of $1V$:$5H$ and $1V$:$2H$ are common.

8.4.9   Bendway weirs

A bendway weir is a low sill located in a navigation channel bend. Bendway weirs are usually angled 20° to 30° upstream with approximately equal length and spacing. The level crest is at an elevation low enough to allow normal

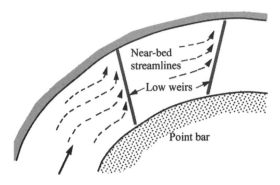

Figure 8.26. Bendway weirs.

river traffic to pass over unimpeded (Fig. 8.26). The weir should be high and long enough to intercept a large percentage of flow at the river cross section. Weirs are typically built in sets (4 to 14 weirs per bend) and are designed to control near-bed flow velocities and current directions through the bend and well into the downstream crossing. Secondary flows are reduced, and water flowing over the weir is redirected at an angle perpendicular to the weir. When the weirs are angled upstream, water is directed away from the outer bank and toward the inner part of the bend. Prototype and model results indicate that construction of a series of bendway weirs in a navigation channel bend results in: (1) improved navigation through the bend; (2) deposition at the toe of the revetment on the outside of the bend and thus an increase in bank stability; (3) more uniform surface-water velocities across any cross section; (4) flow patterns in the bends that are generally parallel with the banks; and (5) the thalweg of the channel being moved away from the toe of the outer-bank revetment.

Case Study 8.2   Bendway weirs of the Mississippi River, United States. A physical movable-bed model study of a 20-mile section of the middle Mississippi River was built at Waterways Experiment Station and operated for the St. Louis District of the Corps of Engineers (Pokrefke and Combs, WES, 1998, personal communication). The model was designed to study and improve bendway navigation and address environmental concerns: (1) high-maintenance dredging costs; (2) need to protect least tern nesting areas; (3) constricted navigation channel; (4) high velocities; (5) detrimental high-flow current patterns; and (6) inadequate navigation channel widths in the crossing downstream of the bend. With this model, the bendway weir concept, developed in January 1988, was refined for Price's Landing and Brown's Bends. Bendway weirs have been

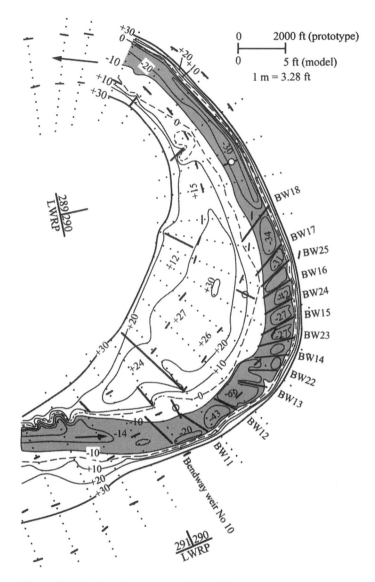

Figure CS.8.2.1. Bendway weirs of the Mississippi River.

tested in 11 models at WES, e.g., Fig. CS.8.2.1, to improve both deep and shallow-draft navigation, align currents through highway bridges, divert sediment, and protect docking facilities. From 1989 to 1995, over 120 weirs have been built in 13 bends of the Mississippi River. Analysis of the five oldest weir installations show that from 1990 to 1995 dredging was reduced by 80%, saving

U.S. $3 million. In addition, towboat accidents were reduced, tow delay times at bends were reduced, sediment and ice management was improved, least tern nesting areas were undisturbed, aquatic habitat area was increased, and fish size and density in the weir fields increased (fivefold in some areas).

8.4.10   Drop structures

Grade-control structures, or drop structures, are used to reduce the slope of a channel. The purpose of grade-control structures is to stabilize the banks and bed of a channel by reducing stream slope and flow velocity. The efficiency of grade-control structures decreases with increasing stream size. Vertical concrete, timbers and sheet pile weirs, riprap sloping sills, and soil–cement or gabion drop structures can be designed, as sketched in Fig. 8.27, after considering the stability of the structure and the depth of the scour hole at the toe of the structure.

Drop structures must be designed to contain the effects of energy dissipation while protecting both the bed and the banks upstream and downstream of the structure. Protection of the approach sections is usually ensured with riprap. Rock protection is efficient in breaking up the remaining turbulence in the stilling basin and in protecting bed and bank materials from direct attack. Filters must also be designed when appropriate.

Log and timber drop structures are appropriate in very small streams and field gullies [Fig. 8.27(a)]. Corrugated pipe grade-control structures consist of a culvert driven or buried vertically into the stream bed, depending on channel-bed width. Riprap is placed within the culvert and for a short distance downstream to prevent scour below the structure. The downstream sidewall of the culvert section is trimmed flush with the bottom. The structures are effective for small drops, and a series of them may be used to control grade in a relatively steep channel over a long distance. Gabions may be useful in constructing weirs or drop structures in areas where an adequate filler material is available. These baskets are versatile in that they may be used in series or singularly to provide various drop heights. They may also be used to protect the streambanks in the approaches and to form a stilling basin. Concrete, because of its durability, is probably the most frequently used material in the construction of larger drop structures. These structures vary in design, particularly in the stilling basin portion of the structure.

Weirs are probably the most widely used form of grade-control structure. This is due to their relative ease of design and construction, low cost in many instances, and versatility and adaptability for other purposes. The simplest form

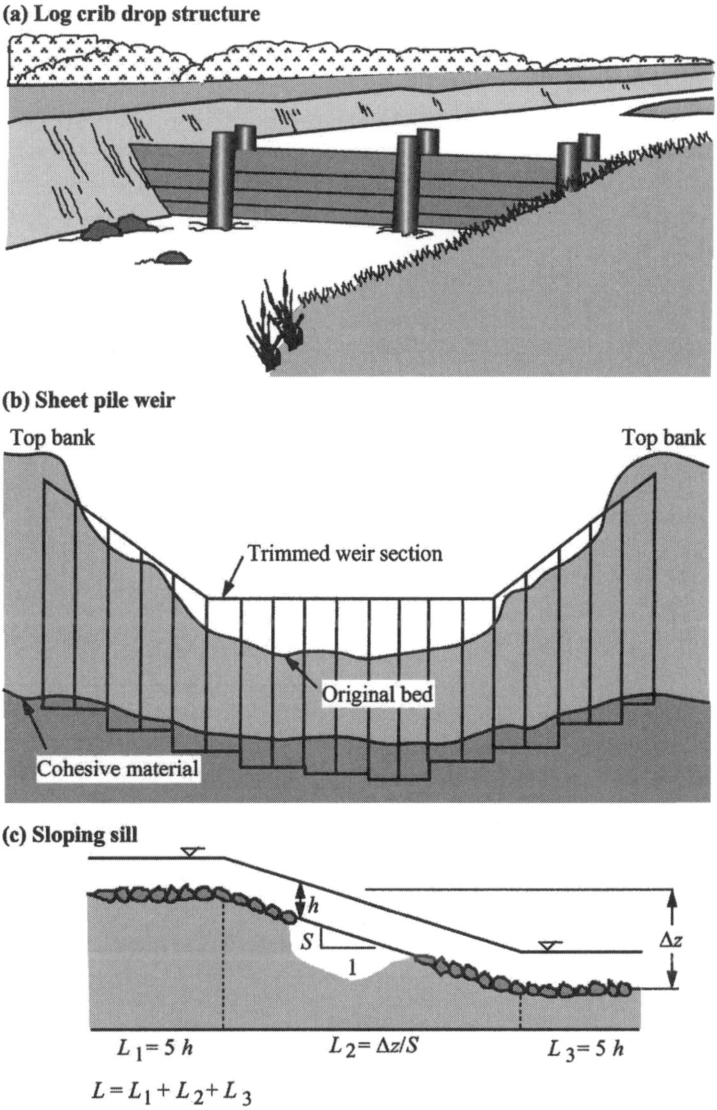

(a) Log crib drop structure

(b) Sheet pile weir

Top bank   Top bank

Trimmed weir section

Original bed

Cohesive material

(c) Sloping sill

$$L_1 = 5h \qquad L_2 = \Delta z/S \qquad L_3 = 5h$$
$$L = L_1 + L_2 + L_3$$

Figure 8.27. Drop structures, weirs, and sills.

is probably that made of sheet piles, the piling serving as both control section and cutoff wall. To form sheet-pile weirs, sheet piles are driven to a depth two to three times the anticipated depth of maximum scour, or to refusal, and are trimmed to a cross section approximately that of the original channel in that area [Fig. 8.27(b)]. If high banks are present, the sheet-pile weir should be tied

nk at both ends. Riprap protection is suggested both upstream and
ι of this type of weir. Sheet piles are particularly useful in wide
ere the required length of concrete or rock structures could be too
ι ne piles may be trimmed at or above the natural channel grade. By
trimming the piles above the channel bottom, the channel will aggrade for a
short distance upstream, resulting in a flatter EGL in that reach. In a stream
requiring reduction of energy over a longer reach, several weirs may be used in
sequence.

Broad-crested weirs, sloping sills, or rock-fill weirs are structures formed
when a pile of stone is placed across the bottom of a stream [Fig. 8.27(c)]. The
amount of scour that could be expected to occur just downstream of the weir,
because of turbulence or a migrating headcut, would determine the volume of
rock needed. Specific methods to determine the local scour depth downstream
of hydraulic structures are presented in Section 9.2.

8.5   Riverbank engineering

The essence of a successful bank-stabilization project is that it should be both
effective and environmentally sound before the economics are considered. The
engineering challenge is to determine the most suitable technique to solve a
specific problem. Environmental and economic factors can then be integrated
into the final selection.

The required project life is short for emergency stabilization during an extreme flow event. Immediate action is required under conditions not permitting
the design and construction of a permanent solution. An intermediate project life
refers to temporary solutions. For instance, an eroding channel can be stabilized
with the expectation that the stream channel will be relocated in the future. The
most common situation refers to a very long project life that typically exceeds
100 yrs.

Operation and maintenance costs can sometimes be traded against initial
construction costs. If the project sponsor has the capability to monitor the
condition of the work and maintain it as required, a less durable technique may
be preferable to a "bomb-proof" method requiring a higher initial investment.

Climatic conditions affect project durability through: (1) the effect of freezing
and thawing on stone; (2) the effect of ice flows on protection structures; (3) the
effect of heaving on slope armoring; (4) the effect of wetting and drying, with the
accompanying damage by bacterial growth and insects, on wooden components;
(5) the effect of sunlight on synthetic materials; and (6) the effect of corrosion on
wire meshes for gabions, mattresses, jacks, and fences. Corrosion and abrasion

can greatly reduce the durability of techniques that rely on metallic components for long-term structural integrity. The critical factors are water chemistry, air quality, and concentration and velocity of coarse sediment impinging on the metallic components.

Debris carried by the flow, usually in the form of uprooted trees, can cause such extensive damage on some streams as to rule out some techniques and may prohibitively raise the cost of other structures designed to withstand debris loads. Debris can make structures more effective in reducing near-bank velocities and accumulating sediment. Debris can also fail structures that accumulate large debris in zones of high velocity.

Vandalism and theft can be reduced by selection of a technique that minimizes temptation. Some materials that are obvious targets for vandals and thieves are posts, boards, concrete blocks and stones of attractive size and shape, small cables and wire, and easily removable fasteners. It is worth considering an increase in the size and the weight of components to make their removal or destruction more tedious. Peeling threaded fasteners and thinly grouting the surface of vulnerable mattresses may suffice to keep the work functional.

Animals can cause problems in a stabilization method. Beavers have a remarkable talent for girdling, felling, or eating vegetation of all sizes and species. This can be disconcerting if the success of the work depends on quickly establishing a strong vegetative cover. Cows, deer, rabbits, and other animals may also find tender young vegetation on new stabilization plantings to their liking. Insect damage can be a problem for wooden components or vegetation. Preservative treatment for wooden components is common practice, and chemical treatment of vegetation at vulnerable stages may be feasible. Water quality and environmental considerations may rule out these options for some projects.

Adjustment to scour or settlement remain important. A stabilization method that has the ability to adjust to scour or bank subsidence has a significant advantage over those that do not. Completely rigid methods must be carefully designed and constructed and perhaps even then supplemented by flexible materials at critical points. The property of flexibility reaches its ultimate application in the design of toe protection. The methods that have this property are stone and other adjustable armor, flexible mattress, and a few types of dikes and retards.

Streambank limitations also need to be considered. It is sometimes expensive or impractical to grade the bank for geotechnical stability or to allow placement of armor protection because of adjacent structures, restricted rights of way, or restrictions on disposing of the excavated material. In such cases, a technique must be selected that leaves at least part of the existing bank intact.

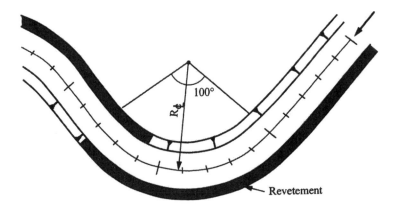

Figure P.8.2.1. Revetment example.

♦**Problem 8.1**

With reference to Example 8.2, determine the rock size that is stable and determine an appropriate gradation curve for the riprap blanket.

Answer:

$d_{50} \simeq 1.25\,\text{ft} = 40\,\text{cm}$, $d_{100} = 2.5\,\text{ft} \simeq 80\,\text{cm}$, $d_{20} = 0.625\,\text{ft} \simeq 20\,\text{cm}$.

♦♦**Problem 8.2**

Use the channel sketched in Fig. P.8.2.1 and the following properties:

  (a) specific weight of stone $= 165\,\text{lb/ft}^3$ $(26\,\text{kN/m}^3)$,
  (b) local depth at toe of outer bank $= 25\,\text{ft}$ $(7.6\,\text{m})$,
  (c) local depth at 20% upslope from toe $= 20\,\text{ft}$ $(6.1\,\text{m})$,
  (d) channel sideslope $= 1V{:}2H$,
  (e) channel downstream slope $= 2\,\text{ft/mile}$ $(3.8\,\text{cm/km})$,
  (f) minimum centerline bend radius $= 1700\,\text{ft}$ $(520\,\text{m})$,
  (g) average velocity $= 7.2\,\text{ft/s}$ $(2.2\,\text{m/s})$,
  (h) water-surface width $= 500\,\text{ft}$ $(150\,\text{m})$

Determine the size, thickness, and geometry of the riprap blanket required for the stabilization of the outer bank of this channel. If the riprap is placed over uniform 0.5-mm sand, determine the characteristics of the filter required for preventing leaching.

Table P.8.3.1. *Riprap gradation*

% Lighter by weight	Weight[a] (lb)	
	Maximum	Minimum
100	292	117
50	86	58
15	43	18

[a]Unit weight of stone is 165 lb/ft^3 (26 kN/m^3).

◆Problem 8.3

Table P.8.3.1 shows the results of a riprap sample that is supposed to meet the requirements for riprap gradation. Determine whether or not this gradation meets the requirements for stability. If d_{50} corresponds to incipient motion, determine what gradation is required.

)

River engineering

This chapter presents a discussion of various types of river engineering structures. The aim of this chapter is to familiarize the reader with solutions to a wide spectrum of river engineering problems. This chapter covers an overview of the following types of river engineering structures: (1) flood control in Section 9.1; (2) river closure and local jet scour in Section 9.2; (3) canal headworks in Section 9.3; (4) bridge crossings in Section 9.4; (5) navigation and waterways in Section 9.5; and (6) dredging in Section 9.6.

9.1   River flood control

This section discusses river engineering solutions to flood control. The methods include flow regulation with reservoirs in Subsection 9.1.1, floodways in Subsection 9.1.2, channel conveyance in Subsection 9.1.3, and levees in Subsection 9.1.4.

9.1.1   Reservoirs

The most direct method of flood control is the storage of surface runoff. Flood control with reservoirs redistributes floodwaters and attenuates floodwaves. A reservoir fills up when the inflow exceeds the outflow, and water storage acts as a buffer to decrease the peak discharges and increase the low discharges. An increase in low discharges is beneficial to hydropower, navigation, and irrigation. It is best to keep the reservoir as full as possible in order to maintain sufficient storage capacity to increase the low discharge reserve in periods of drought. For flood control, however, the reservoir should be kept as low as possible to store unexpected large floods and decrease flood peaks.

In upper watersheds, detention basins are simply equipped with an uncontrolled conduit and a spillway. As long as the spillway is not overtopped, the flood volume is stored, resulting in a considerable attenuation of the floodwave,

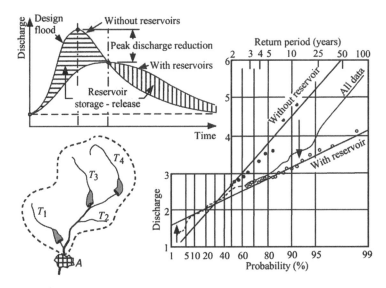

Figure 9.1. Flood storage.

as shown in Fig. 9.1. When the spillway is overtopped, some storage above the spillway crest reduces the outflow discharge. Flood-control reservoirs are most effective when the volume of a reservoir is of the same order of magnitude as that of the upstream floodwave volume.

9.1.2   Floodways

Floodwave attenuation is sometimes possible if some of the floodwater is diverted away from the river. Floodways are used to divert floodwaters into a topographic depression near the river, or into a lake, river, or sea. Floodway outlets consist of spillways or control gates, which are usually located on or near the floodplain to regulate the overbank flow discharge. Normally floodways are not used for long periods of time. It is therefore important to operate the facilities periodically to ensure proper usage in case of emergency. Floodwaters usually carry a significant suspended sediment load, and the possibility of scour around the structures or filling in the floodway needs to be considered. In some cases, sedimentation can be expected downstream of the diversion but can be removed after the flood. Erosion and sedimentation problems may become more serious in the case of possible capture of the river by the distributary, e.g., the possible capture of the Mississippi River by the Atchafalaya River near the Old River control complex.

9.1.3   Channel conveyance

Lowering the flood levels in a river reach by increasing the discharge capacity of the river channel is sometimes possible: (1) by reducing the roughness of the river bed; (2) by enlarging the conveying cross section; and/or (3) by shortening the river channel and thus steepening the channel slope. Each of these improvements results in lowering of the specific stage in the river reach considered. Effective improvements are restricted to clearing the banks and the floodplain from vegetation and other obstacles, eliminating sandbars and islands, and smoothing the banks with revetments. In cold regions, explosives or ice breakers can break up ice jams. In streams in which resistance to flow depends on bedforms, smoothing the bed by removal of sand dunes is rarely effective because bedforms will shortly reappear.

Enlarging the conveying cross section can sometimes be done by deepening or widening the river channel and by lowering or widening the floodplain. This approach is successful only if the sediment load is small and the bed is stable. In most cases, deepening the low-water bed by dredging will result in a temporary improvement at a high cost that is likely to reoccur. In many cases, an enlarged cross section will gradually fill up with sediment until the original bed level and slope have been restored. Deepening of the channel is attractive only if the sediment load is small and the bed is stable. If not, a permanent improvement can be achieved only by continuous dredging, ultimately resulting in degradation of the river bed. Such a solution might be economically justified if the dredged material can be used elsewhere. However, the stability of hydraulic engineering structures might be endangered by headcuts and bed degradation.

Shortening the river channel can be achieved by meander cutoffs, but such a channel rectification should be carried out with great care. If it is not fixed by embankments, the river might start meandering again. Moreover, the locally steep slope will increase the sediment load, causing erosion upstream and sedimentation downstream. In most cases, the new river channel will gradually increase its sinuosity and regain its original slope. Also, flood protection problems are not solved with an increase in stream conveyance; they are merely passed on downstream.

9.1.4   Levees

A levee is an earth embankment constructed along a stream to protect land on the floodplain from being flooded (Fig. 9.2). A floodwall is a concrete structure that serves the same purpose and is found in urban areas where insufficient space prohibits building a levee. For thousands of years, river levees have been built for the

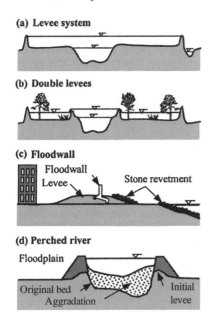

(a) Levee system

(b) Double levees

(c) Floodwall

Floodwall
Levee
Stone revetment

(d) Perched river

Floodplain
Original bed
Aggradation
Initial levee

Figure 9.2. Levees and floodwalls.

protection of people and their property against flooding. It still is the most expedient method for flood control. Levees reduce the storage capacity of the floodplain and thus restrict the flow conveyance of the river. These reductions will induce higher water levels and limit floodwave attenuation. Because of the lateral migration of meandering streams, the levees should be placed at a fair distance from migrating channels, preferably outside the meander belt.

The elevation of levees is primarily determined by the stage of the design flood. River engineers perform a benefit–cost analysis to compare the flood-protection gain against the construction cost of raising the levee above the design level. In general, they measure the benefit of additional protection by considering the exceedance and the cost of flood damages at different water levels. The cost of possible levee breaching and the corresponding extent of flooding need to be considered. Moreover, if flood control is regarded as a purely economical problem, it is seldom justified. Public, environmental, political, and military pressure may justify the large sums needed to protect against flooding. If the increased value of landside property, with improved protection against flooding, is taken into account, the ratio between costs and benefits may become more favorable. Other factors affecting the levee height include tides and wind waves in coastal areas and wide rivers. The design of the levee must prevent breaching as a result of seepage, piping, sliding, slope failure, and revetment erosion. Levees should provide safety until their crests are overtopped.

The choice of elevation of the levees may also depend on anticipated aggradation of the riverbed in coming years. River sedimentation within the levees causes the flow-rating curves to shift upward, thus necessitating the levees' being raised beyond the initial level, e.g., Yellow River and Rio Grande. As sketched in Fig. 9.2(d), the volume of material required for levee protection increases with the square of the levee height, and the cost can become prohibitive in the case of perched rivers. For rivers for which the flood hydrographs rise or fall quickly, the levee section should be analyzed for instability following rapid drawdown. Also, the design of internal filters is particularly critical because of the frequent need to use less than desirable material for embankment construction.

Levee crown elevation is based on the design flood profile plus an allowance for settlement and freeboard. Settlement of either the foundation or the levee embankment, or both, may result in a loss of freeboard, which provides additional levee height for factors that cannot be rationally accounted for in flood-profile computations. Minimum freeboard allowances are usually 2 ft (60 cm) for agricultural levees and 3 ft (1 m) for urban levees. Additional freeboard may be needed at the upstream end of a levee, near drainage structures, bridges, and other constrictions, and for wave action. Levee crown width is primarily dependent on dimensions for the patrol road on the levee crown. Minimum widths of 10 to 12 ft (3 to 9 m) are commonly used with occasional wider turning or passing areas. Where public roads are located on the levee crown, a wider crown width is required.

Compacted levees with sheep foot or rubber-tired rollers require strong foundations of low compressibility and water content of borrow material close to specified range. Semicompacted levees are not commonly used, and uncompacted levees with fill dumped in place with little spreading and compaction is used only for emergency work.

Borrow areas should be on the riverside of the levee, and long, shallow borrow areas along the levee alignment are favored because there are fewer potential problems (Fig. 9.3). Riverside borrow areas should be designed to fill slowly on rising stages and drain fully on fall stages, and the bottom of the excavated area should slope away from the levee.

Underground seepage in pervious foundations beneath levees may result: (1) in the buildup of excessive hydrostatic pressure beneath the impervious top layer on the landside; and (2) in sand boils or in piping beneath the levee itself unless seepage control measures are provided. Principal control measures include riverside impervious blankets, landside seepage berms, pervious toe trenches, pressure-relief wells, cutoff trenches, and sheet piling.

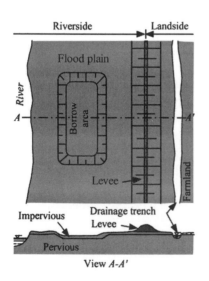

Figure 9.3. Leveed floodplain.

Agricultural use on the landside requires the control of the groundwater table. The efficiency of the drainage system improved by the construction of drains, ditches, and canals may be affected by aggradation or degradation of the riverbed and the floodplain. Local surface drainage of landside agricultural fields through the levees is usually possible with: (1) levees along the tributary channels; (2) culverts with flood gates; or (3) pumping plants. For major tributaries, levees are constructed along the tributaries, tying to the main channel levee and extending upstream to the limits of backwater influence.

Culverts with floodgates are effective only when the main river is at low flow. The culverts can be equipped with control devices or with flap gates that automatically open to permit outflow when the water level on the landside is higher than on the riverside of the levee. Floodgates automatically close to prevent entry of backwater from the main river when the head differential is reversed in ponding areas, and pumping stations are usually required for land drainage when the main river is at high stage. Pumping plants are most frequently effective where a limited amount of temporary ponding is available. Pumping plants may be combined with culverts with floodgates in some situations.

Case Study 9.1   Flood Protection of the Mississippi River, United States.
The Mississippi River has threatened the valley for a long time. In 1543, la Vega, in his history of the expedition by DeSoto, described the first recorded flood of the Mississippi River. The severe flood began on approximately March 10, 1543 and crested 40 days later. The river had returned to its banks by the end of May, having been in flood for ~80 days.

The Mississippi River discharges an average of 520 km^3 of water each year past the cities of Vicksburg and Natchez, Mississippi. Not all parts of the Mississippi River drainage basin contribute water in equal measure; see Fig. CS.9.1.1(a). One-half of the water discharged to the Gulf of Mexico is contributed by the Ohio River and its tributaries (including the Tennessee River) whose combined drainage areas constitute only one-sixth of the area drained by the Mississippi River. By contrast, the Missouri River drains 43% of the area but contributes only 12% of the total water. The Mississippi River now discharges an average of about 200 million metric tons of suspended sediment per year past Vicksburg and eventually to the Gulf of Mexico.

The suspended sediment load carried by the Mississippi River to the Gulf of Mexico has decreased by one-half in the past 200 yrs. The decrease has happened mostly since the 1950s, as the largest natural sources of sediment in the drainage basin were cut off from the Mississippi River main stem by the construction of large reservoirs on the Missouri and Arkansas Rivers. This large decrease in

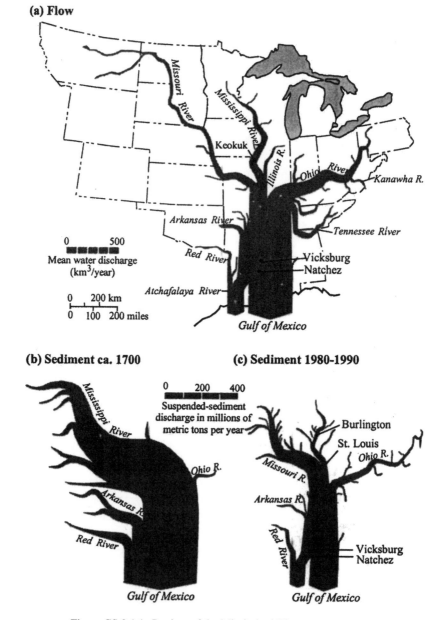

(a) Flow

Mean water discharge
(km³/year)

(b) Sediment ca. 1700

(c) Sediment 1980-1990

Suspended-sediment
discharge in millions of
metric tons per year

Figure CS.9.1.1. Regime of the Mississippi River.

sediment load from the western tributaries was counterbalanced somewhat by a fivefold-to-tenfold increase in sediment load in the Ohio River as a result of deforestation and rowcrop farming.

By the year 1879, the need for improvement of the Mississippi River had become widely recognized. The necessity for coordination of engineering operations through a centralized organization had finally been accepted. Accordingly, in that year, the Congress established the Mississippi River Commission (MRC) with the assignment and the duty to protect the banks of the Mississippi River, improve the navigation thereof, prevent destructive floods, promote and facilitate commerce, trade, and postal service.

Great floods in 1882 and subsequent years plagued the valley as levees were overtopped or crevassed. These disasters and the rising flood heights between the levees caused many to question the total reliance on building levees to contain the river's floodwaters. Other approaches to improving flood protection – reforestation of the floodplain, cutoffs to speed up the river's flow, reservoirs to hold back floodwaters, and floodways to divert flows away from the main channel – were suggested but always rejected by the MRC in favor of a "levee-only" policy.

The role of the MRC grew with each flood, finally culminating in the Flood Control Act of 1917, which authorized the MRC to construct an extensive program of flood protection with cost sharing by states and local interests. The program maintained the levee-only approach and included new levee construction and strengthening of existing levees to standards set 3 ft above the high water of 1912. By the end of 1926, the improved levee system had successfully passed several major high-water events. These successes convinced the MRC and the public that the flood-control problem was nearly solved.

The false sense of security in the Lower Mississippi Valley vanished in the flood of 1927, a natural disaster of great proportions. This tremendous flood extended over nearly 26,000 miles^2, killed more than 500 people, and drove more than 700,000 people from their homes. Thirteen crevasses in the main Mississippi River levees occurred, demonstrating that even the largest and strongest levees would not alone protect from flooding. To prevent a recurrence of the 1927 flood, Congress authorized the Mississippi River and Tributaries (MR&T) project in the Flood Control Act of 1928. The levee-only policy of the past was discarded and the U.S. Army Corps of Engineers adopted a new approach based on improved levees plus floodways, including a spillway to divert water at Bonnet Carré into Lake Pontchartrain above New Orleans.

The four major elements of the MR&T project are: (1) levees for containing flood flows; (2) floodways for the passage of excess flows past critical reaches of

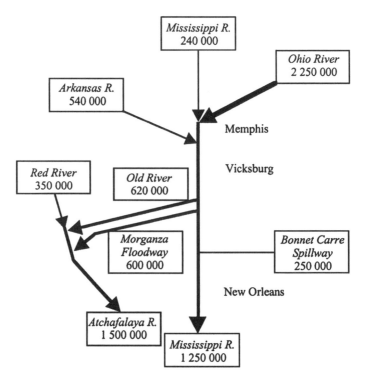

Figure CS.9.1.2. Design of flood-control plan of the Mississippi River.

the Mississippi; (3) channel improvement and stabilization in order to provide an efficient navigation alignment, to increase the flood-carrying capacity of the river, and for protection of the levee system; and (4) tributary basin improvements for major drainage and for control, such as dams and reservoirs, pumping plants, and auxiliary channels. The flood-control plan sketched in Fig. CS.9.1.2 is designed to control the "project flood." It is a flood larger than the record flood of 1927. At Cairo, the project flood is estimated at 2,360,000 ft³/s or 66,820 m³/s. The project flood is 11% greater than the flood of 1927 at the mouth of the Arkansas River and 29% greater at the latitude of Red River Landing, amounting to 3,030,000 ft³/s (85,800 m³/s) at the location ∼120 miles (200 km) below Vicksburg.

The Atchafalaya River. When the first European settlers arrived, they found the Red River emptying into the Mississippi at Turnbull's Bend and the Atchafalaya River a well-defined distributary flowing out of Turnbull's Bend a few miles to the South [Fig. CS.9.1.3(a)]. In 1931, Captain Shreve dug a canal across the narrow neck of Turnbull's Bend. The river accepted the shortcut

(a) 18th Century   **(b) 1831**

(c) 1963

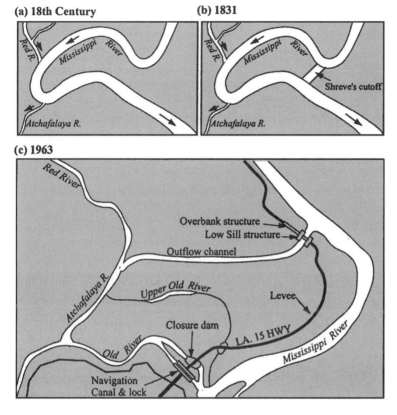

Figure CS.9.1.3. Old River Control.

and abandoned its old channel, the upper part of which eventually silted up, leaving the lower section open, which became known as Old River [Fig. CS.9.1.3(b)].

The Red River no longer flowed into the Mississippi River, but into the Atchafalaya River. Old River connected them to the Mississippi. The current usually flowed west from the Mississippi through Old River into the Atchafalaya; however, during high water on the Red River, the flow sometimes reversed. For years the head of the Atchafalaya River was blocked by a massive "raft" – a 30-mile-long (50-km) log jam – that defied efforts of settlers to remove it. In 1839, the State of Louisiana began to dislodge the raft to open up the river as a free-flowing and navigable stream. The removal of the log jam provided an opportunity for the Atchafalaya to enlarge, become deeper and wider, and carry more and more of the flow from the Mississippi. The Atchafalaya offered the Mississippi River a shorter outlet to the Gulf of Mexico – 142 miles (227 km)

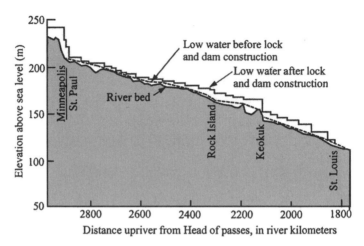

Figure CS.9.1.4. Locks and dams of the Upper Mississippi River.

compared with 315 miles (504 km) – and by 1951 it was apparent that, unless something was done soon, the Mississippi River would take the course of the Atchafalaya River.

The Old River control structures were designed and operated to maintain the distribution of flow and sediments between the Lower Mississippi River and the Atchafalaya River in approximately the some proportions as occurred naturally in 1950. That distribution was determined to be approximately 30% of the total latitude flow (combined flow in the Red River and the Mississippi River above the control structures) passing down the Atchafalaya River on an annual basis [Fig. CS.9.1.3(c)].

The Mississippi River is the main stem of a network of inland navigable waterways that form a system of ~12,350 miles (19,760 km) in length, not including the Gulf Intracoastal Water of 1,173 miles (1,877 km). Waterborne commerce on the Mississippi rose from 30 million tons in 1940 to almost 400 million tons in 1984. In Fig. CS.9.1.4, the entire 1080-km reach of the Upper Mississippi River between Minneapolis, Minnesota, and St. Louis, Missouri, is controlled for barge navigation by a series of 29 lock-and-dam structures. The first structure, completed in 1913 at Keokuk, Iowa, was built to impound water to generate hydroelectric power. The other 28 structures were built, mostly during the 1930s, to maintain a minimum river depth of 9 ft (2.7 m) for barge navigation. Before the dams were built, navigation during low-water periods was extremely hazardous, if not impossible, across rapids such as those at Keokuk and Rock Island, and it was difficult in many other reaches of the upper river.

(a) Stage I

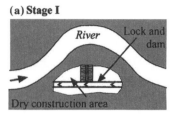

(b) Stage II

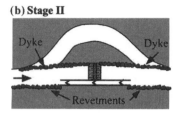

Figure 9.4. Dam construction beside the river.

9.2 River closure

The construction of dams requires closing off a river reach and diverting stream-flow around a dry construction site. There are essentially three types of river closures: (1) dam construction beside the river; (2) complete closure and flow diversion; and (3) partial closure and flow contraction. The first type is illustrated in Fig. 9.4, in which a dry construction site can be found next to the river in the first stage. When the construction is complete, the river is then diverted on the structure in the second stage. This is obviously possible for only low structures such as locks and dams.

The complete closure and flow diversion is discussed in Subsection 9.2.1, along with the methods to calculate local scour from permanent structures such as sills, drop structures, sluice gates, and culverts in Subsection 9.2.2. The partial closure and flow contraction approach by use of cofferdams is covered in Subsection 9.2.3.

9.2.1 River closure and diversion

For the construction of earth dams and high concrete dams in deep, narrow canyons, the entire river channel is generally closed by building upstream and downstream earth and rockfill cofferdams. For large dams on major rivers, the streamflow must be conveyed around the work site through tunnels, as sketched in Fig. 9.5. Such tunnels may serve a dual purpose: (1) flow diversion during construction; and (2) regulated outlet works later. On small streams, a temporary flume or pipeline may be adequate to divert streamflow around the construction site.

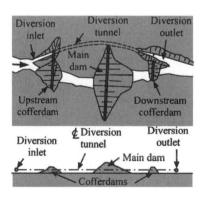

Figure 9.5. River diversion.

The size of a diversion tunnel depends primarily on the diversion design flood and on the upstream cofferdam height. The tunnel inlet must be low enough for flow through the tunnel as soon as the river is closed off. A log boom or trash rack upstream of the inlet may be required for preventing partial blockage of a diversion tunnel by debris. Diversion tunnels are designed to be closed temporarily, usually by gates or stoplogs, at the upstream portal to permit filling of the reservoir. If the tunnel is used solely for diversion, permanent tunnel closure is ensured with a concrete plug.

The construction sequence for channel diversions is as generally follows: (1) construct diversion tunnel; (2) construct temporary upstream cofferdam; (3) construct temporary downstream cofferdam; (4) dewater construction site within the cofferdams; (5) construct permanent upstream cofferdam; and (6) construct main dam.

Initial closure is usually made with temporary cofferdams across the main river channel upstream and downstream of the construction site. The water in the work area between the temporary structures is pumped out for the construction of a permanent cofferdam in the dewatered area. The permanent upstream cofferdam can be constructed as an extension of the temporary upstream cofferdam or as a part of the main dam embankment section.

River closure can be effected in various ways, but the two primary procedures are to: (1) successively narrow and eventually close off the channel by end dumping from one or both banks [Fig. 9.6(a)]; and/or (2) gradually raise a low sill across the channel by dumping uniformly across the gap from a construction bridge or barge or by use of a dragline or moveable dredge disposal line [Fig. 9.6(b)].

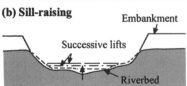

Figure 9.6. River closure.

During a river closure, the upstream water level is raised from the backwater effect of choking the flow. The difference in water levels upstream and downstream of the closure causes an increase in flow velocity and a greater potential for scour in the closure section. In end dumping, the flow velocity increases. The finest material in the fill will move downstream rather than deposit in the gap. As velocity increases, all the fill material being dumped into the closure section will be transported downstream and will not contribute to closing the channel. Just before complete closure, velocities through the gap may be so high that only large stones and prefabricated concrete blocks may be required for making the final closure.

In gradually raising the crown of a low sill across the opening, a mound is built up uniformly across the channel by dumping from a bridge or suspension cable. As material is added, the flow is choked and forces backwater effects upstream of the sill, as described in Example 4.4. The increasing water-level difference upstream and downstream of the closure section increases the flow velocity, tractive force, and sediment-transport capacity. As fill material continues to be added and the tractive force becomes sufficiently high to transport larger-size fractions of the fill, the crest lengthens in the downstream direction. Obstruction of the channel by the sill raises the upstream water surface and diverts part of the streamflow through the diversion tunnel. The reduced discharge over the sill eventually permits the sill to be raised to the water surface for complete closure.

To determine the stability of a rounded stone deposited in flowing water, Isbash's equation has been modified for river closures and stilling basins. As per Eqs. (8.3), the relationship between critical flow velocity V_c for beginning of motion and diameter of a stone d_s is

$$V_c = K_c \, [2g(G-1)]^{1/2} \, d_s^{1/2}, \qquad (9.1)$$

where K_c is a coefficient equal to 1.2 for river closure and 0.86 for stilling basins, g is the gravitational acceleration, and G is the specific gravity of the stone. The stone diameter d_s that can withstand an average flow velocity V_c is calculated from

$$d_s = \frac{1}{2g(G-1)} \left(\frac{V_c}{K_c} \right)^2. \qquad (9.2)$$

The stone weight F_W corresponding to a spherical stone of diameter d_s is calculated from

$$F_W = \gamma_s \frac{\pi d_s^3}{6}, \qquad (9.3)$$

where γ_s is the specific weight of the stone. A sample calculation for a river closure is presented in Example 9.1.

Example 9.1 Application to a river closure. During the closure of a river, the water-surface elevation drops approximately $\Delta h = 6$ ft (1.82 m) within a short distance. Estimate the stone diameter d_s at a specific weight γ_s of 165 lb/ft^3 or $G = 2.65$ that would be stable.

Assuming conservation of energy within this short reach, the velocity corresponding to a 6-ft drop is calculated as

$$V_c = \sqrt{2g\Delta h} = \sqrt{2 \times 32.2 \times 6} = 19 \,\text{ft/s (or 5.8 m/s)}. \qquad \text{(E.9.1.1)} \blacklozenge$$

In the case of a shallow opening, the velocity may correspond to the critical flow depth, given the unit discharge q as $h_c = (q^2/g)^{1/3}$. The required stone diameter calculated from Eq. (9.2) is

$$d_s = \frac{1}{2 \times 32.2 \times 1.65} \left(\frac{19}{1.2}\right)^2 = 2.35\,\text{ft (or 0.71m)},$$

and the corresponding stone weight from Eq. (9.3) is

$$F_W = 165 \times \frac{\pi(2.35)^3}{6} = 1,121\,\text{lb (or 5 kN)}.$$

These results are comparable with those previously presented in Fig. 8.4. We can substitute Eq. (E.9.1.1) into Eq. (9.2) to find that the required stone diameter increases linearly with Δh. The appropriate stone size is approximately one-fifth of the drop height.

9.2.2   Jet scour

The construction of large dams is associated with the need to release water periodically downstream. The bed scour caused by plunging jets or submerged jets needs to be considered in the stability analysis of hydraulic structures.

With reference to the four cases illustrated in Fig. 9.7, the scour depth Δz below plunging jets in Fig. 9.7(a) can be estimated from the empirical equation of Fahlbusch (1994) as a function of unit discharge q, the jet velocity V_1 entering the tailwater depth h_t at an angle θ_j measured from the horizontal at the water surface, and the gravitational acceleration g:

$$\Delta z = K_p \sqrt{\frac{qV_1}{g}} \sin\theta_j - h_t. \tag{9.4}$$

The coefficient for plunging jet K_p depends on grain size with $K_p \simeq 20$ for silts, $5 < K_p < 20$ for sand, and $3 < K_p < 5$ for gravel.

Submerged jets discharge entirely under the free surface, as sketched in Fig. 9.7(b). The example of flow under a sluice gate downstream of a hydraulic structure has a considerable scour potential. Hoffmans and Verheij (1997) applied Newton's second law to a control volume and found the equilibrium scour depth Δz from

$$\Delta z = K_{sj}y_j\left(1 - \frac{V_2}{V_1}\right), \tag{9.5}$$

where V_2 is the outflow velocity, V_1 is the inflow velocity, y_j is the inflow jet thickness, and K_{sj} is a scour coefficient for submerged jets. The value

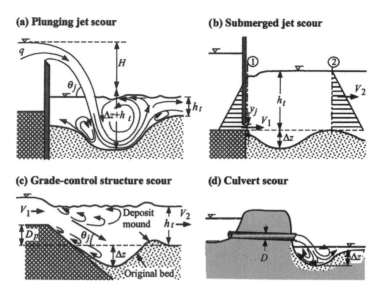

Figure 9.7. Jet scour.

of K_{sj} depends on the particle size and varies from $K_{sj} \simeq 50$ for silts, to $20 < K_{sj} < 50$ for sand, and to $7 < K_{sj} < 20$ for gravel.

As sketched in Fig. 9.7(c), scour below grade-control structures and also sills and drop structures can be estimated from the method of Bormann and Julien (1991) as

$$\Delta z = \left\{ 1.8 \left[\frac{\sin \phi}{\sin(\theta_j + \phi)} \right]^{0.8} \frac{q^{0.6} V_1 \sin \theta_j}{[(G-1)g]^{0.8} d_s^{0.4}} \right\} - D_p, \qquad (9.6)$$

where Δz is the scour depth below the grade-control structure, D_p is the drop height of the grade-control structure, q is the unit discharge, V_1 is the approach velocity, d_s is the particle size, g is the gravitational acceleration, G is the specific gravity of bed material, ϕ is the angle of repose of the bed material, and θ_j is the jet angle measured from the horizontal. Examples 9.2 and 9.3 illustrate how these empirical relationships can be used to estimate the scour depth below hydraulic structures.

As sketched in Fig. 9.7(d), local scour below circular culvert outlets has been studied by Ruff et al. (1982). The scour depth Δz can be predicted as

$$\Delta z = 2.07 \, D \left(\frac{Q}{\sqrt{g D^5}} \right)^{0.45}, \qquad (9.7)$$

where Q is the discharge, D is the culvert diameter, and g is the gravitational acceleration.

The local scour depths predicted from the empirical relationships are subjected to improvements as more field and laboratory data become available. A recent review of different methods showing the analysis of different data sets has been presented by Hoffmans and Verheij (1997).

Example 9.2   Application to scour depth below a sluice gate. A sluice gate, as sketched in Fig. E.9.2.1 is operated at an opening of 0.34 m. The upstream water level is 10 m and the downstream water level is 5 m. The unit discharge in this wide opening is 2 m²/s. The bed material consists of fairly uniform gravel with $d_{50} = 5$ mm and $d_{90} = 7$ mm.

Estimate the scour depth:

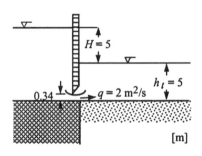

Figure E.9.2.1. Local scour below a sluice gate.

Step 1: The flow velocity in the vena contracta is

$$V_1 = \frac{q}{y_j} = \frac{2}{0.34} = 5.9 \text{ m/s}.$$

Step 2: The outflow velocity is

$$V_2 = \frac{q}{h_t} = \frac{2}{5} = \frac{0.4 \text{ m}}{\text{s}}.$$

Step 3: The scour depth obtained from Eq. (9.5) with $K_{sj} = 15$ is

$$\Delta z = 15 \times 0.34 \left(1 - \frac{0.4}{5.9}\right) \simeq 4.8 \text{ m}.$$

Example 9.3   Application to scour depth below a grade-control structure. A broad-crested weir is built across a 50-m-wide river. The drop height is 2.25 m, and the face angle of the structure is 60°. The scour slope is approximately $1V{:}2H$ in noncohesive material with $d_{50} = 2$ mm and $d_{90} = 2.5$ mm (see Fig. E.9.3.1). Estimate the scour depth when the river discharge is 160 m³/s:

Step 1: Determine $q = (Q/W) = (160/50) = 3.2 \text{ m}^2/\text{s}$, $D_p = 2.25 \text{ m}$, $d_s = 0.002 \text{ m}$, $\theta_j = \tan^{-1} 1/2 = 26°$, and $g = 9.81 \text{ m/s}^2$.

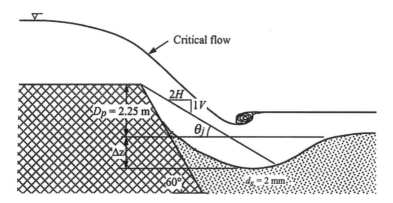

Figure E.9.3.1. Local scour below a drop structure.

Step 2: Estimate $\phi = 40°$ and $G = 2.65$.

Step 3: Assume that the critical flow condition is obtained at the sill crest. The critical flow depth h_c is obtained from $h_c = (q^2/g)^{1/3} = (3.2^2/9.81)^{1/3} = 1$ m.

Step 4: The approach flow velocity is assumed to be critical, or

$$V_1 = \frac{q}{h_c} = \frac{3.2}{1} = 3.2 \,\text{m/s}.$$

Step 5: The scour depth estimated from Eq. (9.6) is

$$\Delta z = \left\{ 1.8 \left[\frac{\sin(40°)}{\sin(66°)} \right]^{0.8} \frac{3.2^{0.6} \, 3.2 \sin 26°}{(1.65 \times 9.81)^{0.8} (0.002)^{0.4}} \right\} - 2.25 \,\text{m} = 2.7 \,\text{m}.$$

9.2.3 Cofferdams

In large rivers in which the entire flow cannot be diverted from the construction site, a partial river closure with flow contraction can be considered. Cellular cofferdams can enclose a part of the streambed that can be dewatered by pumping to provide a dry area for construction. The two-stage construction of a lock and dam is sketched in Fig. 9.8.

Although cofferdams are temporary structures, each stage may be in place for several years, and they should be as cheap as possible while providing the optimum degree of protection. The cofferdam cost is generally one of the major cost items in river construction, particularly when flood stages are frequent or flashy in nature and the construction extends through several years. For preliminary studies, floods with the following recurrence intervals have been

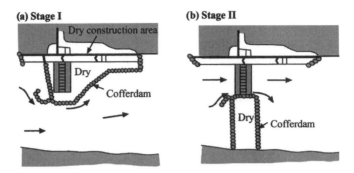

Figure 9.8. Flow contraction with cofferdams.

used for cofferdam design (Petersen, 1986): (1) a 3–5-yr period of return for river construction not exceeding 2 yr; or (2) a 5–25-yr flood when the construction period exceeds 2-yr. Where long streamflow records are available, elevation of the second or the third largest historical flood can be used to determine the cofferdam crest elevation including freeboard. Other important considerations include backwater effects and property damage behind large dams with high cofferdams and potential downstream effects of water release from a sudden cofferdam failure. For final design, the trade-off between risk and potential damage that is due to overtopping and the cost of a higher cofferdam can be assessed. If the risk of loss of life and property damage downstream is high, the standard project flood may be adopted for cofferdam design.

With stage construction on alluvial streams, considerable scour can be expected in the constricted channel cross section at flood stages. Large contraction scour and abutment scour can be calculated from the methods discussed in Subsections 9.4.2 and 9.4.3. If studies indicate that excessive scour might endanger the cofferdam, it may be possible to change the cofferdam configuration to reduce local scour or to armor the bed where maximum scour is expected.

In general, the scour model tests from Franco and McKellar (1968) indicated that: (1) the scour is local rather than general and the point of maximum scour is usually near the upstream corner of the cofferdam; (2) the depth, location, and area of maximum scour can be affected by the alignment of the upper arm of the cofferdam; (3) the extent of scour along the cofferdam is increased as the angle of the upstream arm is increased with respect to the flow direction; and (4) the area of maximum scour can be moved away from the cofferdam with a spur dike or guidebank at the upstream corner of the cofferdam.

Cellular cofferdams are predominately designed either with circular or diaphragm cells (Fig. 9.9). As a rule of thumb, the cell width should

(a) Circular cells

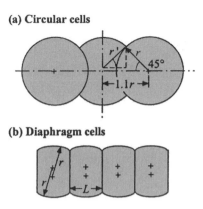

(b) Diaphragm cells

Figure 9.9. Cofferdam cells.

approximately equal the cell height. The circular cells are connected by cells of circular arcs. Circular cells are stable so that each cell can be filled immediately on completion, making it possible for equipment to work from one cell to the next. The diaphragm type consists of two walls of circular arcs connected by straight diaphragm sections, but individual cells are not stable. A template is used for guidance in driving the piles, and when a cell is completed the template is removed and reused for the next cell. According to Petersen (1986), templates are somewhat easier to set for diaphragm cells than for circular cells.

Several diaphragm cells are driven and then carefully filled by keeping the differences in fill elevation in adjacent cells within approximately 4–5 ft (1.2–1.5 m) to avoid distortion of the diaphragms. Circular cells usually require less steel for high structures. However, less steel is usually required for diaphragm cells for low cofferdams. The failure of a circular cell is usually local, with damage limited to one cell, but diaphragm cell failure may extend to adjacent cells.

Sheet-pile cells are usually filled hydraulically with readily available local material such as sand, gravel, rock, or earth for stability. It is particularly important to fill the lower half of each cell with pervious material to facilitate drainage and avoid high hydrostatic forces on the cell walls. A concrete cap (~6 in. or 15 cm) is poured to protect the top of the cell fill against scour in the event the cofferdam is overtopped, to prevent infiltration of rainwater into the fill, and to provide a working surface for the contractor's equipment. The concrete cap also provides a base for sandbagging to provide limited additional height and protection against overtopping beyond the cofferdam design flood elevation and extends the time required for water to saturate the fill in the event of overtopping.

Cellular cofferdams are constructed by driving a wall of interlocking steel sheets through water and saturated pervious material into underlying more impervious clays or rock. Cofferdams are driven to rock whenever possible, and if the material is soft shale, piles are driven 6 in.–1 ft (15–30 cm) into the shale. In pervious materials, the piles are driven as deep as feasible to increase the seepage path and decrease seepage flow into the work area and thus reduce pumping costs. Following the completion of cofferdam construction, the work area inside the cofferdam is dewatered at a limited drawdown rate (usually ∼5 ft in 24 h or 1.5 m per day), depending on riverbank stability and the rate of cell drainage.

9.3   Canal headworks

The regular supply of water to irrigation canals demands adequate control of the water level and sediment transport at the canal intake. Canal headworks control the water level with the construction of a weir to artificially raise the water level at the irrigation canal intake, as sketched in Fig. 9.10. The weir located downstream of the headworks rai-

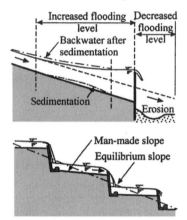

Figure 9.10. Weirs and grade-control structures.

ses the upstream water level, thus artificially decreasing the channel slope and flow velocity near the canal headworks. The artificial slope is therefore less than the equilibrium slope of the channel under natural conditions. Finally, sedimentation problems upstream of the weir and local scour downstream of the weir need to be addressed (e.g., Section 9.2).

There are a number of problems inherent to fixed weirs in streams with variable discharge. Raising the water level upstream of the weir is required for water supply during periods of low river discharge. In periods of high discharge, however, the raised water level may increase the risk of flooding. The erosive forces near the weir at high discharges may also necessitate expensive bed and bank stabilization.

Canal headworks also control the sediment intake into the canal. Fixed weirs usually reduce the sediment concentration at the intake for a period of time. The decreased sediment-transport capacity upstream of the weir results in aggradation that will eventually reach the canal intake. Canal headworks may be located on the outside of a stable river bend. In river bends, the surface water (low sediment concentration) seeks the outside of the bend whereas the sediment-laden streamlines near the bed are deflected away from the canal headworks (e.g., refer to Fig. 6.6). Therefore, headworks on the concave bank benefit from the natural river flow curvature and reduce the necessity for extensive sediment exclusion and sediment ejection at the canal headworks, as sketched in Fig. 9.11. The concave bank approach is possible

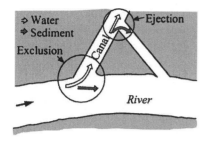

Figure 9.11. Sediment exclusion and ejection.

when the river is stable, without lateral migration, and when the river does not carry a significant load of floating debris. Subsection 9.3.1 focuses on sediment exclusion followed by sediment ejection in Subsection 9.3.2.

9.3.1   Sediment exclusion

Sediment exclusion is intended to prevent sediment from entering the canal by deflecting sediment away from the canal headworks, as sketched in Fig. 9.11. Natural flow curvature causes the heavily sediment laden water to flow away from the canal intake, as shown in Fig. 9.12(a). Guide walls can be designed on both small and large systems to create a favorable flow path, as shown in Fig. 9.12(b). Guide walls increase exclusion efficiency and are effective in continuous sluicing operations.

Guide vanes simply produce favorable secondary currents for sediment removal. Bottom and surface vanes, shown in Fig. 9.13, induce secondary currents that divert the bottom streamlines containing a heavy sediment load away from the canal headworks, and surface water containing a relatively light sediment load can be diverted through the canal headworks. Under steady flow, guide

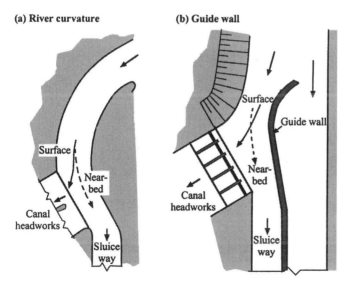

Figure 9.12. Canal headworks with river curvature.

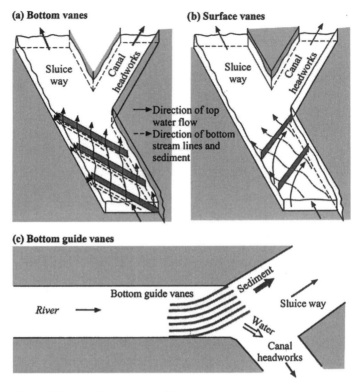

Figure 9.13. Canal headworks with vanes.

vanes can exclude practically the entire bedload. Conversely, they should be avoided when the flow is unsteady.

9.3.2   Sediment ejection

As sketched in Fig. 9.11, sediment ejectors remove sediment that should not be transported through the canal system. Ejection methods do not preclude other sediment control methods, but provide a factor of safety should the exclusion devices fail to perform according to design. The sediment ejector sketched in Fig. 9.14 should be close to the head regulator where settling is predominant; otherwise the sediment deposited in the canal may not reach the ejector.

The tunnel ejector shown in Fig. 9.14(a) can eject coarse bedload from straight canals. The bed of the canal forms a ramp sloping down toward the ejector. The ramp height may sometimes be level with the roof of the ejector. The incoming bed material rolls over the ramp and is ejected through the tunnels. The roof over the guide vanes prevents coarse material from passing over the vanes because

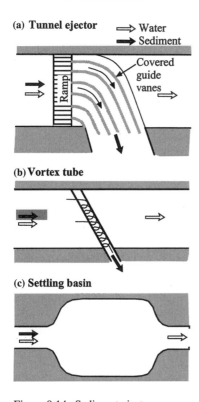

(a) Tunnel ejector ⟹ Water
⟹ Sediment

Covered
guide
vanes ⟹

(b) Vortex tube

(c) Settling basin

Figure 9.14. Sediment ejectors.

of the turbulence resulting from the interaction of the bottom guide vanes and large canal discharges. For small discharges the roof can be omitted and the guide vanes alone will deflect the bedload.

The vortex tube in Fig. 9.14(b) is extremely efficient at removing coarse bedload-like gravels in small canals. The main feature of the vortex tube is a pipe with a slit opening along the top side. As water flows over the tube, the shearing action sets up a vortex motion that catches the bedload as it passes over the lip of the opening. The sediment is carried to the outlet at the downstream end of the vortex tube.

Settling basins sketched in Fig. 9.14(c) remain popular for the removal of both bedload and suspended load at canal headworks. The underlying principle is simply to provide a section wide and long enough to reduce flow velocity and allow sediment to settle out. The basic relation for the design of settling basins is the trap efficiency T_E, which represents the ratio of the weight of sediment settling in the basin to the weight of sediment entering the basin:

$$T_E = 1 - e^{\frac{-\omega W L_{sb}}{Q}}, \qquad (9.8)$$

where W is the settling-basin width, ω is the settling velocity of a particle, L_{sb} is the length of settling basin, and Q is the flow discharge. At a given discharge and particle size, the efficiency depends on the surface area of the settling basin. A calculation example is detailed in Example 9.4.

Example 9.4 Calculation of a settling-basin length. Determine the length of a 30-m-wide settling basin designed to trap 90% of $d_s = 0.05$-mm silt entering a canal at a discharge of $Q = 3$ m^3/s. The unit discharge $q = Q/W = 3/30 = 0.1$ m^2/s. The settling velocity of 0.05-mm particles is approximately 2 mm/s.

ling-basin length is calculated from Eq. (9.8), solved for L as

$$L_{sb} = \frac{q}{\omega} [-\ln (1 - T_E)] = \frac{0.1\,\text{s}}{0.002\,\text{m}} \frac{\text{m}^2}{\text{s}} [-\ln (1 - 0.9)] = 115\,\text{m}.$$

Repeating the application to different basin widths shows that identical results are obtained as long as the settling-basin area $A_{sb} = L_{sb}\,W = 3{,}450\,\text{m}^2$.

9.4   Bridge scour

River engineers are concerned with bridge crossings in the following regards: (1) careful selection of the bridge site to minimize the total bridge costs; and (2) protection against possible structural failure from scour undermining the embankments and piers. The depth of scour during a flood has to be determined in order to design stable foundations for the bridge.

The river crossing should preferably be located in straight river reaches or in stable bends without lateral migration. Sites with narrow floodplain width, rock outcrops, or high bluffs are good locations for bridge crossings. Straight river reaches are often selected in order to avoid problems with lateral migration and deep scour holes in bends.

Protecting embankments and piers against scour requires consideration of the following items: (1) general scour (Subsection 9.4.1); (2) contraction scour (Subsection 9.4.2); (3) abutment scour (Subsection 9.4.3); and (4) pier scour (Subsection 9.4.4). The total scour depth is obtained from the sum of all components. Live-bed scour occurs when there is transport of upstream bed material into the scour hole. Conversely, clearwater scour is without upstream bed sediment transport. Selected methods are presented below, and a detailed treatment of scour at bridges can be found in Richardson and Davis (1995). It should be remembered that all scour estimation procedures serve as approximations and engineering judgement should be exercised.

9.4.1   General scour

Progressive degradation or aggradation can be associated with changes in the river regime caused by natural processes or human activities on the stream or watershed. Factors that affect long-term bed changes are: (1) dams and reservoirs both upstream and downstream of the bridge site; (2) changes in watershed land use like urbanization, deforestation, etc.; (3) channel stabilization and rectification; (4) natural or artificial cutoff of a meander bend; (5) changes in the downstream base level of the bridge reach, including headcuts; (6) gravel

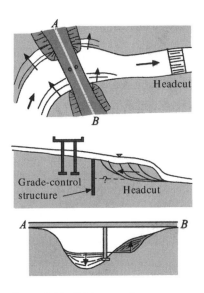

Figure 9.15. Bridge crossing.

mining from the streambed; (7) diversion of water into or out of the stream; (8) lateral migration of a river bend; and (9) thalweg shifting in a braided stream. The engineer must assess the present state of the stream and watershed and consider planned future changes in the river system.

Rivers are free to change cross-section geometry, and the location of the thalweg may also shift considerably in time, e.g., Fig. CS.7.3.2. The bed level may still vary considerably as a consequence of traveling of sand waves, bedforms, riffles, and pools in relatively straight river reaches and changes in pools and crossings during flood conditions. If bed-level measurements are available from previous years, an analysis of specific-gauge data may yield some insight into the trend to be expected in the future, e.g., see Figs. 4.7 and 7.7.

When general scour is expected to be significant, as sketched in Fig. 9.15, it is often advisable to construct a grade-control structure downstream of the bridge crossing. The purpose of the structure is to control the bed elevation at the bridge site and protect the abutment and piers.

9.4.2   Contraction scour

Contraction scour results from flow acceleration in river contractions. The approach flow depth h_1 and average approach flow velocity V_1 result in the sediment transport rate q_{S_1}. The total transport rate to the contraction is $W_1 q_{S_1}$, in which W_1 is the width of the approach. If the water flow rate $Q_1 = W_1 q_1$ in the upstream channel is equal to the flow rate at the contracted section, then, by continuity,

$$q_2 = \frac{W_1}{W_2} q_1, \qquad (9.9)$$

where $q_1 = h_1 V_1$, $q_2 = h_2 V_2$, and the subscript 2 refers to conditions in the contracted section. The sediment-transport rate at the contracted section after

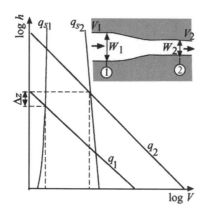

Figure 9.16. Contraction scour (after Nordin, 1971).

equilibrium is established must be

$$q_{S_2} = \frac{W_1}{W_2} q_{S_1}. \qquad (9.10)$$

With reference to Fig. 9.16, the depth of scour Δz that is due to the contraction is then

$$\Delta z = h_2 - h_1. \qquad (9.11)$$

Here h_1, V_1, and W_1 are the depth, velocity, and width of the approach flow, respectively, and h_2 is the contracted flow depth at the bridge. The contracted scour depth represents an average over the channel width, and symmetry is assumed in the calculation. When only one side is being contracted, the abutment scour equation of the following subsection should be considered.

9.4.3   Abutment scour

Abutments, as well as spur dykes, can have different shapes, and they can be set at various angles to the flow. As sketched in Fig. 9.17, the tip of earth and

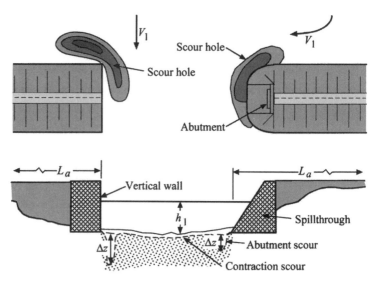

Figure 9.17. Abutment scour (after Richardson et al., 1990).

rockfill abutments will generally have a spillthrough shape when compared with sheet piles standing as vertical walls. Local scour at abutments depends on the amount of flow intercepted by the bridge abutments. The equilibrium scour depth for local live-bed scour in sand near a spillthrough abutment under subcritical flow is

$$\frac{\Delta z}{h_1} = 1.1 \left(\frac{L_a}{h_1}\right)^{0.4} Fr_1^{0.33}, \tag{9.12a}$$

where Δz is the equilibrium depth of abutment scour, h_1 is the average upstream flow depth in the main channel, L_a is the abutment and embankment length measured at the top of the water surface and normal to the side of the channel from where the top of the design flood hits the bank to the outer edge of the abutment, and $Fr_1 = \{V_1/[(gh_1)^{0.5}]\}$ is the upstream Froude number.

If the abutment terminates at a vertical wall and the wall on the upstream side is also vertical, then the scour hole in sand calculated by Eq. (9.12a) nearly doubles (Liu et al., 1961, and Gill, 1972):

$$\frac{\Delta z}{h_1} = 2.15 \left(\frac{L_a}{h_1}\right)^{0.4} Fr_1^{0.33}. \tag{9.12b}$$

Field data for scour at the end of rock dikes on the Mississippi River indicate that the equilibrium scour depth for spillthrough abutment scour when $L_a/h_1 > 25$ can be estimated by

$$\frac{\Delta z}{h_1} = 4 Fr_1^{0.33}. \tag{9.13}$$

This method can be used for very long abutments when flood waters extend onto very wide floodplains. Example 9.5 illustrates abutment scour calculations.

9.4.4 Pier scour

Pier scour is caused by the horseshoe vortex induced by secondary flow at the pier base [Fig. 9.18(a)]. The horseshoe vortex removes bed material away from the base region in front of and along the side of the pier. The strength of the vortex decreases as the depth of scour is increased. Equilibrium scour is reached when the transport rates entering and leaving the scour hole are equal.

The Colorado State University (CSU) equation calculates the pier scour as

$$\frac{\Delta z}{h_1} = 2.0 K_1 K_2 \left(\frac{a}{h_1}\right)^{0.65} Fr_1^{0.43}, \tag{9.14}$$

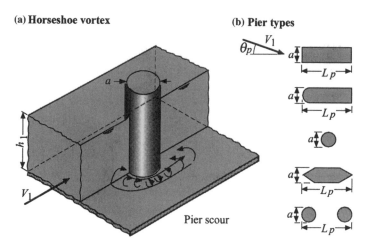

(a) Horseshoe vortex

(b) Pier types

Pier scour

Figure 9.18. Pier scour (after Richardson and Davis, 1995).

where Δz is the scour depth, h_1 is the flow depth just upstream of the pier, K_1 is the correction for pier shape from Table 9.1 corresponding to Fig. 9.18(b); K_2 is the correction for the flow angle of attack from Table 9.2, a is the pier width and $\mathrm{Fr}_1 = V_1/(gh_1)^{0.5}$ is the upstream Froude number.

The extent to which a pier footing or pile cap affects local scour at a pier is not clearly determined. Under some circumstances the footing may serve to impede the horseshoe vortex and reduce the depth of scour. In other cases in which the footing extends above the streambed into the flow, it may increase the effective width of the pier, thereby increasing the local pier scour. In the calculations, the pier width can be used if the top of the pier footing is slightly

Table 9.1. *Pier-type correction factor K_1 (after Richardson et al., 1990)*

Pier type	K_1
Square nose	1.1
Round nose	1.0
Circular cylinder	1.0
Sharp nose	0.9
Group of cylinders	1.0

Note: See Fig. 9.18(b).

Table 9.2. *Flow-angle correction factor K_2*
(after Richardson et al., 1990)

Attack Angle θ_p (deg)	Correction factor K_2		
	$L_p/a = 4$	$L_p/a = 8$	$L_p/a = 12$
0	1.0	1.0	1.0
15	1.5	2.0	2.5
30	2.0	2.5	3.5
45	2.3	3.3	4.3
90	2.5	3.9	5.0

Note: θ_p = skew angle of flow, L_p = pier length, a = pier width.

above or below the streambed elevation. The footing width is used when the pier footing projects well above the streambed or when general scour is expected in the river reach.

As a rule of thumb, the scour depth is approximately two times the pier width. In debris-laden streams or when ice jams can be expected at the bridge site, the actual local scour may be considerably larger than that determined by the equation. The extra scour resulting from debris or ice accumulation against the pier must be given due consideration. Ice and debris can both produce static and dynamic pressure up to 200 Pa on bridge piers. These forces are over and above the static and dynamic forces caused by the traffic on the bridge and the forces exerted on the piers by the flowing water.

When bedforms travel along the riverbed, the maximum scour depth will be one-half the dune height, greater than the mean scour depth. The CSU equation has been slightly modified to account for dune bedforms and for scour reduction from bed armoring (Richardson and Davis, 1995). Scour may also be reduced if riprap is placed around bridge piers. A riprap layer twice the design diameter thick should extend between 1.5 and 6 times the pier width. Example 9.5 illustrates how to calculate the local scour depth around a bridge pier.

Example 9.5   Application to abutment and pier scour. A 200-m-long bridge is to be constructed over a sand-bed channel with 300-m-long spillthrough abutments $1V:2H$. Six rectangular bridge piers measuring 1.5 m thick and 12 m long are aligned with the flow. At a design 100-yr flow discharge of 850 m³/s, the upstream flow velocity is 3.75 m/s and the flow depth is 2.8 m upstream of the piers. Estimate (a) the abutment scour depth and (b) the pier scour depth.

(a) Abutment scour

Step 1: The approach Froude number is

$$Fr_1 = \frac{V_1}{\sqrt{gh_1}} = \frac{3.75}{\sqrt{9.81 \times 2.8}} = 0.71.$$

Step 2: The ratio of abutment length L_a to flow depth h_1 is $(L_a/h_1) = (300/2.8) = 107$.

Step 3: The abutment scour depth is calculated with Eq. (9.13) because $L_a > 25h$:

$$\Delta z = 4\, h_1 Fr_1^{0.33} = 4 \times 2.8 \times 0.71^{0.33} = 10\,\text{m}.$$

Guidebanks in Subsection 8.4.3 could be designed to reduce abutment scour.
(b) Pier scour

Step 1: The Froude number is $Fr_1 = 0.71$.

Step 2: The rectangular pier shape corresponds to $K_1 = 1.1$ from Table 9.1.

Step 3: The pier length/width ratio $(L/a) = (12/1.5) = 8$ and $K_2 = 1.0$ from Table 9.2 as long as the pier is aligned with the flow, $\theta_p = 0$.

Step 4: The pier scour depth calculated from the CSU equation (9.14) is

$$\Delta z = 2\, h_1 K_1 K_2 \left(\frac{a}{h_1}\right)^{0.65} Fr_1^{0.43} = 2 \times 2.8 \times 1.1 \times 1 \times \left(\frac{1.5}{2.8}\right)^{0.65} (0.71)^{0.43}$$
$$= 3.5\,\text{m}.$$

9.5   Navigation waterways

Channel improvement for navigation is justified on the basis of savings in commercial shipping costs. The reduced cost of moving commodities by means of waterways instead of other modes of transportation is compared with the costs of construction, operation, and maintenance. Navigation requirements are discussed in Subsection 9.5.1, followed by waterway alignment and cutoffs in Subsection 9.5.2, and locks and dams in Subsection 9.5.3.

9.5.1   Navigation requirements

General requirements for channel depth, channel width, and lock dimensions for commercial navigation are governed by a number of factors, including type and volume of probable future tonnage, types and sizes of vessels, and tows in general use on connecting waterways. Tows on the Lower Mississippi frequently have more than 40 barges and transport 50,000 to 60,000 tons (450–540 MN) of cargo (Fig. 9.19). Higher-powered towboats have an average of ~3,000 hp (2.2 MW), with three 10-ft- (3-m-) diameter, five-bladed stainless steel propellers housed in Kort nozzles. Three barge types are common: (1) open-hopper barges for transporting coal, sand and gravel, and sulfur; (2) covered-hopper barges for grain and mixed cargo; and (3) tank barges for petroleum and chemicals (Fig. 9.19). Barge sizes vary around a standard 35 ft (10.7 m) in width and 195 ft (60 m) in length.

Pilots navigate towboats at the stern of the tow with control of the engine thrust and direction of the rudder. The navigation width depends on channel alignment, size of tow, and whether one-way or two-way traffic is planned. One-way traffic may be adequate when the traffic is light if the reach is relatively straight with good visibility and if passing lanes are provided. Two-way traffic permits heavy traffic to move faster except when tows are meeting or passing. Figure 9.20(a) shows the recommended channel widths for commercial navigation in straight channels.

Wider navigation channels are required in bends because tows take an oblique position [shown in Fig. 9.20(b)]. The drift angle α varies with the radius of

Type		Length (ft)	Breadth (ft)	Draft (ft)	Capacity (tons)
	Open hopper barges				
	Standard	175	26	9	1000
	Jumbo	195	35	9	1500
	Super jumbo	250-290	40-52	9	2500-3000
	Covered hopper barges				
	Standard	175	26	9	1000
	Jumbo	195	35	9	1500
	Integrated chemical and petroleum barges	150-300	50-54	9	1900-3000
	Towboats	65-160	24-50	5-9	300-7000 hp

Note : 1 m = 3.28 ft ; 1 ton = 2000 lb = 8.96 kN

Figure 9.19. Barge types (after Petersen, 1986).

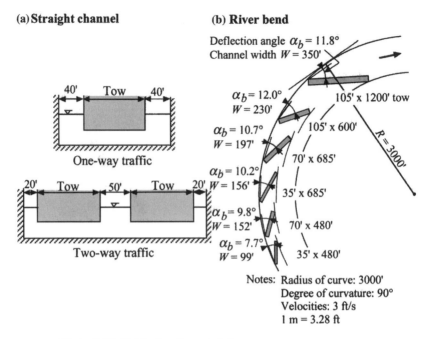

Figure 9.20. Navigation characteristics.

curvature of the channel; the speed, power, and design of the craft, and wind forces; whether the tow is empty or loaded; whether the traffic is going up or down the river; and the flow pattern. The navigation channel width is a direct function of the drift angle, which is larger for downbound tows than for upbound tows. For example, the drift angle for downbound tows of various sizes in a bend with a 3,000-ft (914-m) radius is shown in Fig. 9.20(b).

9.5.2   *Waterway alignment and cutoffs*

Navigation is preferable in fairly straight channels or bends with long radii of curvature. In wide channels, the waterway alignment can be controlled with revetments, spur dykes, and longitudinal dykes. Once the radius of curvature has been determined, the design of river alignment structures is based on the methods discussed in Chap. 8, considering the cost and the effectiveness of different structures.

Navigation in meandering channels usually requires straightening the river. Cutoffs across the neck of long meander bends improve the alignment of the river, reduce sinuosity, and shorten the river length, thus increasing the channel

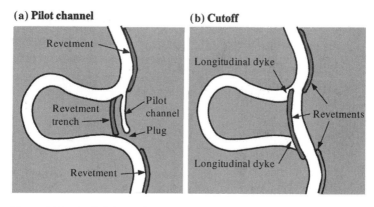

Figure 9.21. Artificial cutoff.

slope. For instance, the natural cutoffs of the Mississippi River shortened the river by ~50 miles (80 km). Artificial cutoffs were constructed to improve river alignment for navigation and for flood control. Indeed, straight river reaches are easier to navigate than long meandering rivers, e.g., Fig. 9.21. Also, steeper gradients reduce flood stages. Historically, structural cutoffs of the Mississippi River were prevented because they (1) disrupted the river regime, (2) aggravated bank erosion upstream, (3) increased downstream shoaling stages, and (4) produced currents difficult to navigate.

Pilot channels are excavated from the downstream end for the construction of artificial cutoffs. Pilot channels have $1V:3H$ sideslopes with bottom widths from 50 to 200 ft (15 to 60 m) and bottom elevations from 6 to 12 ft (2 to 4 m) below the low-water reference plane. When the river length is long compared with the pilot channel, the slope ratio if favorable for natural enlargement of the cut and narrow pilot channels are adequate. Wider and deeper pilot channels are initiated when there is little length or slope advantage. A parallel trench is excavated for the placement of riprap revetment. Earth plugs near the upstream end of the excavated pilot channels are left in place to block low flows. Plugs are designed to be overtopped and washed out during floods after the revetment has been completed.

An example of river alignment is shown in Fig. 9.22. A radius of curvature of 8,000 ft (2.7 km) is maintained throughout the reach. The construction of riprap revetments, spur dykes, and longitudinal dykes provides flow control as long as dyke overlaps and tiebacks prevent flanking of the structures. Note the increase in slope that is due to the reduced reach length. Case Study 9.2 illustrates the changes in the Rhine and the Waal Rivers in The Netherlands.

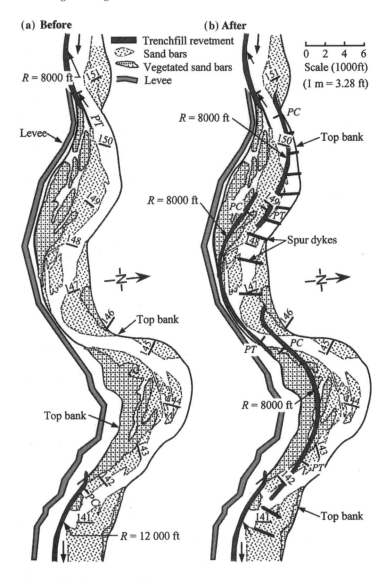

Figure 9.22. River control for navigation.

**Case Study 9.2   Control of the Rhine and the Waal Rivers, The Nether-
lands.** According to RIZA (1999), people living near the Rhine River in past
centuries had to cope with flood disasters. Levees failed repeatedly during
floods. In winter, drifting ice often led to the formation of ice jams that raised

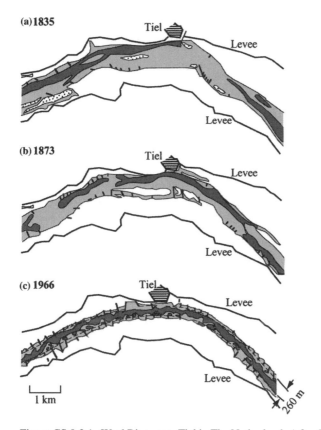

Figure CS.9.2.1. Waal River near Tiel in The Netherlands (after Jansen et al., 1979).

the water levels and increased the risk of levee overtopping. Levees, built on the floodplain and groynes or spur dykes, have been built since the eighteenth century to channelize the river, as shown in Fig. CS.9.2.1. The width of the main channel has been reduced from more than 500 to 260 m. Islands and sandbanks were removed, and the banks were protected from erosion with a series of groynes spaced approximately every 200 m on both sides of the river. As a result, the main channel of the Rhine River has eroded considerably over the past 100 years. In the past 60 years, the Rhine's channel at Lobith has degraded an average of 1–2 cm each year. This in turn threatens the stability of the groynes and the banks.

Approximately one-third of the sediment transported by the Rhine reaches the North Sea. The rest is deposited in the river's lower reaches. During floods, material deposits between the groynes and on the floodplain, particularly behind the summer levees, which is the primary reason for the double-levee system. Up to 10 cm of sediment can deposit along the riverbanks during extended periods of flooding.

The main navigable channel of the Waal River is currently maintained at 150 m wide with at least 2.5 m of water 95% of the time. The river engineering challenge arises from increasing the navigation width and depth without detrimental effects on the environment. Measures include the use of bottom vanes and bendway weirs to widen bends and extension of groynes and construction of longitudinal dykes to increase the flow depth locally.

Stream rehabilitation projects do not necessarily imply restoring the stream to natural conditions that existed before humanity appeared. Natural conditions that existed several thousands of years ago can be looked at as reference conditions. In many instances, stream restoration to ancient conditions is not a feasible proposition. Stream rehabilitation offers a compromise between reference conditions and the present situation. The present policy for stream rehabilitation may include the vital components for mankind development and offer adequate habitat for aquatic and riparian species as well as ensure survival of endangered species. Figure CS.9.2.2 shows the present policy reflects a compromise solution between present and referenced floodplain situations.

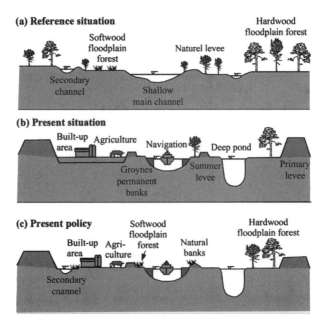

Figure CS.9.2.2. Floodplain management of the Rhine-Waal in The Netherlands (after RIZA, 1999).

Target situations are often idealistic, and the completion of a stream rehabilitation project does not guarantee that the designed habitat is fit for the intended species. Efforts to restore aquatic habitat and develop wetlands have become an integral part of river engineering practice.

9.5.3 Locks and dams

Dams are required to sustain an adequate navigation depth, and locks allow ships to navigate in a river steepened by artificial cutoffs. Locks and dams must be spaced to maintain the navigation depth throughout the pools. Spacing the dams farther apart may eliminate dams and reduce the initial cost of construction. However, maintenance dredging and channel contraction with dikes are also required for maintaining the sediment-transport capacity of the natural channel.

Straight reaches are desirable locations for locks because they are easier to navigate than bends. However, straight reaches tend to be unstable with inadequate flow depth in the downstream approach channel. Wide straight reaches are also prone to sedimentation. Adverse cross currents from spillway discharges may also affect traffic in the lock approaches. In river navigation projects the lock is usually located near the bank at one end of the dam to minimize the adverse effects of spillway discharge on traffic. A typical lock and dam layout is illustrated in Fig. 9.23.

A lock is a structure designed to enable vessels to gain access to lower or higher water levels on either side of the dam. It is an open chamber with gates at both the upper and the lower pool. Locks admit water to fill the chamber from the upper pool and discharge to the lower pool to empty the lock, as illustrated schematically in Fig. 9.24. A downstream-bound tow approaches a lock, the emptying valves and lower lock gates are closed, and the water surface in the lock chamber is brought to the same elevation as that of the upper pool by opening the filling valves. The upper lock gates are then opened for the tow to move into the lock chamber. The upper lock gates and filling valves are then closed, and the emptying valves are opened to bring the water surface in the lock down to the level of the lower pool. The lower lock gates are then opened, and the tow moves out of the lock chamber into the lower pool. The procedure is reversed for a tow moving upstream. Locks are designed to provide adequate depth for navigation. Undersized locks may force large tows to be very slow and cautious when entering a lock chamber. Large locks are not only more expensive, but they require longer filling and emptying times. Filling and emptying times for a lock are designed to be as short as possible without excessive turbulence,

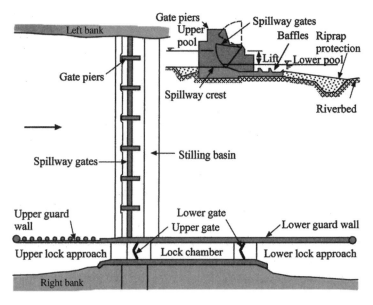

Figure 9.23. Lock and dam (after Petersen, 1986).

surges, or crosscurrents in the lock chamber that might damage the tow or cause the tow to damage the lock.

Locks fill and empty through culverts and ports in or on the chamber floor or at the base of the lock walls. Deep locks provide a cushion of water to dampen turbulence so that tows are not damaged and stresses in the hawsers (the cables that secure tows to the lock walls) are within acceptable limits. In the lock chamber, shown in Fig. 9.25, flow from the wall culverts pass into a crossover culvert across the center of the lock chamber. The splitter wall at the crossover culvert distributes flow equally into two longitudinal floor culverts with ports. Ports in the upper guard wall reduce crosscurrents by permitting the flow intercepted by the lock to pass through the wall to the spillway. The total cross-sectional area of port openings in the guard wall should be equivalent to the cross-sectional area of the approach channel. Upper guard wall ports should be designed as low as possible to increase bottom velocities and reduce shoaling. Currents and velocities from a lock-emptying system can be dangerous to tows approaching the lower lock. The lock emptying outlet should be outside the lower lock approach. However, when the discharge outlet is located outside the lock approach, the water-surface elevation at the outlet sometimes may be higher than that in the lower lock approach, resulting in difficulty in opening the lower lock gates. The outlet discharge manifold in the lower lock approach can induce turbulence to reduce shoaling.

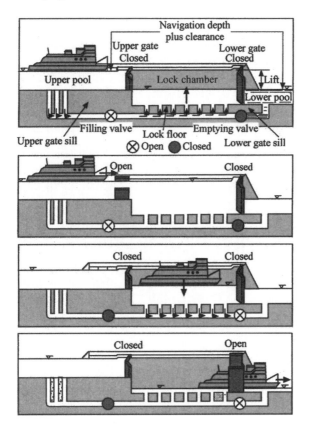

Figure 9.24. Navigation through a lock (after Petersen, 1986).

9.6   Dredging

Dredging is the process of removing material from the bed or the banks of a waterway for the purpose of deepening or widening navigation channels or to obtain fill material for land development. Dredging is a very costly operation (in excess of U.S. $1.00 per cubic yard) that requires heavy equipment and long pipelines. Dredging equipment can be classified as either mechanical or hydraulic. Mechanical dredges (Fig. 9.26) lift the dredged material by means of diggers or buckets. Hydraulic dredges pick up the dredged material by means of suction pipes and pumps. Mechanical dredges remove bed material by a dipper or bucket, and the excavated material is dumped into disposal barges for unloading at the disposal site. Mechanical dredges are usually not self-propelled and must be towed to the work site.

Dipper dredges have considerable digging power to excavate hard compacted material and blasted-rock fragments. It can operate with very little maneuvering

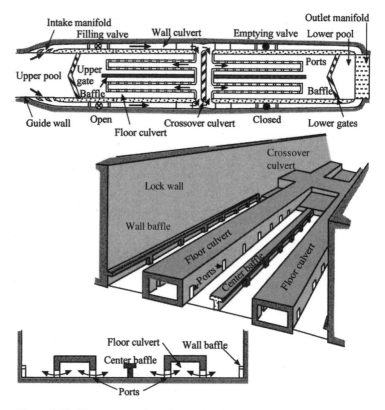

Figure 9.25. Flow pattern through a lock (after Petersen, 1986).

space and can be accurately controlled in the vicinity of bridges and other structures. Bucket dredges use interchangeable buckets (clamshell, orange peel, dragline) for different operational purposes. An open bucket digs into 12 yd³ (9 m³) of bed material, and then closes to be raised and emptied. A modified system in which buckets are fixed on a conveyor belt has also been used in Europe. Considerable fine material is lost from the bucket as it is raised from depths up to 100 ft (30 m), and the maximum concentration of the suspended turbidity plume is typically less than 1,000 ppm.

Hydraulic suction dredges, shown in Fig. 9.27, can be categorized by means of picking up the dredged material (cutterhead, plain suction, and dustpan dredges) and by means of disposal of the dredged material (hopper, pipeline, and sidecasting dredges). Hopper dredges, sketched in Fig. 9.27(a), are self-propelled seagoing vessels used primarily for maintenance dredging and progressive deepening by successive passes. Hopper dredges draw concentrated material in contact with the channel bottom through suction pipes and store it in hoppers in the

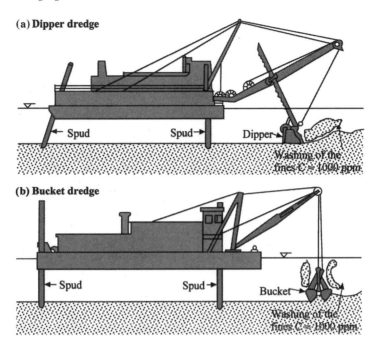

Figure 9.26. Mechanical dredges (after U.S. Army Corps of Engineers, 1983).

dredge. Sediment resuspension occurs near the suction lines at concentrations of a few parts per thousand, and up to several tens of parts per thousand when there is overflow from the hoppers. When fully loaded, the dredge moves to the disposal area under its own power. Hopper dredges are emptied by opening of the bottom doors and dumping of the entire contents in a few seconds. A well-defined plume of dredged material entrains water as it settles to the bottom. Most of the material forms a mound on the bottom, and some spreads horizontally.

Sidecasting dredges are self-propelled seagoing vessels designed to remove material from shallow coastal harbors. A sidecasting dredge picks up bottom material through two suction pipes and pumps it directly overboard beside the dredge. The dredge operates back and forth across the bar, successively deepening the channel on each pass. Sidecasting dredges are not suitable for dredging contaminated material.

Hydraulic pipeline dredges loosen the bottom material with a cutterhead or with water jets (dustpan dredges). The slurry is then pumped through a floating discharge line to the disposal site. Dustpan dredges [Fig. 9.27(b)] are self-propelled vessels suitable for dredging only noncohesive material in waters without significant wave action. Dustpan dredges are equipped with: (1) pressure water jets that loosen the bottom material; and (2) a wide-flared and flat

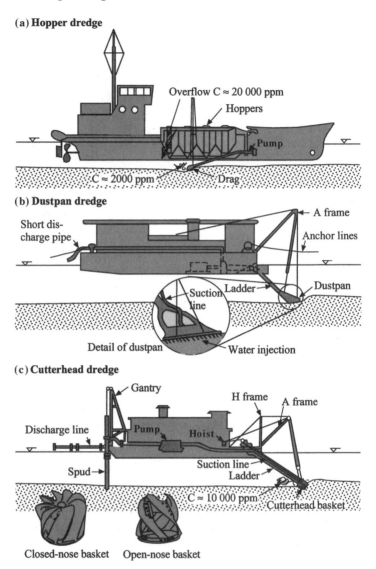

Figure 9.27. Suction dredges (after U.S. Army Corps of Engineers, 1983).

suction line intake for sediment removal. It normally discharges into open water through a relatively short pipeline, up to 100 ft (30 m) long; a longer disposal line requires a booster pump.

Cutterhead dredges are the most efficient and versatile, and thus the most widely used. The cutterhead dredge shown in Fig. 9.27(c) has a rotating cutter around the suction pipe intake and can dig and pump alluvial material including

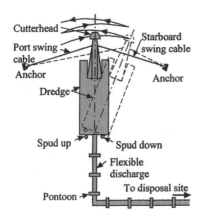

Figure 9.28. Cutterhead dredge with pipeline (after U.S. Army Corps of Engineers, 1983).

compacted clays and hardpans. Suction pipe diameters range from 8 to 30 in. (20 to 90 cm). The cutterhead dredge sketched in Fig. 9.27(c) consists generally of a cutterhead, baskets, ladder, suction line, A-frame, H-frame, pumps, gantry, spuds, and a pipeline up to 30 in. (90 cm) in diameter.

A cutterhead dredge operates by circling about one anchored spud with the cutterhead describing an arc. As the swing is completed, the second spud is anchored and the first one raised to circle in the opposite direction and move forward, as illustrated in Fig. 9.28.

The floating discharge line is made up of 30–50-ft- (9–16-m-) long pipe sections, each supported by pontoons. They are connected together by flanges, ball joints, or rubber sleeves to give the dredge some flexibility in moving. Land pipeline disposal can range up to 3 miles (5 km) without additional sections of shore pipe. For longer transport distances, booster pumps are required on the discharge line. Slurries of 10%–20% sediment concentration by weight are typically pumped in pipelines up to 30 in. (90 cm) in diameter at velocities ranging from 15 to 20 ft/s (5 to 8 m/s).

According to Petersen (1986), the disposal of dredged material has received considerable environmental consideration since the 1960s. Not all dredged material is contaminated, and only a small percentage of the sites are highly contaminated. Clay content and organic matter in dredged material tend to retain many contaminants, and sands easily release contaminants through mixing, resuspension, and transport. Adverse water quality effects are unlikely unless the dredged material is highly contaminated. The physical impacts of open-water disposal are likely to be of greater potential consequence than the chemical and the biological impacts. Adverse biological effects are often unlikely because many organisms usually recolonize disposal sites. Except during fish migration and spawning, turbidity in river waters is more likely to be an aesthetic problem than a biological problem.

Confined land disposal is sometimes the only possible alternative for disposing of some highly polluted sediments. The dredge material is ponded until the suspended solids have settled out. The long-term storage capacity of land disposal sites depends on consolidation of the dredged material, compressibility of foundation soils, effectiveness in dewatering the dredged material, and management of the site. Case Study 9.3 illustrates how recurring dredging problems can be alleviated with flow-control structures.

Case Study 9.3   Dredging at Choctaw Bar of the Mississippi River, United States. In this river engineering case study from Phil Combs (USACE WES, 1998), Choctaw Bar is an important location along the Mississippi River. The Choctaw Bar area is located between river mile 557–565 AHP (above head of passes) on the Mississippi River. It is located ∼17 river miles upstream of Greenville, Mississippi. The small town of Arkansas City, Arkansas, is located adjacent to Choctaw Bar and is protected by the Mississippi River Mainline levee. Choctaw Bar is located on the right descending bank of the river, and the Eutaw-Mounds revetment is located on the left descending bank. Stabilization of the left bank began as early as 1947 and has continued until the present, costing more than $10 million. Stabilization of the river in this reach was critical as levee crevasses had occurred many times in the past and the mainline levee is relatively close to the river. In addition to the stabilization of the left bank, dikes have been constructed on the right bank. Dikes in the vicinity of Chicot Landing were constructed in 1968 for approximately $1.2 million.

Stabilization of the river is for flood protection and to ensure a 9-ft-deep navigable channel. Channel maintenance for navigation has become a significant problem within the reach. Table CS.9.3.1 lists the channel maintenance dredging for the years 1971–1973.

Review of the 1972 hydrographic surveys indicates that a bar in the vicinity of mile 559 AHP essentially blocked the navigation channel. Table CS.9.3.1 of maintenance dredging reflects the attempt to maintain the channel by mechanical means. The 1972 hydrographic surveys also indicate that there is a considerable chute cutoff development on the right bank, as shown in Fig. CS.9.3.1(a).

Regular discharge measurements have been taken to establish the flow distribution in the channel and the right chute. The flow in the chute has increased to the point that, in 1972, ∼50% of the total river flow was passing through the chute.

Table CS.9.3.1. *Dredging data (after Combs, USACE WES, 1998)*

Year	Days	Cubic Yards
1971	15	404,000
1972	27	1,531,000
1973	71	4,121,000

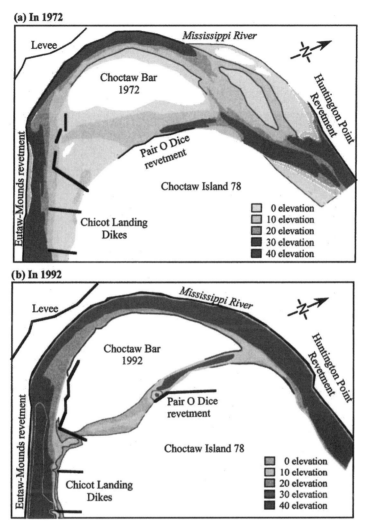

Figure CS.9.3.1. Dredging at Choctaw bar of the Mississippi River (after Combs, USACE, WES, 1998).

Additional information includes the facts that: (1) the left bank has been stabilized; (2) rock dikes are located generally on the right-hand side of the channel; and (3) contours on plan sheets refer to elevations in feet relative to the low water reference plane (LWRP). The LWRP is the water-surface elevation plane corresponding to the discharge exceeded 97% of the time on the flow duration curve. The elevation of the LWRP shown on the plan sheets is varying from 97–100 ft NGVD (National Geodetic Vertical Datum). NGVD elevations are identical to above mean-sea-level (MSL) elevations.

The assignment has been to develop an engineering plan to provide dependable navigation through the reach. The plan should utilize and enhance the existing stabilization infrastructure in the reach, if at all possible. The plan should consider that the aquatic ecosystem in the right chute is a valuable resource. Additionally, the plan should not conflict with the overall flood carrying capacity of the river. If the existing infrastructure cannot be used, sufficient engineering and environmental justification has to be provided to alter the plan.

Figure. CS.9.3.1(b) shows the structural solution proposed by the U.S. Army Corps of Engineers and the topography at Choctaw Bar in 1992. Note the effect of reduced channel width on flow depth and ease of navigation.

♦Problem 9.1

During a river closure by the end-dumping method, the water-surface elevation drops $\sim$15 ft, or 4.5 m, within a short distance. Estimate the stone diameter and the stone weight required for closure.

Answer: $d_s \simeq 3$ ft (90 cm) and $F_W \simeq 2,200$ lb (10 kN).

♦Problem 9.2

Estimate the scour depth below a plunging jet at a velocity of 18 ft/s (5.5 m/s) at an angle of 60° from the horizontal. The jet thickness is 1 ft (30 cm) and the tailwater depth is 6 ft (1.8 m). Sand covers a thick layer of gravel $\sim$5 ft (1.5 m) below the surface.

Answer: $\Delta z \simeq 5.8$ ft (1.8 m).

♦♦Problem 9.3

Estimate the scour depth below the sluice gates of Example 9.2 when the opening is 1 m high. Determine the flow velocity, assuming conservation of energy on both sides of the gate.

Answer: $V_1 \simeq \sqrt{2gh} = 9.9$ m/s, $q = Vh = 9.9$ m²/s,

$V_2 \simeq 2$ m/s, $\Delta z \simeq 12$ m.

♦♦Problem 9.4

Estimate the scour depth below the grade-control structure in Example 9.3 when

(a) the discharge is 320 m³/s

(b) for scour in medium sand ($Q = 160$ m³/s).

♦♦Problem 9.5

Estimate the scour depth below a 2-ft-diameter culvert flowing full at a velocity of 15 ft/s.

Answer: The discharge $Q = 47$ ft³/s (1.33 m³/s) and $\Delta z \simeq 5$ ft (1.5 m).

Problem 9.6

Repeat the length calculations for the settling basin in Example 9.4 for a width of 60 m and show that the basin area is 3,450 m². Also determine the trap efficiency of a basin area of 2,000 m².

Answer: $T_E = 74\%$.

♦♦Problem 9.7

Calculate the local scour depth at the end of a 100-m-long spillthrough abutment in a river flowing at a velocity of 3 m/s and a flow depth of 4 m. If the water spreads 200 m farther out the floodplain during floods, estimate the scour depth, considering that the Froude number does not change significantly.

Answer: The scour depth for Eq. (9.12a) is 12.5 m and does not change during floods because $L_a > 25\, h_1$.

♦♦Problem 9.8

Estimate the scour depth around a rectangular pier that is 1 m wide and 5 m long in a river that is 4 m deep. The flow velocity is 3 m/s at an angle of 30° from the pier alignment.

♦Problem 9.9

Reevaluate the pier scour depth from Example 9.5 should the flow alignment against the pier change to 20° in the coming years.

Answer: At $L_p/a = 8$, $\theta_p = 20°$, and $K_2 \simeq 2.2$ from Table 9.2, the scour depth would more than double at $\Delta z = 7.7$ m.

10

Physical river models

Physical models of rivers have existed at least since 1875, when Louis Jerome Fargue built a model of the Garonne River at Bordeaux. Physical models are usually built to test various river engineering structures and to carry out experiments under controlled laboratory conditions as opposed to costly field programs. The main purposes of physical models include: (1) a small-scale laboratory duplication of a flow phenomenon observed in a river; (2) the examination of the performance of various hydraulic structures or alternative countermeasures to be considered in the final design; and (3) investigation of the model performance under different hydraulic and sediment conditions.

This chapter first describes hydraulic similitude in Section 10.1 in terms of geometric, kinematic, and dynamic similitude. Hydraulic models can be classified into two categories: (1) rigid-bed models, discussed in Section 10.2 and (2) mobile-bed models, discussed in Section 10.3. The analysis leads to the definition of model-scale ratios, and several examples and case studies are presented.

10.1 Hydraulic similitude

The prototype conditions, denoted by the subscript p, refer to the full-scale field conditions for which a hydraulic model, subscript m, is to be built in the laboratory. Model scales, subscript r, refer to the ratio of prototype to model conditions. For instance, the gravitational acceleration in the prototype is g_p, the gravitational acceleration in the model is g_m, and the scale ratio for gravitational acceleration is defined as $g_r = g_p/g_m$. Hydraulic models usually have the same gravitational acceleration in the model and the prototype; thus $g_r = 1$.

For all scale models, the following considerations are relevant: (1) the model length must be large enough to ensure the accuracy of the measurements, e.g., a flow-depth measurement error of 1 mm in a model at a scale of 1:100, or $z_r = 100$, represents an error of 10 cm in the prototype flow-depth measurement; (2) we must consider the physical limitations on space, water discharge, and instrumentation accuracy, e.g., we cannot realistically model the Mississippi River in a 100-m-long hydraulics laboratory; and (3) we must appropriately simulate the boundary conditions, e.g., the stage and the discharge of inflow tributaries, and the possible tidal effects at the downstream end must be properly accounted for.

Hydraulic models use water and require that the scales of mass density ρ_r and kinematic viscosity ν_r be unity. Because the scale ratio for gravitational acceleration $g_r = 1$, the scale for specific weight γ_r and dynamic viscosity μ_r are also unity; thus $\rho_r = g_r = \nu_r = \gamma_r = \mu_r = 1$ in hydraulic models.

Geometric similitude describes the relative size of two Cartesian systems of coordinates (x, y, z). The vertical z_r length scale is defined as the ratio of the prototype vertical length z_p to the model vertical length z_m such that $z_r = z_p/z_m$. For instance, a length scale $z_r = 100$ indicates that a model length of 1 m corresponds to a prototype of 100 m. The horizontal length scales are defined in a corresponding manner in the downstream x and lateral y directions as x_r and y_r. Exact geometric similitude is obtained when the vertical and the horizontal length scales are identical, i.e., $L_r = x_r = y_r = z_r$. The corresponding area and volume scales are respectively $A_r = L_r^2$ and $\mathrm{Vol}_r = L_r^3$. If accurate quantitative data are to be obtained from a model study, there must be exact geometric similitude in every linear dimension. Model *distortion* implies that the vertical z_r and the lateral y_r scales are not identical. The distortion factor is obtained from the ratio of y_r/z_r. Model *tilting* results from different vertical z_r and downstream horizontal x_r scales. The downstream model slope $S_r = z_r/x_r$ is effectively tilted when the horizontal length scale is different from the vertical scale. In a distorted model, the surface area scales for horizontal and cross-sectional surfaces are respectively $A_r = x_r y_r$ and $A_{xr} = z_r y_r$. The volume scale for a distorted model corresponds to $\mathrm{Vol}_r = x_r y_r z_r$.

Kinematic similitude refers to parameters involving length and time, e.g., velocity V, acceleration a, kinematic viscosity ν, etc. For instance, the velocity scale V_r is defined as the ratio of prototype to model velocities as $V_r = V_p/V_m$. The time scale $t_r = t_p/t_m$ appropriately describes kinematic similitude when fluid motion in the model and the prototype are similar. With the kinematic relationships that $L = Vt$ and $V = at$, time can be canceled from these two relationships to obtain $V^2 = aL$. Any experiment in which the gravitational acceleration is the same in the model and the prototype requires that $a_r = g_r = 1$. When applied to the model and the prototype, this relationship yields one of the most important kinematic relationships in physical modeling:

$$\frac{V_r}{z_r^{0.5}} = 1. \tag{10.1} \blacklozenge\blacklozenge$$

This is known as the Froude similitude criterion. Accordingly, the time scale and the velocity scales for exact kinematic similitude are identical, $t_r = V_r = z_r^{0.5}$. It is important to consider that the time scale for distorted and tilted models varies with direction. Distorted and tilted models are restricted to simulate 1D flows and time scales for flow velocities in the y and the z directions are irrelevant. Distorted and tilted models therefore cannot appropriately account for 2D and 3D convective and turbulent accelerations and should not be used to model vorticity, diffusion, turbulent mixing, and dispersion.

Dynamic similitude implies a similarity in the dynamic behavior of fluids. Dynamic similitude refers to parameters involving mass, e.g., mass density ρ, specific weight γ, and dynamic viscosity μ. For instance, the mass-density scale ρ_r is defined as the ratio of prototype to model mass densities as $\rho_r = \rho_p/\rho_m$. The mass scale $M_r = M_p/M_m$ appropriately describes dynamic similitude, besides the readily defined length and time scales. The basic concept of dynamic similitude is that individual forces acting on corresponding fluid elements must have the same force ratio in both systems. Individual forces acting on a fluid element may be due either to a body force such as weight in a gravitational field, or surface forces resulting from pressure gradients, viscous shear, or surface tension. The resulting inertial force necessitates that the force polygon be geometrically similar.

Gravitational and viscous effects are respectively described by the Froude number $\mathrm{Fr}_r = V_r/(g_r z_r)^{0.5}$ and the Reynolds number $\mathrm{Re}_r = V_r z_r/\nu_r$. In hydraulic models, the Froude and the Reynolds numbers, i.e., $\mathrm{Fr}_r = \mathrm{Re}_r = 1$, can be simultaneously satisfied only when $V_r^2/z_r = V_r z_r$, which is the trivial full scale $V_r = z_r = 1$. We thus conclude that exact similitude of all force ratios in hydraulic models is strictly impossible except at full scale. Of course, forces that are negligible compared with others will not affect the force polygon. Therefore the art of hydraulic modeling is to center the analysis around the force components that are dominant in the system.

The art of hydraulic modeling thus consists of determining whether gravity or viscosity is the predominant physical parameter and to determine the scale parameters accordingly. This approach is reasonable as long as either gravitational or viscous terms can be neglected. In open channels, gravitational effects are typically predominant and resistance to flow does not depend on viscosity as long as flows are hydraulically rough. In most river models, however, the weight force ratio $F_{gr} = \rho_r L_r^3 g_r$ should balance the inertial or hydrodynamic force $F_{ir} = \rho_r L_r^2 V_r^2$. The ratio of inertial to weight forces implies that $F_{ir}/F_{gr} = V_r^2/L_r g_r = 1$, which is simply the Froude similitude criterion. The Froude similitude criterion thus satisfies similarity in the ratio of inertial to weight forces. We also recognize that the Froude number also properly scales the ratio of velocity head to flow depth. Consequently Froude similitude describes similarity conditions in specific-energy diagrams for the model and the prototype. The Froude similitude is therefore useful in describing rapidly varied flow conditions.

10.2   Rigid-bed model

Rigid-bed models are built to simulate flow around river improvement works and hydraulic structures. A rigid boundary implies that the bed is fixed, i.e., no sediment transport. This is the case when the Shields parameter of the bed material is $\tau_* < 0.03$. Rigid-bed model scales can be determined in either one of two cases: (1) exact geometric similitude, in which resistance to flow can be neglected; and (2) distorted/tilted models, in which resistance to flow is important. Exact geometric similitude and Froude similitude can be simultaneously maintained in rigid-bed models only when resistance to flow can be neglected (Subsection 10.2.1). Such models are well suited to the analysis of 3D flow around hydraulic structures, in which sediment transport is not important. When long river reaches are considered and resistance to flow cannot be neglected, both the Froude and the resistance similitude can be simultaneously satisfied in tilted/distorted models (Subsection 10.2.2).

10.2.1   Exact Froude similitude

The model scales for hydraulic models with exact geometric similitude can be determined from the Froude similitude criterion. Exact geometric similitude is required when the flow is 3D and when vertical accelerations are not negligible. This type of model is particularly well suited for modeling the flow near hydraulic structures. The scale ratios for hydraulic models with exact geometric similitude reduce to $V_r = t_r = L_r^{0.5}$ and $M_r = L_r^3$. All other scale ratios can be derived from the length, time, and mass scales by use of the fundamental dimensions of any considered variable. The scale ratio for several variables is listed in the third column of Table 10.1.

The exact geometric similitude imposes constraints that are usually difficult to work with in modeling river reaches. Scaling the size of roughness elements according to the length scale will maintain the same resistance parameter as long as $Re_* > 70$ for both the model and the prototype. In practice, this is possible for only very coarse bed channels such as cobble and boulder bed streams. The difficulty with the undistorted (untilted) Froude similitude criterion is that the near-bed conditions are drastically changed. Indeed, the laminar sublayer thickness δ in hydraulic models is relatively too thick. For instance, it requires a scale $\delta_r = (u_{*_r})^{-1} = (L_r)^{1/2}$. A strict geometrical similitude cannot be maintained because the length similitude requires very small particles whereas scale

Table 10.1. *Scale ratios for hydraulic models*

		Rigid-bed (Froude)		Mobile-bed			
	Scale	Exact	Tilted	Complete General	$(d_{*r}=\tau_{*r}=1)$ $m=1/6$	$d_{*r}\neq 1$	Incomplete $Fr_r\neq 1$
Geometric							
Depth	h_r	L_r	z_r	z_r	z_r	z_r	z_r
Width	W_r	L_r	y_r	y_r	y_r	y_r	y_r
Length	x_r	L_r	x_r	$z_r\left(\frac{1+4m}{1+m}\right)$	$z_r^{1.43}$	$z_r^{1+2m}d_{sr}^{-2m}$	$z_r^2 d_{sr}^2$
Particle diameter	d_{sr}	L_r	$z_r^4 x_r^{-3}$	$z_r\left(\frac{2m-1}{2+2m}\right)$	$z_r^{-0.286}$	d_{sr}	d_{sr}
X-section area	$W_r h_r$	L_r^2	$z_r y_r$	$y_r z_r$	$y_r z_r$	$y_r z_r$	$z_r y_r$
Volume	$X_r W_r h_r$	L_r^3	$x_r y_r z_r$	$y_r z_r\left(\frac{2+5m}{1+m}\right)$	$y_r z_r^{2.43}$	$y_r z_r^{2+2m}d_{sr}^{-2m}$	$y_r z_r^3 d_{sr}^2$
Kinematic							
Time (flow)	t_r	$L_r^{1/2}$	$x_r z_r^{-1/2}$	$z_r\left(\frac{1+7m}{2+2m}\right)$	$z_r^{0.928}$	$z_r^{0.5+2m}d_{sr}^{-2m}$	$z_r^{2-m}d_{sr}^{3+m}$
Time (bed)	t_{br}	—	—	$z_r\left(\frac{2+5m}{1+m}\right)$	$z_r^{2.428}$	$z_r^{1.5+3m}d_{sr}^{-1-3m}$	$z_r^m d_{sr}^2$
Velocity	V_r	$L_r^{1/2}$	$z_r^{1/2}$	$z_r^{1/2}$	$z_r^{1/2}$	$z_r^{1/2}$	$z_r^m d_{sr}^{-1-m}$
Shear velocity	u_{*r}	$L_r^{1/2}$	$z_r x_r^{-1/2}$	$z_r\left(\frac{1-2m}{2+2m}\right)$	$z_r^{0.286}$	$z_r^{0.5-m}d_{sr}^m$	d_{sr}^{-1}
Settling velocity	ω_r	—	—	$z_r\left(\frac{1-2m}{2+2m}\right)$	$z_r^{0.286}$	—	d_{sr}^{-1}
Discharge	Q_r	$L_r^{1/2}$	$y_r z_r^{3/2}$	$y_r z_r^{3/2}$	$y_r z_r^{1.5}$	$y_r z_r^{1.5}$	$y_r z_r^{1+m}d_{sr}^{0.5-m}$
Unit bedload discharge	q_{br}	—	—	1	1	$d_{sr}^{1+m}z_r^{0.5-m}$	1

Dynamic

Mass	M_r	L_r^3	$x_r y_r z_r$	$y_r z_r \left(\dfrac{2+5m}{1+m}\right)$	$y_r z_r^{2.43}$	$y_r z_r^{2+2m} d_{sr}^{-2m}$	$y_r z_r^3 d_{sr}^2$
Pressure	p_r	L_r	z_r	z_r	z_r	z_r	z_r
Shear stress	τ_r	L_r	$z_r^2 x_r^{-1}$	$z_r\left(\dfrac{1-2m}{1+m}\right)$	$z_r^{0.57}$	$z_r^{1-2m} d_{sr}^{2m}$	d_{sr}^{-2}
Force	F_r	L_r^3	$x_r y_r z_r$	$y_r z_r\left(\dfrac{2+5m}{1+m}\right)$	$y_r z_r^{2.43}$	$y_r z_r^{2+2m} d_{sr}^{-2m}$	$y_r z_r^3 d_{sr}^2$

Dimensionless

Slope	S_r	1	$z_r x_r^{-1}$	$z_r\left(\dfrac{-3m}{1+m}\right)$	$z_r^{-0.43}$	$d_{sr}^{2m} z_r^{-2m}$	$z_r^{-1} d_{sr}^{-2}$
Darcy–Weisbach	f_r	1	$z_r x_r^{-1}$	$z_r\left(\dfrac{-3m}{1+m}\right)$	$z_r^{-0.43}$	$d_{sr}^{2m} z_r^{-2m}$	$z_r^{-2m} d_{sr}^{2m}$
Froude	Fr_r	1	1	1	1	1	$z_r^{m-0.5} d_{sr}^{-1-m}$
Reynolds	Re_r	$L_r^{3/2}$	$z_r^{3/2}$	$z_r^{3/2}$	$z_r^{1.5}$	$z_r^{3/2}$	$z_r^{1+1m} d_{sr}^{-1-m}$
Shields	τ_{*r}	—	—	1	1	1	1
Grain Reynolds	Re_{*r}	$L_r^{3/2}$	$z_r^5 x_r^{-3.5}$	1	1	$d_{sr}^{1+m} z_r^{0.5-m}$	1
Dimensionless diameter	d_{*r}	—	—	1	1	$d_{sr}^{\frac{2+2m}{3}}\, z_r^{\frac{1-2m}{3}}$	1
Sediment density	$(G-1)_r$	—	—	$z_r\left(\dfrac{3-6m}{2+2m}\right)$	$z_r^{0.857}$	$z_r^{1-2m} d_{sr}^{2m}-1$	d_{sr}^{-3}

modeling produces a very large laminar sublayer thickness in the model. In the hydraulically smooth regime, resistance to flow increases as the Reynolds number decreases, and resistance to flow will be larger for the model than for the prototype. Exact similitude of near-bedflow conditions cannot be preserved when the same fluid is used because the viscous effects cannot be neglected. Exact similitude in these cases would require different fluids for the model and the prototype.

10.2.2   Froude similitude for tilted river models

Model distortion and tilting is viewed as a feasible practical alternative. Model distortion and tilting are acceptable only when vertical and lateral accelerations of the water can be neglected with respect to the gravitational acceleration. This practical solution allows the use of different scales for flow depth and sediment size. The model is distorted when $y_r \neq z_r$ and tilted when $x_r \neq z_r$, which should be appropriate for near 1D flow conditions. Rigid-bed hydraulic models require that resistance to flow be the same for the model and the prototype. In the hydraulically rough regime, $Re_* > 70$, resistance to flow depends on relative submergence h/d_s. The governing equation to be preserved in gradually varied flow models with rigid boundaries is the resistance relationship $S_r = f_r \mathrm{Fr}_r^2$ whereby tilting is required because $\mathrm{Fr}_r = 1$ and $d_{sr} \neq 1$. In general terms, resistance to flow can be defined as $\sqrt{8/f} = a(h/d_s)^m$ where $m = [1/(\ln 12.2\frac{h}{d_s})]$. The governing equation of similitude for resistance to flow can be written as

$$\mathrm{Fr}_r^2 = \frac{z_r}{x_r}\left(\frac{z_r}{d_{sr}}\right)^{2m}. \tag{10.2}$$

As a particular case, according to Strickler's relationship between Manning coefficient n and bed roughness diameter $n \sim d_s^{1/6}$, the Manning–Strickler equation corresponds to $m = 1/6$.

The model scales in distorted Froude models must simultaneously satisfy the Froude and the Manning–Strickler similitude criteria. The Manning–Strickler similitude criterion in a distorted model is defined as $(z_r/d_{sr})^{1/6}[(z_r^{1/2}S_r^{1/2})/V_r] = 1$. A tilted hydraulic model $S_r = z_r/x_r \neq 1$ that satisfies the Froude similitude $\mathrm{Fr}_r = 1$ implies that $d_{sr} = z_r^4/x_r^3$. According to this relationship, the user has 2 degrees of freedom in selecting two of the three scale parameters, x_r, z_r, or d_{sr}. During the calibration of rigid-boundary models, model roughness is typically increased when disproportionately large blocks and sticks are used to reproduce a stage–discharge relationship comparable with that of the prototype. The modeling of design structures with distorted rigid-boundary models thus requires the intuition and judgement of experienced engineers. The scale ratios for a distorted rigid-bed model are presented in column 4 of Table 10.1 and a calculation example is presented in Example 10.1.

Model distortion is often encountered in engineering practice whereby the flow depth is increased compared with that of exact similitude. A distorted model with different horizontal and vertical scales allows different scales for the bed material and for flow depth. The practical interest in distorted models is that in increasing flow depth and decreasing resistance to flow, and the model user can empirically increase the size of roughness elements until the model results compare with field measurements. The model is then said to be calibrated. Because the kinematic similarity is not exact, however, any attempt to determine kinematic properties such as streamlines and turbulent mixing cannot be properly scaled in distorted models.

Example 10.1   Application to a tilted-rigid-bed model. A 2,000-m gravel-bed river reach has a flow depth of 2 m, a width of 50 m, and a mean velocity of 0.3 m/s. If the prototype Manning coefficient $n = 0.025$ and $d_{50} = 5$ mm, determine the model scales. The maximum length in the laboratory is 20 m.

The prototype Froude number is

$$\text{Fr}_p = \frac{V_p}{\sqrt{gh_p}} = \frac{0.3}{\sqrt{9.81 \times 2}} = 0.0677,$$

and the prototype slope is

$$S_p \cong \frac{n^2 V^2}{h^{4/3}} = \frac{0.025^2 \times 0.3^2}{2^{4/3}} = 2.23 \times 10^{-5}.$$

The Shields parameter is

$$\tau_{*_p} = \frac{h_p S_p}{(G-1)d_{s_p}} = \frac{2 \times 2.23 \times 10^{-5}}{1.65 \times 0.005} = 0.005 < 0.03.$$

Therefore a rigid-bed model is appropriate.

The length scale $x_r = (x_p/x_m) = (2{,}000/20) = 100$ or a model depth of $h_m = h_p/h_r = 2/100 = 0.02$ m and an exact Froude similitude gives $V_r = \sqrt{x_r} = 10$, and the model velocity $V_m = V_p/V_r = 0.03/10 = 0.03$ m/s. The corresponding model Reynolds number is $\text{Re}_m = [(V_m h_m)/\nu] = [(0.03 \times 0.02)/(1 \times 10^{-6})] = 600$ and the model flow would be laminar.

Model distortion is necessary in order to increase the accuracy of flow-depth and velocity measurements and to guarantee turbulent flow in the model. The user has 1 degree of freedom in selecting either the vertical scale, the slope, or the size of roughness elements. The vertical scale is arbitrarily set as $z_r = 25$

or a flow depth of 8 cm. The model Reynolds number would then be

$$\mathrm{Re}_m = \frac{\mathrm{Re}_p}{\mathrm{Re}_r} = \frac{V_p\,h_p}{v z_r^{3/2}} = \frac{2 \times 0.3}{1 \times 10^{-6} \times 25^{1.5}} = 4,800.$$

Model scales are calculated from column 4 of Table 10.1 for tilted rigid beds. For example, planform geometry is similar when $y_r = x_r = 100$. The model velocity is $V_m = (V_p/V_r) = (V_p/\sqrt{z_r}) = 0.3/\sqrt{25} = 0.06$ m/s, and the time scale is $t_r = x_r z_r^{-1/2} = (100/\sqrt{25}) = 20$. The model discharge is $Q_m = Q_p/Q_r = V_p W_p h_p/y_r z_r^{3/2} = [(0.3 \times 50 \times 2)/(100 \times 25^{3/2})] = 0.0024$ m³/s. The tilted model slope is $S_m = S_p/S_r = S_p/z_r x_r^{-1} = 2.23 \times 10^{-5} \times 100/25 = 8.9 \times 10^{-5}$. The size of the roughness elements should be $d_{sm} = d_{sp}/d_{sr} = d_{sp}\,x_r^3/z_r^4 = 0.005\,100^3/25^4 = 13$ mm. In this type of model, resistance blocks or plates would be placed on a smooth surface until the field stage–discharge conditions can be adequately duplicated. This is a trial-and-error calibration procedure.

10.3   Mobile-bed river models

Mobile-bed models are useful when sediment transport is significant, e.g., when $\tau_* > 0.06$. Typical examples include drop structures, local scour, erosion below spillways, sills, locks and dams, reservoir sedimentation, etc. The bed mobility provides 1 additional degree of freedom in selecting the mass density of sediment. Similitude in sediment transport is obtained when the Shields parameter τ_* and the dimensionless grain diameter d_* are similar in both systems, i.e., $\tau_{*_r} = 1$ and $d_{*_r} = 1$. Of course, these conditions also imply that $\mathrm{Re}_{*_r} = 1$ because $\tau_* d_*^3 = \mathrm{Re}_*^2$.

There are four similitude criteria for mobile-bed models: (1) Froude similitude; (2) resistance, e.g., Manning–Strickler; (3) dimensionless grain diameter; and (4) bed-material entrainment or Shields parameter. These four similitude criteria must be simultaneously satisfied in river reaches with rapidly varied flow and sediment transport. The governing criteria involve seven parameters: V_r, g_r, z_r, d_{sr}, S_r, $(G-1)_r$, and v_r. Hydraulic models with $g_r = v_r = 1$ and four equations of similitude leave only 1 degree of freedom, e.g., the model length scale z_r, besides the lateral scale y_r, which is not specified by the equations. The model is tilted because the equations impose $S_r = z_r/x_r$. For instance, the user may prefer a model that is undistorted but tilted, $y_r \neq z_r$, to a distorted and tilted model that preserves planform geometry $x_r = y_r$.

The mobile-bed similitude is said to be complete, with 1 degree of freedom, when the four equations of similitude are simultaneously satisfied (Subsection 10.3.1). When complete similitude is impossible, it is sometimes possible to sacrifice one of the governing equations for an additional degree of freedom (Subsection 10.3.2).

10.3.1   Complete mobile-bed similitude

It is important to acknowledge that complete mobile-bed similitude implies that the downstream direction is dominant and the accelerations in the lateral and the vertical directions are negligible. Complete mobile-bed similitude is therefore essentially suitable for 1D sediment-transport processes. Similitude in dimensionless particle diameter $d_{*_r} = 1$ in hydraulic models imposes the following relationship between the particle diameter and the particle density:

$$d_{sr}^3 = \frac{1}{(G-1)_r}. \tag{10.3}$$

It is clear from this relationship that hydraulic models require very light sediment when large particles are used in the model.

The properties of light material commonly used in practice are listed in Table 10.2. It is interesting to note that, for prototype sediment at a specific gravity of 2.65, any lightweight material corresponds to a specific scale ratio for the particle diameter.

Table 10.2. *Lightweight sediment properties for mobile-bed models*

Material	Specific gravity G	Typical size d_s (mm)	Comment
Polystyrene	1.035–1.05	0.5–3	Durable but difficult to wet and tends to float
Gilsonite	1.04		
Nylon (polyamidic resins)	1.16	0.1–5	
Lucite	1.18		
PVC	1.14–1.25	1.5–4	Hydrophobic
Perspex	1.18–1.19	0.3–1	Dusty
Acrylonitrile butadiene styrene	1.22	2–3	Adds detergent against air-bubble adherence
Coal	1.2–1.43 (up to 1.6)	0.3–4	Possible inhomogenity in specific gravity and sorting
Ground walnut shells	1.33	0.15–0.41	Deteriorate in 2–3 months, color water (dark brown)
Bakelite	1.38–1.49	0.3–4.0	Porous, tends to rot, changes diameter, and floats
Pumice	1.4–1.7		
Loire sand	1.5	0.63–2.25	Dusty
Lytag (fly-ash)	1.7	1–3	Porous
Quartz sand	2.65	0.1–1	

Similitude in Shields parameter $\tau_{*_r} = 1$ imposes the following relationship between the particle diameter and the slope similitude S_r:

$$\tau_{*r} = z_r S_r / [(G - 1)_r d_{sr}] = 1. \tag{10.4}$$

It is clear from Eqs. (10.3) and (10.4) that in order to simultaneously satisfy $d_{*_r} = 1$, $S_r = z_r / x_r$, and $\tau_{*_r} = 1$, the condition $d_{sr} = x_r^{1/2} / z_r$ must be satisfied. The scale ratios are obtained from simultaneously satisfying Eqs. (10.1)–(10.4). In a nutshell, the particle-diameter scale d_{sr} is directly obtained as a result of the Manning–Strickler relationship and the Froude similitude $d_{sr} = z_r^4 / x_r^3$. The density of the sediment is then obtained from the dimensionless particle diameter such that $(G - 1)_r = x_r^9 / z_r^{12}$. Finally, by substitution into the Shields equation, the condition that satisfies the same Shields parameter imposes that $z_r = x_r^{7/10}$. This condition further simplifies previous requirements as $d_{sr} = x_r^{-0.2}$ and $(G - 1)_r = x_r^{0.6}$.

The criterion for sediment suspension based on settling velocity ω can be defined from $\omega = 8(\nu / d_s)[(1 + 0.0139 d_*^3)^{0.5} - 1]$. The settling velocity scale in water is thus $\omega_r = 1 / d_{sr}$ as long as $d_{*_r} = 1$. The criterion for sediment suspension is defined from identical values of the ratio of shear velocity to settling velocity, or $\omega_r / u_{*_r} = 1$. This leads directly to $d_{sr} = x_r^{1/2} / z_r$, which is identical to the condition previously obtained from the Shields parameter. We can thus conclude that similitude in Shields parameter is equivalent to similitude in the ratio of bedload to sediment suspension, provided that $d_{*_r} = 1$. This strengthens the requirement that $d_{*_r} = 1$ and $\tau_{*_r} = 1$ for similitude in sediment transport.

Similitude in bedload sediment transport can be determined from the Einstein–Brown relationship as $q_{bv} / \omega d_s = f(\tau_*)$, in which, with $\omega_r = 1 / d_{sr}$ when $d_{*_r} = 1$, we obtain directly $q_{bvr} = f(\tau_{*_r})$. It is interesting to note that $q_{bvr} = 1$ when $\tau_{*_r} = 1$, which again strengthens the requirement for similitude in Shields parameter through $\tau_{*_r} = 1$.

Bed aggradation and degradation relates to the sediment continuity relationship applied to bedload discharge written in 1D form as $\partial q_b / \partial x = -p_0 (\partial z_0 / \partial t_s)$, where q_b is the unit bedload discharge, p_0 is the porosity of the bed material, z_0 is the bed-elevation, and t_s refers to time. The time scale for bedload motion t_{sr} that describes bed-elevation changes is then obtained as $t_{sr} = [(p_{0r} z_r x_r) / q_{br}]$. It can be assumed that the porosity ratio $p_{0r} = 1$. This time ratio refers to the sedimentation time scale that is useful in the analysis of local bed-elevation changes through local scour, bedforms, and changes in bedload transport. The time scale for bed-elevation changes is different from the time scale obtained from the Froude similitude criterion.

In diffusion–dispersion studies, the time scale for vertical mixing can be estimated from $t_{vr} = z_r / u_{*_r}$ comparatively with the time to lateral mixing given by $t_{tr} = y_r^2 / z_r u_{*_r}$. Of course these two scales are equivalent only as long as the model is not distorted. Also, the length scale for vertical mixing is $x_{vr} = [(z_r V_r)/u_{*_r}]$ is comparable with the length for lateral mixing given by $x_{tr} = V_r y_r^2/z_r u_{*_r}$. These length scales are compatible for only undistorted models.

The scale ratios for rapidly varied mobile-bed models are listed in column 5 of Table 10.1. Example 10.2 illustrates how to calculate the scale ratios for mobile-bed models. In practice, complete mobile-bed similitude is somewhat restricted to model scales that are not too small (approximately $z_r < 25$) because larger scale models necessitate unreasonably light material.

Example 10.2 Calculation of complete mobile-bed similitude. Consider the model of a large sand bed at a flow depth of 8 m and a velocity of 2 m/s. The slope is 7×10^{-5}, and the discharge is 40,000 m³/s. Determine the scale ratios for complete similitude at $z_r = y_r = 100$. The prototype Shields parameter is

$$\tau_{*p} = \frac{h_p S_p}{(G-1)_p d_{sp}} = \frac{8 \times 7 \times 10^{-5}}{1.65 \times 0.0002} = 1.7 > 0.03.$$

The bed is mobile and the value of $m = [\ln(\frac{12.2 \times 8}{0.0002})]^{-1} = 0.076$. The scale ratios for the following parameters are obtained from column 5 of Table 10.1:

$m = 0.0763$

downstream distance $x_r = 100^{1.21} = 266$

model particle diameter $d_{sm} = \dfrac{d_{sp}}{d_{sr}} = \dfrac{0.2\,\text{mm}}{100^{-0.393}} = 1.2\,\text{mm}$

time scale $t_r = 100^{0.71} = 26.6$

time scale of bed $t_{sr} = 100^{2.21} = 26{,}300$

model velocity $V_m = \dfrac{V_p}{V_r} = \dfrac{2\,\text{m/s}}{10} = 0.2\,\text{m/s}$

model discharge $Q_m = \dfrac{Q_p}{Q_r} = \dfrac{40{,}000\,\text{m}^3/\text{s}}{100^{5/2}} = 0.4\,\text{m}^3/\text{s}$

model slope $S_m = \dfrac{S_p}{S_r} = \dfrac{7 \times 10^{-5}}{100^{-0.21}} = 1.9 \times 10^{-4}$

model sediment density $(G-1)_m = \dfrac{(G-1)_p}{100^{1.18}} = \dfrac{1.65}{230} = 7 \times 10^{-3}.$

$G_m = 1.007$, which is lower than the density of polystyrene.

An incomplete mobile-bed model will unfortunately be indicated at this model scale.

10.3.2 Incomplete mobile-bed similitude

When the conditions for complete similitude are not practically possible, one constraint can sometimes be sacrificed in order to benefit from an additional degree of freedom. As the model further deviates from complete similitude, there is a greater risk that the model may yield incorrect results. There are nevertheless a number of possibilities, depending on which conditions of similitude should be preserved in both the model and in the prototype. Two types of models are considered here: (1) non-Froudian similitude $Fr_r \neq 1$; and (2) quasi similitude in sediment transport $d_{*_r} \neq 1$. First, near-equilibrium streams in which the flow is gradually varied can be simulated with different values of the Froude number as long as the flow is subcritical, i.e., fine-grained alluvial rivers with low Froude numbers can be simulated with the same bed material at higher, yet subcritical, Froude numbers. Second, coarse bed material in which bedload transport is predominant can be simulated with smaller values of d_* as long as the flow is hydraulically rough. The scale values for these two cases of incomplete mobile-bed similitude are presented in columns 7 and 8 of Table 10.1.

In non-Froudian models, gradually-varied flow in large alluvial rivers does not impose large changes on the specific-energy diagram. In cases in which the Froude number remains fairly constant along the study reach, we may use different values of the Froude number for the model and the prototype as long as the flow remains subcritical in both systems. In this case only the three governing equations, (10.2), (10.3), and (10.4), are simultaneously satisfied. The Froude similitude criterion would be replaced with an additional degree of freedom, e.g., y_r, z_r, and d_{sr}.

When the same density of sediment is used, the similitude in dimensionless particle diameter d_* implies that $(G - 1)_r = 1/d_{sr}^3$; thus the sediment must have the same density and particle diameter. The Shields similitude condition can be rewritten as $d_{sr} = x_r^{1/2}/z_r = 1$ and Manning–Strickler equation (10.2) can be rewritten in terms of the Froude number as $Fr_r = z_r^{-0.5+m}$. The Froude scale is not unity but changes slightly with model scale and the scale ratios for gradually varied alluvial flow models are listed in column 8 of Table 10.1.

This type of model allows the use of the same sediment in the model and the prototype. Undistorted cross-sectional geometry can be simulated with $y_r = z_r$, and similitude in planform geometry is obtained when $x_r = y_r$. This type of model offers similitude in bedload and suspended sediment transport. The drawbacks of this approach are that the model does not obey the Froude similitude and should be limited to near-uniform subcritical flows. This type of model should appropriately simulate sediment transport and resistance to flow. However, the force diagrams and the lateral/vertical accelerations are not appropriately simulated.

Models with quasi similitude in sediment transport, $d_{*_r} \neq 1$, may be used from Table 10.1, column 7, in which bedload transport is dominant, as detailed in Example 10.3. It is noteworthy that the similitude in total sediment transport is questionable when $\tau_{*_r} = 1$ and $d_{*_r} \neq 1$. Scale effects can be found in other sediment-transport processes such as suspended load and bedform prediction. Case Study 10.1 and Problem 10.3 illustrate how the scales of a model that uses the same material for both systems can be determined. Case Study 10.2 relates physical modeling with protection against scour at highway bridges.

Example 10.3   Application to bedload transport in coarse-gravel-bed streams. A gravel-bed river with $d_{50} \approx 30$ mm and $d_{90} = 100$ mm has a natural bed slope $S = 3 \times 10^{-3}$. If the flow depth reaches 4 m during floods, determine the model scales that would allow a 10-km reach to be modeled within 40 m in the hydraulics laboratory.

The downstream length scale $x_r = 10{,}000/40 = 250$ and the Shields parameter of the prototype is

$$\tau_{*p} = \frac{h_p S_p}{(G-1)_p d_{s_p}} = \frac{4\,\text{m} \times 3 \times 10^{-3}}{1.65 \times 0.03\,\text{m}} = 0.24 > 0.06;$$

hence there is sediment transport. Complete mobile-bed similitude would require a model sediment size of $d_{s_r} = (d_{s_p}/d_{s_m}) \cong z_r^{-0.286} = 250^{-0.286} = 0.208$ or $d_{s_m} \gg 30$ mm, which is impractical.

The modeler must therefore resort to incomplete mobile-bed modeling. In this case, a model that does not satisfy the Froude condition would not be indicated because the prototype Froude number is certainly quite high. Because the bed material is very coarse and the flow is hydraulically rough,

$$\text{Re}_{*_p} = \frac{\sqrt{gh_p S_p d_{s_p}}}{\nu} = \frac{\sqrt{9.81 \times 4 \times 3 \times 10^{-3}} \times 0.03}{1 \times 10^{-6}} = 10{,}300,$$

it is considered that $d_{*_r} \neq 1$ would be appropriate because finer bed material would still be in the hydraulically rough regime. The scale ratios for incomplete mobile-bed similitude with $d_{*_r} \neq 1$ leave 2 degrees of freedom. From prototype values, $\text{m} = [\ln(12.2 \times \frac{h_p}{d_{s_p}})]^{-1} = (\ln 12.2 \times \frac{4}{0.03})^{-1} = 0.135$, and the 2 degrees of freedom in $z_r = 175$ and $d_{s_r} = 20$ are arbitrarily selected, along with $y_r = 250$. There is no similitude in planform geometry because $y_r = 250$ and $x_r = z_r^{1+2m} d_{sr}^{-2m} = 314$. The resulting scale ratio for grain shear Reynolds number from Table 10.1, column 7 is $\text{Re}_{*_r} = (\text{Re}_{*_p}/\text{Re}_{*_m}) = d_{sr}^{1+m} z_r^{0.5-m} = 20^{1+0.135} 175^{0.5-0.135} = 197;$ then $\text{Re}_{*_m} = \text{Re}_{*_p}/\text{Re}_{*_r} = 10{,}300/197 = 52$, which is considered hydraulically rough. Other model scales calculated from Table 10.1 are $Q_r = y_r z_r^{1.5} = 5.8 \times 10^5$ and $V_r = z_r^{1/2} = \sqrt{175} = 13.3$. The model sediment diameter is $d_{s_m}/d_{s_r} = 0.03\,m/20 = 1.5$ mm at a particle density of $(G-1)_m = (G-1)_p/(G-1)_r = 1.65/z_r^{1-2m}d_{sr}^{2m-1} = 1.65/175^{1-2\times0.135}20^{2\times0.135-1} = 0.34$ or $G_m = 1.34$, in which coal or Bakelite would be appropriate from Table 10.2. The model will also require tilting and $S_m = S_p/S_r = S_p/d_{s_r}^{2m}z_r^{-2m} = 3 \times 10^{-3}/20^{2\times0.135}175^{-2\times0.135} = 0.0054$. The scale for unit bedload discharge is calculated assuming similitude in $q_{br}/u_{*_r}d_{sr} = 1$ or $q_{br} = d_{sr}^{1+m}z_r^{0.5-m} = 20^{1+0.135}175^{0.5-0.135} = 197$. Finally, the time scale for bed-elevation changes is obtained from $t_{sr} = z_r^{1.5+3m}d_{s_r}^{-1-3m} = 175^{1.5+3\times0.135}20^{-1-3\times0.135} = 278$, which means that bed changes over 1 min of model time correspond to 4.6 h of prototype time. Suspended sediment transport is not in similitude. Therefore the similitude in total sediment discharge is questionable, but aggradation and degradation features from bedload discharge should be well represented by the model.

Case Study 10.1   Mobile-bed model of the Jamuna River, Bangladesh.
Klaassen (1990, 1992) reported on a physical scale model of the Jamuna River in Bangladesh. The Jamuna River is a flat sand-bed river, with, on the average, three braids upstream of Sirajganj (near the proposed bridge site). The purpose of this movable-bed scale model investigation is: (1) to study the effect of the river training works on channel patterns; (2) to identify "worst" channel patterns (during both floods and low-flow conditions) for the river training works; and (3) to study the upstream river shifts on the channel patterns near the bridge and the training works.

For the present model investigation the limitations include: (1) model dimensions at a maximum of approximately 50×20 m^2 because of costs; (2) model material sand, because lightweight material (which would be preferable) is too expensive in large quantities; and (3) model material size, which should be $\sim$200 μm because it is readily available in large quantities.

The main characteristics of the related model are listed in Table CS.10.1.1. The scaling procedure resulted in a considerably distorted model. For a model discharge of $\sim$100 l/s, a fair reproduction of the channel pattern is probably obtained for a model slope of approximately 8×10^{-3}. If the Chézy coefficient C_m in the model is assumed to be $\sim$25 m$^{1/2}$/s, the average velocity in the channels is estimated to be $\sim$0.4 m/s. This implies that the Froude number in the model is of the order of 0.5 whereas in the prototype Fr $\cong$ 0.2. Locally in the model even higher values of the Froude number may be found. Such high Froude numbers will certainly affect the flow pattern reproduction and consequently induce scale effects.

Very good morphological results were obtained with the model, and the objectives have been satisfactorily met. One of the main concerns about the model relates to local scour. Local scour is reproduced correctly in only undistorted models that satisfy the Froude condition. Because both conditions are

Table CS.10.1.1. *Characteristics of the Jamuna River model (after Klaassen, 1990)*

Parameter	Prototype	Scale factor	Model
Particle size	$d_{sp} = 0.2$ mm	$d_{sr} = 1$	$d_{sm} = 0.2$ mm
Density	$G_p = 2.65$	$(G-1)_r = 1$	$G_m = 2.65$
Slope	$S_p = 7 \times 10^{-5}$	$S_r = 0.01$	$S_m = 7 \times 10^{-3}$
Discharge	$Q_p = 10{,}000 \, \text{m}^3/\text{s}$	$Q_r = 10^6$	$Q_m = 0.01 \, \text{m}^3/\text{s}$
	$= 90{,}000 \, \text{m}^3/\text{s}$		$= 0.09 \, \text{m}^3/\text{s}$
Bankfull width	$W_p = 3{,}000 \, \text{m}$	$y_r = 1{,}000$	$W_m = 3.3 \, \text{m}$
Total width	$15{,}000 \, \text{m}$	$y_r = 1{,}000$	$15 \, \text{m}$
Flow depth	$h_p = 5.8 \, \text{m}$	$z_r = 200$	$h_m = 0.032 \, \text{m}$
	$h_{mp} = 8 \, \text{m}$		$h_{mm} = 0.04 \, \text{m}$
Sediment transport	$q_{sp} = 1.4 \times 10^{-3} \, \text{m}^2/\text{s}$	$q_{sr} = 80$	$q_{sm} = 1.34 \times 10^{-5} \, \text{m}^2/\text{s}$
Flood duration	$T_p = 78$ days	$t_r = 2{,}500$	$T_m = 0.03$ day
Froude number	$\text{Fr}_p = 0.1{-}0.2$	$\text{Fr}_r = 0.25$	$\text{Fr}_m = 0.4 - 0.8$

not fulfilled, local scour cannot be scaled. Another concern is the local presence of supercritical flow.

Through comparisons with the conditions in Example 10.2, the reader will note that complete mobile-bed similitude requires excessively light material. It is also given as Problem 10.3 that the reader can verify that the model scales are quite comparable with those obtained with incomplete mobile-bed similitude with $z_r = 100$, $y_r = 1{,}000$, and $d_{sr} = 1$.

Case Study 10.2   Bridge scour model for Schoharie Creek, United States.
Richardson and Lagasse (1996) report that there are more than 575,000 bridges in the U.S. National Bridge Inventory. Approximately 84% of these bridges span over water. On April 5, 1987, the center span of the 165-m-long bridge over Schoharie Creek, New York, collapsed during a near-record flood of $\sim 1{,}750$ m³/s. Each rigid-frame pier was supported on a spread footing bearing on a glacial till just below the streambed. The bridge designers assumed that the glacial till substrate was nonerodible.

A physical model of the Schoharie Creek bridge was built at a model scale of 1:50 in a 20 ft long × 100 ft flume. The following scale ratios were obtained from an exact Froude similitude with $z_r = 50$: (1) the velocity scale was 7.07; (2) the time scale was also 7.07; and (3) the discharge scale was 17,678. The horseshoe vortex near the base of the bridge pier was at the origin of the scour underneath the pier and the spread footing.

The 1987 bridge failure resulted in establishing a National (U.S.) Program that requires each state to evaluate all bridges over water for vulnerability to failure from erosion of the foundations. This evaluation is to be carried out by an interdisciplinary team consisting of hydraulic, structural, and geotechnical engineers. This evaluation is in addition to the National Bridge Inspection Program that requires the states to inspect all bridges every two years to determine their structural integrity. As a design philosophy, bridges should be designed by use of the flow of a 100-yr flood and to withstand the effects of scour from a flow exceeding the 100-yr flood.

Scour monitoring devices include sonic fathometers for scour data measurements that can be stored in data loggers. Also, magnetic sliding collars have been designed to slide freely over a small stainless steel pipe driven into the streambed at the expected scour location. An audible signal can be heard at the bridge deck when the magnet comes close to the sensor.

◆Exercise 10.1

Derive the scale ratios for tilted-rigid-bed models from similitude in Froude number and Manning–Strickler. Compare the results with those of column 4 in Table 10.1. Also determine the results of the particular case in which $x_r = y_r = z_r = L_r$.

Exercise 10.2

Demonstrate that, in a hydraulic model, the scale parameters that satisfy the same Reynolds number and the same Shields parameter also satisfies the same dimensionless particle diameter.

◆Exercise 10.3

In Example 10.2, show that the grain Reynolds number of the model and the prototype are quite similar despite the model distortion.

◆◆Exercise 10.4

Derive the scale ratios for complete mobile-bed similitude from simultaneously solving for similitude in: (1) Froude number $Fr_r = 1$; (2) resistance equation; and (3) sediment transport $\tau_{*_p} = d_{*_r} = 1$. Compare the results with those of Table 10.1, column 5, and check the values for the Manning–Strickler relationship ($m = 1/6$).

◆◆Exercise 10.5

Derive the scale ratios for incomplete mobile-bed similitude after considering similitude in: (1) the resistance equation; and (2) sediment transport with $\tau_{*_r} = d_{*_r} = 1$. Compare the results with those of Table 10.1, column 8, for $Fr_r \neq 1$.

♦♦Exercise 10.6

In Case Study CS.10.1, determine whether the scale factors selected in Table CS.10.1.1 satisfy: (1) the Froude similitude; (2) the Manning–Strickler similitude; and (3) sediment-transport similitude in $d_{*_r} = \tau_{*_r} = 1$.

♦♦Problem 10.1

A clearwater open-channel model is to be designed such that the maximum laboratory discharge is 2 l/s for a stream discharge of 300 m³/s. If the laboratory space allows a maximum length of 60 m to model a river reach of 10 km, determine a suitable scaling length for the model. From this, determine the scaling ratios for time, discharge, and hydropower.

Problem 10.2

The filling and emptying gates of a canal lock extend the full height of the lock chamber. When a vessel is lowered in the prototype lock, the gates at the outlet end are programmed to open at a fixed rate. The waves and the currents produced by the outflow cause the vessel to pull at its moorings. In a 1/25 scale model of the system that uses water, the maximum tension in the moorings is 1.6 lb (0.36 N) when the gates are opened at the proper rate. Determine the maximum mooring-line tension in the prototype.

Answer: $F_p = F_m F_r = (1.6 \text{ lb}) L_r^3 = 25{,}000 \text{ lb} (112 \text{ kN})$.

♦♦Problem 10.3

With reference to Case Study 10.1, demonstrate that the model scales in Table CS.10.1 are comparable with those obtained from Table 10.1, column 8, with $y_r = 1{,}000$, $z_r = 100$, $d_{sr} = 1$, and $m = 0.076$.

Answer: $t_r = 100^{2-m} = 7050$, $Q_r = y_r z_r^{1+m} = 142{,}000$,
$S_r = z_r^{-1} = 0.01$, and $Fr_r = 100^{m-0.5} = 0.14$.

♦♦Problem 10.4

With reference to Example 10.3, calculate the ratio u_*/ω for both the model and the prototype and verify that most of the sediment transport is bedload in both cases.

Answer: $u_{*_p}/\omega_p = 0.57$ and $u_{*_m}/\omega_m \simeq 0.5$, thus predominantly bedload.

11

Mathematical river models

Numerous river engineering problems can be conveniently investigated by means of mathematical models. Mathematical models must properly describe the physical processes and provide a numerical solution to a system of differential equations that are solved together with suitable boundary conditions and empirical relationships that describe resistance to flow and turbulence.

The differential equations describing river mechanics problems are usually simplified forms of the equations of conservation of mass and momentum, leading to a set of partial differential equations involving two independent variables (time and space or two spatial variables). Examples that use the finite-difference method are presented in this chapter. The finite-element method also provides useful solutions to river engineering problems but is beyond the scope of this chapter.

The algorithms to be used in the finite-difference method depend on the type of differential equation to be solved. Table 11.1 provides a simple classification of river engineering problems. The information propagates at a celerity c in hyperbolic equations, and the celerity is effectively infinite in parabolic equations.

Once a river engineering problem has been defined and a mathematical model chosen, field data need to be gathered to describe initial and boundary conditions, geometrical similitude, material properties, and design conditions. Additional data are also required for calibration and verification. The governing equations can be simplified to preserve the main features of the physical problem; the time and the space increments are determined at this stage. A schematization can be made of the design conditions to be investigated.

Model calibration is usually necessary because empirical parameters are involved to describe resistance to flow and because of the implications to the governing equations. Parameters can be adjusted to obtain good correspondence between numerical results and continuum values. Of course, the adjustment should not be extended beyond physically acceptable values. The

352

Table 11.1. *Differential equation types in river engineering*

Equation type	Equation	River engineering problem
Hyperbolic	$\dfrac{\partial \phi}{\partial t} + v \dfrac{\partial \phi}{\partial x} = 0$	Advection (v constant)
	$\dfrac{\partial^2 \phi}{\partial t^2} = c^2 \dfrac{\partial^2 \phi}{\partial x^2}$	Floodwave propagation (c^2 constant)
Parabolic	$\dfrac{\partial \phi}{\partial t} = K_d \dfrac{\partial^2 \phi}{\partial x^2}$	Diffusion–dispersion (K_d constant)
Elliptic	$\dfrac{\partial^2 \phi}{\partial x^2} + \dfrac{\partial^2 \phi}{\partial y^2} = 0$	Flow net

precision of a model refers to the error margin of the numerical calculations. Model accuracy usually refers to the comparison of the model with field measurements. For instance, a model that calculates the floodstage to the nearest centimeter but is 1 m off from the field measurements is precise but not accurate. The method of adjusting parameters by running the model at different values until a satisfactory result is obtained is called hindcasting. It is a very useful way to determine the sensitivity of the model results to changes in the model parameters. The calibration phase should also comprise a check of the numerical accuracy by varying numerical parameters such as the time step.

Model verification involves simulation for a different set of prototype data with the coefficients previously obtained during the calibration. If a model run satisfactorily reproduces the measured prototype conditions without further adjustment, a reasonable confidence is gained in the application of the model to design conditions that have never occurred in the prototype. It is often possible to calibrate a model with the first half of a field data set and to verify the model with the second half. This chapter describes finite-difference approximations in Section 11.1, followed by some typical 1D models in Section 11.2, and a brief discussion of multidimensional models in Section 11.3.

11.1 Finite-difference approximations

Let us consider a function $h(x, t)$ defined in space x and time t. We may divide the x–t plane into a grid, as shown in Fig. 11.1(a). The grid spacing along the x axis is Δx and the time interval along the t axis is Δt.

(a)

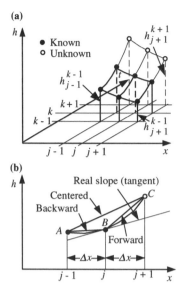

(b)

Figure 11.1. Finite-difference grid and approximation.

The value of the variable h will use the spatial location as a subscript and the time as a superscript, e.g., h_j^k refers to the value of flow depth at the jth spatial grid point and kth time grid point. By the known time level, we mean that the values of different dependent variables are known at the time level h^k and we want to compute their values at the unknown time level h^{k+1}.

If the computations progress from one step to the next, then the procedure is referred to as a marching procedure. Most of the phenomena described by hyperbolic partial differential equations are solved with marching procedures. The conditions specified at time $t = 0$ are referred to as the initial conditions. The conditions specified at the channel ends are called the end, or boundary, conditions.

Finite-difference approximations are first introduced before a presentation on consistency and convergence (Subsection 11.1.1), a linear stability analysis (Subsection 11.1.2), higher-order approximations (Subsection 11.1.3), and boundary conditions (Subsection 11.1.4). The finite-difference method is based on a Taylor series expansion of the variable h_{j+1} written as a function of h_j as

$$h_{j+1}^k = h_j^k + \Delta x \left(\frac{\partial h}{\partial x}\right)_j^k + \frac{\Delta x^2}{2!}\left(\frac{\partial^2 h}{\partial x^2}\right)_j^k + \frac{\Delta x^3}{3!}\left(\frac{\partial^3 h}{\partial x^3}\right)_j^k + 0(\Delta x^4), \quad (11.1)$$

where the derivative $(\partial h/\partial x)_j^k$ is evaluated at grid point j and time level k and $0(\Delta x^m)$ indicates m-order terms. The Taylor series could be similarly expanded to define h_{j-1}^k from h_j^k as

$$h_{j-1}^k = h_j^k - \Delta x \left(\frac{\partial h}{\partial x}\right)_j^k + \frac{\Delta x^2}{2!}\left(\frac{\partial^2 h}{\partial x^2}\right)_j^k$$
$$- \frac{\Delta x^3}{3!}\left(\frac{\partial^3 h}{\partial x^3}\right)_j^k + 0(\Delta x^4). \quad (11.2)$$

Rearranging Eqs. (11.1) and (11.2) and dividing by Δx gives, respectively,

$$\left(\frac{\partial h}{\partial x}\right)_j^k = \underbrace{\frac{h_{j+1}^k - h_j^k}{\Delta x}}_{\text{forward difference}} - \underbrace{\frac{\Delta x}{2!}\left(\frac{\partial^2 h}{\partial x^2}\right)_j^k - \frac{\Delta x^2}{3!}\left(\frac{\partial^3 h}{\partial x^3}\right)_j^k}_{\text{truncation error}}$$

$$+ 0(\Delta x^3) \simeq \underbrace{\frac{h_{j+1}^k - h_j^k}{\Delta x}}_{\text{downwind}} + 0(\Delta x), \tag{11.3}$$

$$\left(\frac{\partial h}{\partial x}\right)_j^k = \underbrace{\frac{h_j^k - h_{j-1}^k}{\Delta x}}_{\text{backward difference}} + \underbrace{\frac{\Delta x}{2!}\left(\frac{\partial^2 h}{\partial x^2}\right)_j^k - \frac{\Delta x^2}{3!}\left(\frac{\partial^3 h}{\partial x^3}\right)_j^k}_{\text{truncation error}}$$

$$+ 0(\Delta x^3) \simeq \underbrace{\frac{h_j^k - h_{j-1}^k}{\Delta x}}_{\text{upwind}} + 0(\Delta x). \tag{11.4}$$

The first approximation of the partial derivative in Eq. (11.3) is written in the form of a forward (downwind) difference and a first-order truncation error $0(\Delta x)$ that will approach zero as Δx becomes very small. Similarly, Eq. (11.4) includes a backward (upwind) difference and a truncation error $0(\Delta x)$. The truncation error approaches zero as Δx approaches zero as long as the high derivatives remain continuous. Therefore both forward and backward finite-differences are first-order approximations.

A central finite-difference approximation can then be obtained from taking half of the sum of Eqs. (11.3) and (11.4), or

$$\left(\frac{\partial h}{\partial x}\right)_j^k = \frac{h_{j+1}^k - h_{j-1}^k}{2\Delta x} + 0(\Delta x^2). \tag{11.5}$$

The truncation error is of the order of $(\Delta x)^2$ because the terms in Δx in Eqs. (11.3) and (11.4) cancel. The central difference is thus said to be a second-order approximation.

Figure 11.1(b) shows a geometrical representation of the forward, backward, and central finite-difference approximations. The real slope is the tangent of the function at B. The forward finite-difference approximation uses the slope of the secant curve line BC, the backward finite-difference approximation uses the slope of line AB, and the central finite-difference approximation uses the slope

Table 11.2. *Explicit and implicit finite differences*

Finite difference	Explicit	Implicit
Backward upwind	$\dfrac{\partial h}{\partial x} \simeq \dfrac{h_j^k - h_{j-1}^k}{\Delta x}$	$\dfrac{\partial h}{\partial x} \simeq \dfrac{h_j^{k+1} - h_{j-1}^{k+1}}{\Delta x}$
Forward downwind	$\dfrac{\partial h}{\partial x} \simeq \dfrac{h_{j+1}^k - h_j^k}{\Delta x}$	$\dfrac{\partial h}{\partial x} \simeq \dfrac{h_{j+1}^{k+1} - h_j^{k+1}}{\Delta x}$
Central	$\dfrac{\partial h}{\partial x} \simeq \dfrac{h_{j+1}^k - h_{j-1}^k}{2\Delta x}$	$\dfrac{\partial h}{\partial x} \simeq \dfrac{h_{j+1}^{k+1} - h_{j-1}^{k+1}}{2\Delta x}$

of the line AC, although all three approximations become exact as Δx goes to zero (first-order approximation). It is clear from these figures that the central finite-difference approximation is more accurate (second-order approximation) than the forward or the backward finite-difference approximations.

Explicit formulations refer to partial derivatives at the known level k whereas implicit formulations refer to the unknown level $k + 1$. Table 11.2 lists some typical explicit and implicit finite-difference approximations for the spatial partial derivative, $\partial h / \partial x$, at the grid point (j,k).

11.1.1   Consistency and convergence

Four properties of consistency, stability, convergence, and accuracy are important in numerical analysis. The following formulation of the advection equation, or floodwave propagation problem, is used to illustrate these properties. Hence

$$\frac{\partial h}{\partial t} + c\frac{\partial h}{\partial x} = 0, \tag{11.6}$$

where c is the celerity, is approximated with a forward difference in time and a backward difference in space (FTBS) to give

$$\frac{h_j^{k+1} - h_j^k}{\Delta t} + c\frac{h_j^k - h_{j-1}^k}{\Delta x} = 0. \tag{11.7}$$

Rearranging to find the flow depth at the unknown level $k + 1$ as a function of the flow depth at the known level k, we obtain

$$h_j^{k+1} = h_j^k - \frac{c\Delta t}{\Delta x}\left(h_j^k - h_{j-1}^k\right), \tag{11.8a}$$

$$h_j^{k+1} = C_c h_{j-1}^k + (1 - C_c) h_j^k, \tag{11.8b}$$

where $C_c = (c\Delta t / \Delta x)$ is the Courant number.

To get started, the initial condition of flow depth needs to be known for all j values at $k = 0$. The algorithm can then march in time given the boundary condition of flow depth for all k values at $j - 1 = 0$.

Consistency is the property of a finite-difference scheme to reduce to the partial differential equation as the truncation error disappears. In our example, the values of h_{j+1}^k and h_{j-1}^k from Eqs. (11.1) and (11.2) are substituted back into Eq. (11.8b) to give

$$
h_j^k + \Delta t \left(\frac{\partial h}{\partial t} \right)_j^k + \frac{\Delta t^2}{2!} \left(\frac{\partial^2 h}{\partial t^2} \right)_j^k + 0(\Delta t^3) = \left(1 - \frac{c\Delta t}{\Delta x} \right) h_j^k
$$

$$
+ \frac{c\Delta t}{\Delta x} \left[h_j^k - \Delta x \left(\frac{\partial h}{\partial x} \right)_j^k + \frac{\Delta x^2}{2!} \left(\frac{\partial^2 h}{\partial x^2} \right)_j^k + 0(\Delta x^3) \right]. \quad (11.9)
$$

Rearranging the equation after canceling the terms in h_j^k and dividing by Δt results in

$$
\underbrace{\left(\frac{\partial h}{\partial t} \right)_j^k + c \left(\frac{\partial h}{\partial x} \right)_j^k}_{\text{original equation}} + \underbrace{\left[\frac{\Delta t}{2!} \left(\frac{\partial^2 h}{\partial t^2} \right)_j^k - \frac{c\Delta x}{2!} \left(\frac{\partial^2 h}{\partial x^2} \right)_j^k + 0(\Delta t^2) + 0(\Delta x^2) \right]}_{\text{truncation error}} = 0,
$$

$$
(11.10)
$$

where the first part is the original equation. This numerical scheme is unconditionally consistent with the partial differential equation because the truncation error vanishes regardless of how Δt and Δx approach zero.

A method is said to be convergent when the difference between the solutions of the differential and difference equations tends to zero as the time step goes to zero. It has been shown that a consistent method, if stable, is also convergent and vice versa. Consequently it is generally sufficient to check consistency and stability to ensure convergence. It is therefore indicated to examine the stability of numerical schemes.

11.1.2 Linear stability analysis

The stability of a difference method is concerned with the propagation of an error, introduced for example by inaccurate initial or boundary data or rounding in the numerical calculations. Such errors will be propagated by the difference method. If they do not grow, the method is called stable.

The linear stability analysis, also referred to as the von Neumann procedure, examines the property of the response of the finite-difference scheme to input perturbations written as a Fourier series in complex form as

$$h_j^k = \sum_{\tilde{n}=1}^{N/2} \zeta_{\tilde{n}}^k \, e^{\frac{i\tilde{n}2\pi j \Delta x}{L}}, \tag{11.11}$$

where N is the number of points per wavelength $L = N\Delta x$, $i = \sqrt{-1}$, and $\tilde{n}$ is the wave-number index. The complex function $e^{i\tilde{\alpha}j\tilde{n}}$ can be separated into a real and an imaginary part according to Euler's relation $e^{i\tilde{\alpha}j\tilde{n}} = \cos(\tilde{\alpha}j\tilde{n}) + i\sin(\tilde{\alpha}j\tilde{n})$. When viewed in the complex plane, the Fourier coefficients $\zeta_{\tilde{n}}^k$ exhibit the amplitude and the angle $\tilde{\alpha} = 2\pi/N$ represents a phase angle. The linear stability analysis method examines how each Fourier coefficient changes in time for any wave-number index $\tilde{n}$. Looking at any wave number (say $\tilde{n} = 1$), we obtain

$$h_j^k = \zeta^k e^{i\tilde{\alpha}j}. \tag{11.12a}$$

In the following example from Abbott and Basco (1989), the stability analysis of the algorithm in Eq. (11.8) is examined after h_j^{k+1} and h_{j-1}^k are defined from Eq. (11.12a) as

$$h_j^{k+1} = \zeta^{k+1} e^{i\tilde{\alpha}j}, \tag{11.12b}$$

$$h_{j-1}^k = \zeta^k e^{i\tilde{\alpha}(j-1)}. \tag{11.12c}$$

The terms in Eqs. (11.12) are substituted back into Eq. (11.8b):

$$\zeta^{k+1} e^{i\tilde{\alpha}j} = C_c \zeta^k e^{i\tilde{\alpha}(j-1)} + (1 - C_c) \zeta^k e^{i\tilde{\alpha}j}. \tag{11.13}$$

After canceling the common term $e^{i\tilde{\alpha}j}$, we obtain

$$\zeta^{k+1} = \frac{[(1 - C_c) + C_c e^{-i\tilde{\alpha}}] \zeta^k}{|\ \text{amplification factor } A_\alpha\ |} \tag{11.14}$$

The term in the brackets of Eq. (11.14) is the amplification factor A_α, which is a complex number. As sketched in Fig. 11.2(a), in the complex plane, the factor A_α is simply a circle of magnitude C_c centered at $1 - C_c$, and the modulus $|A_\alpha|$ determines whether the Fourier coefficient grows (when $|A_\alpha| > 1$), stays constant (when $|A_\alpha| = 1$), or decays (when $|A_\alpha| < 1$) as a function of time.

Of course, growth in the Fourier coefficients means that perturbations grow as the calculations progress in time, even when $\Delta x \to 0$ and $\Delta t \to 0$. Therefore a finite-difference scheme is stable as long as $|A_\alpha| \leq 1$. In this example, shown in Fig. 11.2, $A_\alpha = (1 - C_c + C_c \cos \tilde{\alpha}) + iC_c \sin \tilde{\alpha}$ and $|A_\alpha| = \sqrt{[1 - C_c(1 - \cos \tilde{\alpha})]^2 + C_c^2 \sin^2 \tilde{\alpha}}$. Numerical stability $|A_\alpha| \leq 1$ corresponds to $C_c \leq 1$, which is referred to as the Courant–Friedrich–Levy (CFL)

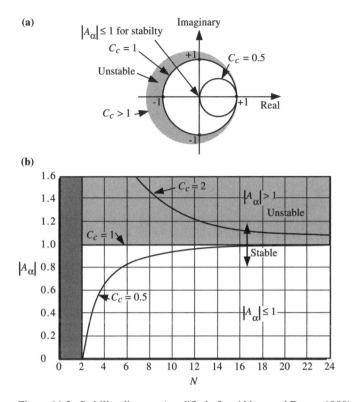

Figure 11.2. Stability diagram (modified after Abbott and Basco, 1989).

condition of stability for this numerical scheme. It is interesting to note that the CFL stability condition implies that the celerity c of the analytical solution must be less than the celerity $(\Delta x / \Delta t)$ of the numerical solution, or

$$C_c = \frac{c}{\Delta x / \Delta t} \leq 1. \tag{11.15a}$$

Accordingly, once the space size Δx has been determined, the time increment Δt that satisfies the stability requirement is specified as

$$\Delta t < \frac{\Delta x}{c}. \tag{11.15b} \blacklozenge\blacklozenge$$

The physical interpretation of the CFL condition is that the numerical time step Δt may not exceed the characteristic time step $\Delta x / c$. Otherwise, all the physical information does not have sufficient time to propagate to the next time step, and this will manifest itself as an instability. In practice, the CFL condition often imposes an important restriction on the time step because the spatial step

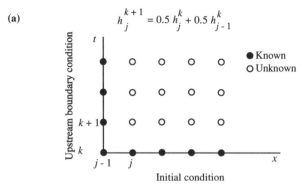

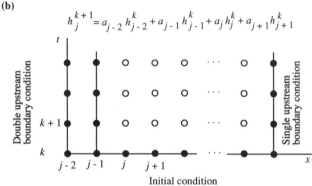

Figure 11.3. Examples of initial and boundary conditions.

Δx is generally determined by geometric considerations. This restriction has led many investigators to prefer implicit methods for several decades. With the development of very fast computers, explicit numerical schemes are nowadays gaining popularity.

11.1.3   Higher-order approximations

Finite differences with higher-order approximations are obtained by the introduction of differences at additional nodes in order to cancel out higher-order terms that would otherwise appear in the truncation error. For instance, a Taylor series expansion between nodes j and $j + 2$ gives

$$h^k_{j+2} = h^k_j + 2\Delta x \left(\frac{\partial h}{\partial x}\right)^k_j + \frac{4\Delta x^2}{2!}\left(\frac{\partial^2 h}{\partial x^2}\right)^k_j + \frac{8\Delta x^3}{3!}\left(\frac{\partial^3 h}{\partial x^3}\right)^k_j + 0(\Delta x^4).$$

$$(11.16)$$

The second derivative is approximated after Eq. (11.1) is doubled and subtracted from Eq. (11.16):

$$\left(\frac{\partial^2 h}{\partial x^2}\right)_j^k = \frac{h_{j+2}^k - 2h_{j+1}^k + h_j^k}{\Delta x^2} - \Delta x \left(\frac{\partial^3 h}{\partial x^3}\right)_j^k + 0(\Delta x^2). \quad (11.17)$$

The truncation error of this approximation is only first-order accurate. However, substituting Eq. (11.17) back into the right-hand side of Eq. (11.3) yields, after rearranging,

$$\left(\frac{\partial h}{\partial x}\right)_j^k = \frac{-h_{j+2}^k + 4h_{j+1}^k - 3h_j^k}{2\Delta x} + \left(\frac{\Delta x^2}{2!} - \frac{\Delta x^2}{3!}\right)\left(\frac{\partial^3 h}{\partial x^3}\right)_j^k \quad (11.18a)$$

or

$$\left(\frac{\partial h}{\partial x}\right)_j^k \simeq \frac{-h_{j+2}^k + 4h_{j+1}^k - 3h_j^k}{2\Delta x} + 0(\Delta x^2). \quad (11.18b)$$

As a result, the first derivative is turned into a second-order approximation after an additional grid point is added into the numerical scheme. Higher-order finite-difference approximations can be obtained in a similar fashion, and several useful finite-difference schemes of the first and the second derivatives are listed in Table 11.3. For instance, Relation (11.18b) corresponds to the second-order forward-difference scheme of the first derivative in Table 11.3.

Table 11.3. *First- and second-order finite-difference schemes*

	ϕ_{j-3}	ϕ_{j-2}	ϕ_{j-1}	ϕ_j		ϕ_j	ϕ_{j+1}	ϕ_{j+2}	ϕ_{j+3}
(a) Backward differences $0(\Delta x)$					(b) Forward differences $0(\Delta x)$				
$\Delta x \dfrac{\partial \phi}{\partial x} =$			-1	1	$\Delta x \dfrac{\partial \phi}{\partial x} =$	-1	1		
$\Delta x^2 \dfrac{\partial^2 \phi}{\partial x^2} =$		1	-2	1	$\Delta x^2 \dfrac{\partial^2 \phi}{\partial x^2} =$	1	-2	1	
(c) Backward differences $0(\Delta x)^2$					(d) Forward differences $0(\Delta x)^2$				
$2\Delta x \dfrac{\partial \phi}{\partial x} =$		1	-4	3	$2\Delta x \dfrac{\partial \phi}{\partial x} =$	-3	4	-1	
$\Delta x^2 \dfrac{\partial^2 \phi}{\partial x^2} =$	-1	4	-5	2	$\Delta x^2 \dfrac{\partial^2 \phi}{\partial x^2} =$	2	-5	4	-1
(e) Central differences $0(\Delta x^2)$									
$2\Delta x \dfrac{\partial \phi}{\partial x} =$			$-1\phi_{j-1} + 0\phi_j + 1\phi_{j+1}$						
$\Delta x^2 \dfrac{\partial^2 \phi}{\partial x^2} =$			$1\phi_{j-1} - 2\phi_j + 1\phi_{j+1}$						

11.1.4   Boundary conditions

Elliptic equations require that boundary conditions be specified over a completely closed boundary. The boundary data may consist of the value (Dirichlet type of boundary condition) or its normal derivative (Neumann type of boundary condition). The solution to hyperbolic and parabolic equations usually requires both initial and boundary conditions. Starting from given initial conditions at $t = 0$, a marching method finds the values at successive unknown time intervals from the boundary conditions.

The zone of influence of boundary conditions propagates through the domain at each time step. Lower-order approximations require simple boundary conditions. For instance, a scheme $h_j^{k+1} = 0.5\, h_{j-1}^k + 0.5\, h_j^k$ requires only the upstream boundary condition h_0^k and the initial condition h_j^0 for calculations over the entire domain. A downstream boundary condition is not required.

As sketched in Fig. 11.3, higher-order approximations require additional information about the boundary conditions. For instance, consider the scheme

$$h_j^{k+1} = a_{j-2} h_{j-2}^k + a_{j-1} h_{j-1}^k + a_j h_j^k + a_{j+1} h_{j+1}^k$$

with nonzero coefficients. Besides the initial condition at all nodes h^0, this algorithm requires a double upstream boundary condition at h_{-1} and h_0 and also requires a single downstream boundary condition because the term h_{j+1} implies that the information propagates in the upstream direction. The term h_{j+1} also implies that the information contained at the upstream boundary condition does not entirely propagate in the downstream direction. Indeed, some information contained in the upstream boundary condition propagates outside of the solution domain, in the upstream direction. This becomes particularly important in advection–dispersion problems in which conservation of mass is expected. Higher-order approximations thus provide improved calculation results at the expense of requiring detailed boundary conditions. Example 11.1 illustrates the concepts of numerical stability and convergence discussed in this section.

Example 11.1   Application to Wave Routing. Apply the finite-difference scheme that is FTBS from relations (11.7) and (11.8b) to wave routing at a constant celerity c at 2 m/s. Consider $\Delta x = 2,000$ m and examine calculations of the triangular wave propagation for $C_c = 0.5$, 1.0 and 2.0, as shown in Table E.11.1.1. Comparing the results at $t = 2,000$ s shows that the wave is correctly propagated when $C_c = 1.0$ whereas the peak is numerically decreased to half its value when $C_c = 0.5$ and increased to twice its value when $C_c = 2.0$.

Table E.11.1.1. *Wave propagation at* $\Delta x = 2,000$ *m*

Distance (km)	0	2	4	6	8	10	12
Initial depth (m)	0	0.5	1	0	0	0	0
Routing time (s)	$C_c = 0.5$ or $\Delta t = \dfrac{C_c \Delta x}{c} = 500$ s; $h_j^{k+1} = 0.5\,h_{j-1}^{k} + 0.5\,h_j^{k}$						
0	0	0.5	1	0	0	0	0
500	0	0.25	0.75	0.5	0	0	0
1,000	0	0.125	0.5	0.625	0.25	0	0
1,500	0	0.0625	0.3125	0.5625	0.4375	0.125	0
2,000	0	0.03125	0.1875	0.4375	0.50	0.28125	0.0625
Routing time (s)	$C_c = 1.0$ or $\Delta t = 1,000$ s; $h_j^{k+1} = h_{j-1}^{k}$						
0	0	0.5	1	0	0	0	0
1,000	0	0	0.5	1	0	0	0
2,000	0	0	0	0.5	1	0	0
Routing time (s)	$C_c = 2.0$ or $\Delta t = 2,000$ s; $h_j^{k+1} = 2h_{j-1}^{k} - h_j^{k}$						
0	0	0.5	1	0	0	0	0
2,000	0	−0.5	0	2	0	0	0
4,000	0	0.5	−1	−2	4	0	0

The numerical scheme is stable when $C_c \leq 1$ and unstable when $C_c > 1$, and there is numerical diffusion when $C_c < 1$. The convergence of this numerical scheme is examined at $C_c = 0.5$ and $C_c = 2.0$ after the space increment $\Delta x = 1,000$ m is halved, as shown in Table E.11.1.2. It is most instructive to observe that in the case of $C_c = 0.5$, halving Δx results in a better approximation of the peak value than that calculated in Table E.11.1.1. Specifically, after $t = 2,000$ s, the peak value of h_{max} calculated with $\Delta x = 1$ km is 0.61 m compared with $h_{max} = 0.5$ at $\Delta x = 2$ km from a correct value of $h_{max} = 1$ m. The improvement shows that the truncation error decreases as $\Delta x \to 0$, which is a simple example of convergence in which accuracy is gained at the expense of more extensive calculations. In comparison, the case with $C_c = 2$ shows that decreasing Δx by one-half does not increase the accuracy of the calculation. Indeed, the maximum flow depth h_{max} calculated at $t = 4,000$ s is 8 m for $\Delta x = 1,000$ m compared with $h_{max} = 4$ m at the same time when $\Delta x = 2,000$ m. This illustrates the fact that decreasing Δx and Δt does not necessarily improve convergence, even if the numerical scheme is consistent. A consistent numerical scheme can be convergent only when it is stable.

Table E.11.1.2. *Wave propagation at* $\Delta x = 1,000\ m$

Distance (km)	0	1	2	3	4	5	6	7	8	9	10
Initial depth (m)	0	0.25	0.5	0.75	1	0.5	0	0	0	0	0
Routing times (s)	$C_c = 0.5$ or $\Delta t = 250$ s, $h_j^{k+1} = 0.5\,h_{j-1}^k + 0.5\,h_j^k$										
0	0	0.25	0.5	0.75	1	0.5	0	0	0	0	0
250	0	0.125	0.375	0.625	0.875	0.75	0.25	0	0	0	0
500	0	0.0625	0.025	0.5	0.75	0.81	0.5	0.125	0	0	0
750	0	0.031	0.156	0.375	0.625	0.78	0.656	0.31	0.06	0	0
1,000	0	0.016	0.094	0.266	0.5	0.7	0.72	0.484	0.19	0.03	0
1,250	0	0.008	0.0547	0.18	0.383	0.6	0.71	0.6	0.336	0.11	0.01
1,500	0	0.004	0.031	0.127	0.28	0.49	0.655	0.655	0.468	0.22	0.06
1,750	0	0.002	0.016	0.07	0.20	0.39	0.57	0.655	0.56	0.344	0.14
2,000	0	0.001	0.009	0.05	0.145	0.29	0.46	0.61	0.61	0.45	0.24
Routing times (s)	$C_c = 2.0$ or $\Delta t = 1000$ s, $h_j^{k+1} = 2\,h_{j-1}^k - h_j^k$										
0	0	0.25	0.5	0.75	1	0.5	0	0	0	0	0
1,000	0	−0.25	0	0.25	0.5	1.5	1	0	0	0	0
2,000	0	+0.25	−0.5	−0.25	0	−0.5	2	2	2	0	0
3,000	0	−0.25	1	−0.75	−0.5	0.5	−3	2	4	0	0
4,000	0	+0.25	−1.5	2.75	−1	−1.5	4	−8	0	8	0

11.2   One-dimensional river models

There are numerous one-dimensional models available for the simulation of steady backwater flow, unsteady floodwave propagation, advection–dispersion of sediment and contaminants, and aggradation–degradation in alluvial rivers. The foregoing presentation focuses on a few numerical schemes: (1) explicit schemes in Subsection 11.2.1; (2) the Leonard scheme in Subsection 11.2.2; (3) the MacCormack scheme in Subsection 11.2.3, an explicit predictor–corrector scheme; and (4) the Preissmann scheme in Subsection 11.2.4, an implicit scheme. Subsection 11.2.5 discusses aggradation–degradation simulations in alluvial rivers.

11.2.1   Explicit scheme

Consider the combination of advection and dispersion as a mechanism for the transport of sediment, or contaminant, at a concentration ϕ. The substance is said to be conservative in the sense that the total mass remains constant and therefore without sedimentation or chemical reaction that would cause the decay of the substance; the 1D advection–dispersion equation is written as

$$\frac{\partial \phi}{\partial t} + v \frac{\partial \phi}{\partial x} = K_d \frac{\partial^2 \phi}{\partial x^2}, \tag{11.19}$$

where v is the mean flow velocity in the downstream x direction and $K_d \simeq 250\, hu_*$ is the dispersion coefficient. Note that a similar approach could be used for turbulent diffusion; however, it is assumed here that the substance is well mixed and the concentration is uniform at a given cross section. It can also be noted that when K_d is very small, Eq. (11.19) becomes quite similar to the floodwave propagation analysis of Section 11.1, as described by Eq. (11.6).

The mathematical interest in Eq. (11.19) arises from the fact that the equation is a hybrid between a hyperbolic equation when $K_d = 0$ and a parabolic equation when $v = 0$. In rivers, the flow velocity is usually important and the system should be centered around a solution to hyperbolic equations, and the numerical scheme developed in Section 11.1 can serve as a basis for further analysis. The FTBS scheme can thus be adopted for the advective term and a second-order central difference approximation used for the dispersion term. The resulting finite-difference scheme is written as

$$\frac{\phi_j^{k+1} - \phi_j^k}{\Delta t} + v \frac{\phi_j^k - \phi_{j-1}^k}{\Delta x} = K_d \frac{\phi_{j+1}^k - 2\phi_j^k + \phi_{j-1}^k}{\Delta x^2}. \tag{11.20}$$

The terms are rearranged in an explicit form after $C_u = v \Delta t / \Delta x$ and $C_k = K_d \Delta t / \Delta x^2$, are defined; thus

$$\phi_j^{k+1} = C_k \phi_{j+1}^k + (1 - C_u - 2C_k) \phi_j^k + (C_u + C_k) \phi_{j-1}^k. \qquad (11.21)$$

Although the dispersion term is a second-order approximation, the advection term is approximate to only the first order. The truncation error contains terms on a higher order than the advection term and may thus induce numerical dispersion. Indeed, with reference to the consistency analysis in Subsection 11.1.2, advection equation (11.10) can be rewritten, after ϕ is substituted for h and v for c, as

$$\left(\frac{\partial \phi}{\partial t}\right)_j^k + v \left(\frac{\partial \phi}{\partial x}\right)_j^k + \left[\frac{\Delta t}{2!} \left(\frac{\partial^2 \phi}{\partial t^2}\right)_j^k - v \frac{\Delta x}{2!} \left(\frac{\partial^2 \phi}{\partial x^2}\right)_j^k + 0(\Delta t^2) + 0(\Delta x^2)\right] = 0.$$

$$(11.22)$$

The advection scheme can be written as

$$\frac{\partial \phi}{\partial t} = -v \frac{\partial \phi}{\partial x} + 0(\Delta t, \Delta x). \qquad (11.23)$$

After taking space and time derivatives with constant v, we obtain

$$\frac{\partial^2 \phi}{\partial t \partial x} = -v \frac{\partial^2 \phi}{\partial x^2} + 0(\Delta t, \Delta x), \qquad (11.24a)$$

$$\frac{\partial^2 \phi}{\partial t^2} = -v \frac{\partial^2 \phi}{\partial x \partial t} + 0(\Delta t, \Delta x), \qquad (11.24b)$$

which can be recombined to obtain the wave equation

$$\frac{\partial^2 \phi}{\partial t^2} = v^2 \frac{\partial^2 \phi}{\partial x^2} + 0(\Delta t, \Delta x). \qquad (11.25)$$

Substituting Eq. (11.25) back into Eq. (11.22) gives

$$\underbrace{\left(\frac{\partial \phi}{\partial t}\right)_j^k + v \left(\frac{\partial \phi}{\partial x}\right)_j^k}_{\text{advection term}} = \underbrace{\left(\frac{-v^2 \Delta t}{2} + \frac{v \Delta x}{2}\right) \left(\frac{\partial^2 \phi}{\partial x^2}\right)_j^k + 0(\Delta t^2, \Delta x^2)}_{\text{truncation error}}.$$

$$(11.26)$$

We obtain the fact that the truncation error of the advection scheme includes a numerical diffusion term, which, with $C_u \Delta x = v \Delta t$, can be rearranged as

$$K_{\text{num}} = \frac{-v^2 \Delta t}{2} + \frac{v \Delta x}{2} = \frac{v \Delta x}{2} (1 - C_u). \qquad (11.27)$$

It is now becoming clear that the proposed scheme in Eq. (11.20) really solves the following equation:

$$\frac{\partial \phi}{\partial t} + \frac{v \partial \phi}{\partial x} = (K_d + K_{num})\frac{\partial^2 \phi}{\partial x^2}. \tag{11.28}$$

The numerical dispersion determined in Eq. (11.27) vanishes as C_u approaches unity, hence the interest in running computer models with values of C_u close to unity. This is a challenge, as it has been shown that this numerical scheme becomes unstable as $C_u > 1$. This now explains why the simulations in Example 11.1 were flawless when $C_c = 1$, whereas diffusion was observed for $C_c = 0.5$.

The grid Peclet number $P_\Delta = C_u/C_k = v\Delta x/K_d$ is a measure of the ratio between advection $C_u = v\Delta t/\Delta x$ and dispersion $C_k = K_d\Delta t/\Delta x^2$. As dispersion gradually tends toward a normal distribution, the standard deviation σ_d increases with time Δt as $\sigma_d = \sqrt{2K_d\Delta t}$, and 95% of the dispersed material is contained in a plume of length $\pm 2\sigma_d$. When $C_u = 1$, the advection length is simply given by $\Delta x = v\Delta t$, the grid Peclet number then corresponds to $2\Delta x^2/\sigma_d^2$. The physical significance, as sketched in Fig. 11.4, is that $\Delta x = \sigma_d$

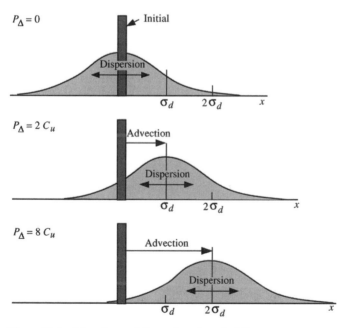

Figure 11.4. Advection and dispersion characteristics.

when $P_\Delta = 2C_u$, or $C_k = 1/2$, and dispersion is dominant when $P_\Delta < 2C_u$. A substance cannot propagate upstream when $\Delta x > 2\sigma_d$, which corresponds to $P_\Delta = 8C_u$, or $C_k = 1/8$; dispersion is dominant when $P_\Delta > 8C_u$. The physical dispersion is sufficiently large to counteract local convection instabilities in regions of large velocity gradients when $P_\Delta \leq 2C_u$.

In practice, modelers must guard against simulations in which $K_{num} \gg K_d$. For instance, consider a model with $v = 1$ m/s, $K_d = 100$ m^2/s, $\Delta x = 2,000$ m, and $\Delta t = 1,000$ s. This model is stable, $C_u = v\Delta t/\Delta x = 0.5 < 1$, and essentially advective, $C_k = K_d\Delta t/\Delta x^2 = 0.025 \ll C_u$. However, $K_{num} = (v\Delta x/2)(1 - C_u) = 500$ is five times larger than the physical dispersion K_d, and the model is physically meaningless as far as the dispersion calculations are concerned. A better approach would be to use, for instance, half the original K_d value, or $K_d/2 = 50$ m^2/s, and set $K_{num} = K_{d/2}$ to determine the grid spacing Δx and time increment Δt as functions of C_u and K_d as

$$\Delta x = K_d/v(1 - C_u), \tag{11.29a}$$
$$\Delta t = K_d C_u/v^2(1 - C_u). \tag{11.29b}$$

For instance, modeling this case with $v = 1$ m/s and $K_d = 100$ m^2/s would require setting the model for $C_u = 0.9$ with $\Delta x = 1,000$ m and $\Delta t = 900$ s to obtain $K_{num} = 50$ m^2/s. A finer grid spacing Δx would be required for reducing K_{num}. It is also clear from Eqs. (11.29) that both Δx and Δt become excessively small when K_d decreases. This approach thus becomes quite restrictive when we are solving for diffusion problems (lower K_d values). Also, when this algorithm is applied to rivers with variable flow velocity, the most restrictive conditions on Δx and Δt are obtained when v is large. This practically means that explicit models are most sensitive to instabilities where and when v is the largest, i.e., near the peak discharge of a flood hydrograph. Peak flood conditions should therefore be used to determine the maximum velocity used in the calculations of Δx and Δt.

11.2.2   Leonard scheme

Leonard (1979) developed a third-order approximation to solve advection–dispersion equation (11.19). The algorithm eliminates the numerical diffusion term contained in the truncation error. Canceling higher-order space derivative terms provides a higher accuracy for the advection term. The explicit algorithm for ϕ_j^{k-1} is given, without derivation, as a function of $C_u = v\Delta t/\Delta x$ and

$C_k = K_d \Delta t / \Delta x^2$, and values of several modes at the known level ϕ^k:

$$\phi_j^{k+1} = \phi_j^k + \left[C_k(1 - C_u) - \frac{C_u}{6} \left(C_u^2 - 3\,C_u + 2 \right) \right] \phi_{j+1}^k$$

$$- \left[C_k(2 - 3C_u) - \frac{C_u}{2} \left(C_u^2 - 2\,C_u - 1 \right) \right] \phi_j^k$$

$$+ \left[C_k(1 - 3\,C_u) - \frac{C_u}{2} \left(C_u^2 - C_u - 2 \right) \right] \phi_{j-1}^k$$

$$+ \left[C_k\,(C_u) + \frac{C_u}{6} \left(C_u^2 - 1 \right) \right] \phi_{j-2}^k .$$

(11.30)

The linear stability analysis shows a wide stability range, as in Fig. 11.5. For practical purposes, it is generally sufficient to know that the algorithm is stable

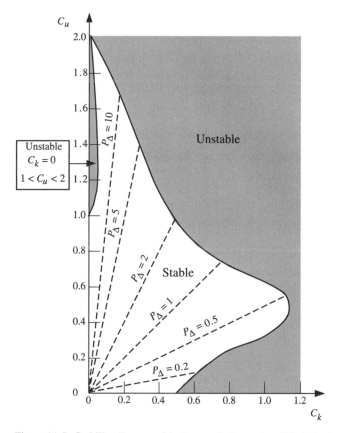

Figure 11.5. Stability diagram of the Leonard scheme (modified after Leonard, 1979).

for $C_u < 1$ and $C_k < 0.4$. We can therefore determine Δx and Δt as functions of K_d and v as

$$\Delta t = \frac{K_d \, C_u^2}{v^2 \, C_k}, \tag{11.31a}$$

$$\Delta x = \frac{K_d \, C_u}{v \, C_k}. \tag{11.31b}$$

For given K_d and v, the values of Δx and Δt are proportional to C_u and inversely proportional to C_k. For instance, when $C_u = 1$ and $C_k = 0.1$, $\Delta t = 10 \, K_d/v^2$, $\Delta x = 10 \, K_d/v$, and $P_\Delta = 10$. The coefficients of Eqs. (11.30) reduce to

$$\phi_j^{k+1} = 0.1 \, \phi_j^k + 0.8 \, \phi_{j-1}^k + 0.1 \, \phi_{j-2}^k. \tag{11.32a}$$

This algorithm should be convenient as long as the changes in ϕ are relatively gradual. Small oscillations may otherwise be amplified because the grid Peclet number $P_\Delta = 10$. To ensure a numerically smooth simulation without oscillations, a value of $P_\Delta = C_u/C_k = 2$ may provide better results, for instance, $C_u = 0.8$ and $C_k = 0.4$. The corresponding grid spacing $\Delta x = 2K_d/v$ and time step $\Delta t = 1.6 \, K_d/v^2$ are smaller and the coefficients of Eq. (11.30) reduce to

$$\phi_j^{k+1} = 0.048 \, \phi_{j+1}^k + 0.376 \, \phi_j^k + 0.304 \, \phi_{j-1}^k + 0.272 \, \phi_{j-2}^k. \tag{11.32b}$$

In the case of pure advection ($K_d = C_k = 0$), the algorithm is stable for $C_u < 1$, which simply imposes $\Delta t < \Delta x/v$ regardless of Δx; this is quite convenient as long as v remains fairly constant within Δx.

Example 11.2   Application to Advection–Dispersion. Consider advection and dispersion in a steep and very rough mountain channel with a slope $S = 4 \times 10^{-3}$. The flow depth h is 3 m, the mean flow velocity is $V = 2$ m/s, and the dispersion coefficient $K_d \simeq 250 \, hu_* \simeq 257$ m^2/s. At the upstream end of a 30-km reach, very fine sediment is released at a concentration of 100,000 mg/l for 15 min with a 3-min pulse at 200,000 mg/l 1 h after the initial release. Consider that the reach is fairly uniform and that the total sediment mass is preserved at all times. Determine the maximum concentration at a distance of 10 km downstream of the release. How long will the concentration exceed 10,000 mg/l at a point located 20 km downstream of the release?

A simple finite-difference model can be used. For instance, with $\Delta t = 180$ s $= 0.05$ h and $\Delta x = 500$ m, the Courant number $C_u = V \, \Delta t/\Delta x = 0.72$

and the model should be stable because $C_u < 1$. The dispersion number $C_k = K_d \Delta t / \Delta x^2 = 0.185$, and the grid Peclet number is $P_\Delta = 3.86$. The user of the Leonard method should check that the coefficients $a_{j-2} = 0.075607$, $a_{j-1} = 0.577656$, $a_j = 0.337868$, and $a_{j+1} = 0.008869$ sum up to unity and that none is negative. The initial sediment concentration can be set at 0, and the upstream boundary condition set as 1×10^5 for the first five time steps and 2×10^5 at the 20th time step; all other values are zero.

These boundary conditions are not sufficient for using the Leonard scheme. An additional upstream boundary condition is required because the algorithm includes both ϕ_{j-1} and ϕ_{j-2}. Additionally, the downstream boundary condition must be specified because the algorithm requires ϕ_{j+1}. To compare the total mass, the second upstream boundary condition could satisfy the requirement that no sediment propagates upstream from the point of release. At the

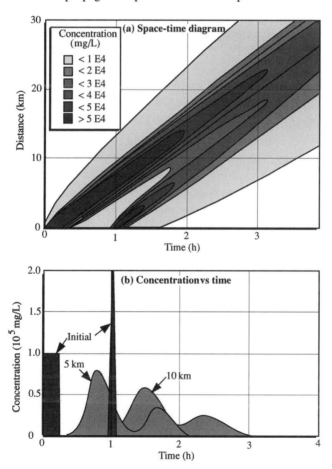

Figure E.11.2.1. Advection–dispersion calculation example.

downstream end, advection is dominant and the concentration could be set identical to the value calculated one grid space upstream at the previous time step. The use of the numerical scheme in Subsection 11.2.1 is comparatively simpler because only one upstream boundary condition must be specified. The numerical results shown in Fig. E.11.2.1 indicate that the second sediment pulse disperses very rapidly and produces a second peak with lower sediment concentration. The maximum concentration 10 km downstream is 61,000 mg/l at a time of 1 h 27 min after release. The Leonard algorithm gives C_{max} at 59,000 mg/l at 1 h 30 min. The results of both algorithms are essentially identical 20 km downstream, and the concentration exceeds 10,000 mg/l between 2 h 25 min until 4 h after release.

11.2.3   MacCormack scheme

The MacCormack scheme is an explicit, two-step predictor–corrector scheme (MacCormack, 1969) that is second-order accurate in both space and time and is capable of capturing the shocks without isolating them. This scheme has been applied for analyzing 1D unsteady open-channel flows by Fennema and Chaudhry (1986).

For 1D flow, two alternatives of this scheme are possible: (1) backward finite differences are used in the predictor part and forward differences are in the corrector part; or (2) forward finite differences are used in the predictor part and backward differences in the corrector part. It is possible to alternate the direction of differencing from one time step to the next. Better results are produced if the direction of differencing in the predictor step is the same as that of the movement of the wave front.

The finite-difference approximations for the first alternative are given below. The predictor algorithm is

$$
\frac{\partial U}{\partial t} \simeq \frac{U_j^* - U_j^k}{\Delta t},
$$
$$
\frac{\partial F}{\partial x} \simeq \frac{F_j^k - F_{j-1}^k}{\Delta x},
$$

(11.33)

in which the superscript * refers to variable U computed during the predictor part for the unknown level $k + 1$.

Substitution of these finite differences into the governing equation in the form $\partial U/\partial t + \partial F/\partial x + R = 0$ and simplification of the resulting equation yield

$$
U_j^* = U_j^k - \frac{\Delta t}{\Delta x}\left(F_j^k - F_{j-1}^k\right) - R_j^k\,\Delta t.
$$

(11.34)

In the continuity equation, the computed value of U_j^* corresponds to flow depth h^* at the level $k + 1$, from which we determine the values of cross-section area A^*, discharge Q^*, and velocity U^* at the unknown level $k + 1$. These values are then used to compute F^* and R^*. The corrector algorithm is implicit in space, as

$$\frac{\partial U}{\partial t} = \frac{U_j^{**} - U_j^k}{\Delta t},$$

(11.35)

$$\frac{\partial F}{\partial x} = \frac{F_{j+1}^* - F_j^*}{\Delta x}.$$

(11.36)

Substituting these finite differences and $R = R_j^*$ into the governing equation yields

$$U_j^{**} = U_j^k - \frac{\Delta t}{\Delta x}(F_{j+1}^* - F_j^*) - R_j^* \,\Delta t,$$

(11.36a)

where the superscript ** refers to the values of the variables after the corrector step. The value of U_j^{k+1} at the unknown time level $k + 1$ is finally given by

$$U_j^{k+1} = \frac{1}{2}(U_j^* + U_j^{**}),$$

(11.36b)

from which all other parameters can be determined at the level $k + 1$. The MacCormack scheme is stable if the CFL condition is satisfied.

11.2.4 Preissmann scheme

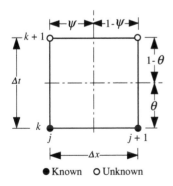

● Known   ○ Unknown

Figure 11.6. Definition sketch of the Preissmann scheme.

The Preissmann scheme has been used since the early 1960s (Preissmann and Cunge, 1961; Liggett and Cunge, 1975). It offers the advantage that a variable spatial grid may be used; steep wave fronts may be properly simulated by varying the weighting coefficients θ and ψ, and the scheme yields an exact solution of the linearized form of the governing equations for a particular value of Δx and Δt.

The partial derivatives are approximated as sketched in Fig. 11.6 with the following weighted four-point scheme

$$
\frac{\partial h}{\partial t} \simeq (1 - \psi)\frac{h_j^{k+1} - h_j^k}{\Delta t} + \psi\frac{h_{j+1}^{k+1} - h_{j+1}^k}{\Delta t}, \tag{11.37a}
$$

$$
\frac{\partial h}{\partial x} \simeq (1 - \theta)\frac{h_{j+1}^k - h_j^k}{\Delta x} + \theta\frac{h_{j+1}^{k+1} - h_j^{k+1}}{\Delta x}, \tag{11.37b}
$$

in which $0 \leq \theta \leq 1$ and $0 \leq \psi \leq 1$ are weighting coefficients. In the partial derivatives, h refers to either velocity V or flow depth h. By selection of a suitable value for θ, the scheme may be made totally explicit ($\theta = 0$) or implicit ($\theta = 1$). The scheme is stable if $0.5 \leq \theta \leq 1$. The truncation errors are of the order of Δt^2 and Δx^2 when $\psi = \theta = 0.5$. Taking $\theta = 0.5$ gives a scheme that is stable without numerical diffusion. Taking values $\theta > 0.5$ introduces truncation errors that produce numerical diffusion. Steep wave fronts are properly simulated for low values of θ, but there are oscillations behind the wave front. These oscillations are eliminated for θ close to unity; however, steep wave fronts are somewhat smeared. For typical applications, $\theta = 0.6$–0.7 may be used.

By substituting these finite-difference approximations into the St. Venant equation, we have $2N$ equations (N is the number of reaches on the channel). We cannot write these equations for the grid points at the downstream end. However, we have $2(N + 1)$ unknowns, i.e., two unknowns for each grid point. Thus, at each time step for a unique solution, we need two more equations. These are provided by two boundary conditions.

Boundary conditions at the upstream and the downstream extremities of the routing reach must be specified in order to obtain solutions to the St. Venant equations. In fact, in most unsteady-flow applications, the unsteady disturbance is introduced at one or both of the external boundaries. At the upstream boundary, either a specified discharge or water-surface elevation time series (hydrograph) can be used. The hydrograph should not be affected by downstream flow conditions. At the downstream boundary, specified discharge or water-surface elevation time series or a tabular relation between discharge and water-surface elevation (single-valued rating curve) can be used. Another downstream boundary condition can be a loop-rating curve based on the Manning equation. The loop is produced with the friction slope S_f rather than the channel bottom slope S_0.

The Preissmann scheme produces a pentadiagonal banded matrix. The set of nonlinear algebraic equations may be solved by an iterative technique. The solution of $2N \times 2N$ simultaneous equations requires an efficient technique such as the double-sweep elimination procedure. Algorithms for subcritical and supercritical flows with velocity or flow depth as the boundary condition are given in Abbott and Basco (1989). The lengthy procedure is beyond the scope of this chapter.

11.2.5 Aggradation–degradation scheme

The analysis of bed-elevation changes in 1D channels involves the combined effects of water-surface calculations and of changes in bed elevation through the sediment continuity relationship. The stability of explicit schemes depends not only on the hydrodynamic conditions, but also on the type of sediment-transport relationship. To illustrate this point, consider a steady 1D flow of unit discharge q in a rectangular canal. Assume a slight perturbation of the bed elevation at point j, as sketched in Fig. 11.7, and let us calculate the change in bed elevation Δz by using a backward difference of the sediment continuity relationship:

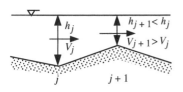

Figure 11.7. Sketch of riverbed.

$$\Delta z_{j+1} = -\frac{T_E}{(1 - p_0)}\frac{(q_{sj+1} - q_{sj})\Delta t_s}{\Delta x}. \tag{11.38}$$

where $T_E = 1 - \exp[(-\Delta x\omega)/q]$ is the trap efficiency, p_0 is the porosity of the bed material, ω is the settling velocity, q is the unit discharge, Δx is the grid spacing, and Δt_s is the time increment for sediment. In practice, the grid spacing is usually sufficiently long that $T_E \simeq 1$. Also, the porosity is usually assumed constant.

The unit sediment discharge q_s is the subject of discussion here because $h_j > h_{j+1}$ and $V_j < V_{j+1}$. Therefore, if the sediment-transport relationship is proportional to the flow depth, $q_s \sim h^b$, then $q_{sj} > q_{sj+1}$ and $\Delta z_{j+1} > 0$ cause aggradation at node $j + 1$ and the scheme is unstable. Conversely, the scheme is stable (as long as Δt_s is reasonable) if $q_s \sim V^b$ because $q_{sj} < q_{sj+1}$ results in degradation. Algorithms based on shear stress, $q_s \sim a\tau^b = a(hS)^b$, can be equivalent to the flow-depth algorithm if the slope is taken as a reach average water-surface slope.

Opposite results in terms of scheme stability are obtained if the backward difference Δz_{j+1} is replaced with a forward difference Δz_j. Also, the case of supercritical flow would also yield different results. The stability of aggradation–degradation algorithms depends not only on the stability of the hydrodynamic scheme but also depends largely on the sediment-transport relationship. In practice, it is often preferable to calculate Δz from Eq. (11.38) and split Δz as $\Delta z_j = \alpha \Delta z$ and $\Delta z_{j+1}(1 - \alpha)\Delta z$, with $0 < \alpha < 1$ as a weighting factor between forward and backward differences.

The time increment for sediment Δt_s is different than the time increment Δt for hydrodynamic calculations. It is clear from Eq. (11.38) that the aggradation–degradation increment Δz is linearly proportional to Δt_s. When the change in bed elevation Δz during $\Delta t_s = \Delta t$ is small compared with the flow depth, the assumption of a rigid boundary calculation for hydrodynamic calculations during Δt is justified. The aggradation–degradation calculations can be performed independently, and the equations are said to be uncoupled. The use of $\Delta t_s > \Delta t$ is thus possible as long as Δz is small compared with h. Conversely, when Δz is large compared with the flow depth when $\Delta t_s = \Delta t$, the sediment and the hydrodynamic equations have to be solved simultaneously, and the equations are said to be coupled. Fortunately, bed-elevation changes are usually sufficiently small during floods to use uncoupled formulations.

An example of aggradation–degradation calculations is shown in Fig. 11.8. The formulation is uncoupled, and the unit discharge is constant throughout the reach. The upstream sediment discharge depends on the local upstream slope, and the upstream bed elevation is assumed to be fixed. The sand-transport capacity is used in the calculations. Over time, aggradation takes place as sediment transport decreases in the downstream direction. The reach becomes increasingly uniform with a new slope that is several times less than the initial reach slope. The sediment-transport capacity also gradually becomes fairly uniform with slightly decreasing unit sediment discharge in the downstream direction. This numerically illustrates the fact that alluvial rivers gradually tend to become quite uniform in the downstream direction with very gradual changes in depth, velocity, slope, and sediment-transport characteristics.

11.3   Multidimensional river models

2D depth-integrated models have been applied to predict surface runoff and sediment-transport rates. A quasi-steady approach can often be used, although some 2D unsteady-flow models are available. A complete description of multidimensional models is beyond the scope of this chapter. A significant number of 2D and 3D codes are commercially available, and some are readily available in the public domain. The fast development of computers make new numerical solutions possible to river engineering problems of increasing complexity. In many cases in which the vertical variation in flow velocity and turbulence are of little interest, vertically averaged horizontal 2D models can be used.

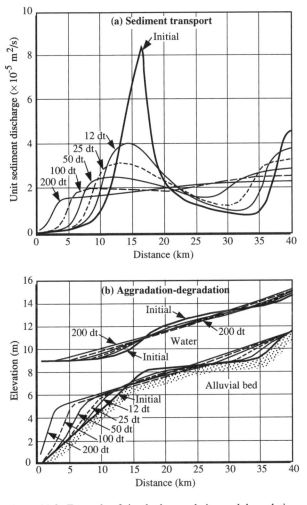

Figure 11.8. Example of riverbed aggradation and degradation.

The following assumptions are usually considered in depth-integrated 2D models: (1) the radius of curvature is usually much larger than the channel width and the mildly curved channel approximation is appropriate; (2) the shallow-water approximation is appropriate, and this assumption is particularly valid when the width–depth ratio of meandering or braiding streams is large enough to neglect sidewall effects; (3) a 2D horizontal flow is considered, the vertical velocity component is disregarded, and this type of model cannot account for secondary currents in bends; (4) hydrostatic pressure distribution is assumed;

(5) the wavelength of the bed deformations is assumed to be longer than the wavelength of ripples, dunes, or antidunes, and bedforms are considered only as roughness elements; (6) the spatial variation in the hydraulic roughness can be neglected; (7) the influence of grain sorting is insignificant, and uniform bed material is considered; and (8) the celerity of the bed disturbances is small, the model assumes a rigid boundary, and hydrodynamic calculations are uncoupled with the aggradation–degradation calculations.

2D models usually solve the continuity and the momentum equations after an empirical relationship is assumed for resistance to flow (e.g., Manning or Chézy). The calculations can be performed in a raster-based or vector-based format, depending on the GIS database available for the calculations. Finite-difference schemes are well suited to raster-based data whereas the finite-element method seems more appropriate for vector-based data and complex geometries. The calculations are repeated at successive time steps for unsteady-flow calculations. Sediment-transport calculations can effectively yield aggradation and degradation results with uncoupled formulations. Examples of models include CASC2D for surface runoff simulation at the watershed scale (e.g., Julien et al., 1995) and FLO-2D for mudflow simulation, typically on alluvial fans (e.g., O'Brien et al., 1993).

Quasi-3D flow models simulate depth-averaged mathematical flow models in combination with the depth-integrated velocity profile. The continuity and the momentum equations are solved with empirical logarithmic velocity profiles. The bedload transport rate in the transverse direction is calculated with empirical formulas. The depth-averaged steady-state equation for the suspended load is described in an orthogonal, curvilinear coordinate system. Models can solve the sediment continuity equation for the 2D bed and suspended sediment transport. Once the depth-averaged 2D velocity calculations are completed, standard logarithmic velocity profiles are considered to determine the vertical velocity profile. Because the model calculations are in two dimensions, secondary currents in river bends cannot be properly simulated with quasi-3D models.

3D models are generally steady-state models used for turbulent-flow simulations. In κ-ε models, the state of turbulence is characterized by the energy and dissipation parameters κ and ε. 3D models typically solve the depth-averaged Reynolds approximation of the momentum equation for velocity. The depth-averaged mass conservation determines the water-surface elevation. The deviation from the depth-averaged velocity is computed for each cell by the solution of the conservation of mass equation in conjunction with a κ-ε closure for vertical momentum diffusion. Sedimentation computations are based on 2D solid mass conservation for the channel bed and the exchange of sediment between bedload and suspended load.

Data required for running multidimensional models include: (1) channel geometry with cross sections; (2) upstream/downstream boundary conditions in terms of discharge and stage as functions of time for unsteady-flow models and flow-velocity profiles for 3D models; (3) particle-size distribution of the bed material; (4) upstream/downstream sediment load (some models require both bedload and suspended load); and (5) suspended sediment concentration profiles for 3D models. The data requirements increase with the number of dimensions in the model. Some model features may require data that are not available, and many times assumptions must be made regarding missing data. For instance, some models can calculate sediment transport by size fractions, and sediment data of the bed material and boundary conditions may not be available for each size fraction.

Models may not necessarily handle the data in the most appropriate manner. For instance, some κ-ε should account for the turbulence generated behind bedforms. Lumped values of κ and ε will be assumed in the model even if longitudinal profile data showing bedforms are available.

Steady 3D models are applied to estimate the initial rate of sedimentation and erosion in a given situation. The reason for this is the vast computer time required for stabilizing the models under steady-state conditions. To run the model over long time periods under different flow conditions to determine aggradation and degradation would be prohibitive. The initial models provide good insight into the short-term effects of a proposed structure (channel diversion, new harbor, closure of a channel, etc.). However, they are of limited value for long-term simulation. It is usually preferable to run long-term 1D simulations in parallel to gain basic knowledge of morphological processes and long-term changes to be expected at the site. Case Study 11.1 illustrates how various models can be used to solve complex river engineering problems.

Case Study 11.1   Lower Mississippi River Sediment Study, United States. The diversion of water and sediment from the Mississippi River into the Atchafalaya River has been closely monitored with physical and mathematical model studies (USACE, 1999). Models of the flow and sediment diversion at the Old River Control Complex required extensive field and laboratory measurements. Among the data needs at the site, the bathymetry and daily water and sediment-discharge records over a period of 50 yr served to

380   *Mathematical river models*

calibrate and test 1D HEC-6 models of the Mississippi and Atachafalaya Rivers. A 3D model, CH3D-SED, also required more accurate field measurements of velocity profiles with acoustic Doppler current profilers, and sediment concentration profiles by size fractions, measured with a P-63 at discharges of ~1,000,000 ft³/s.

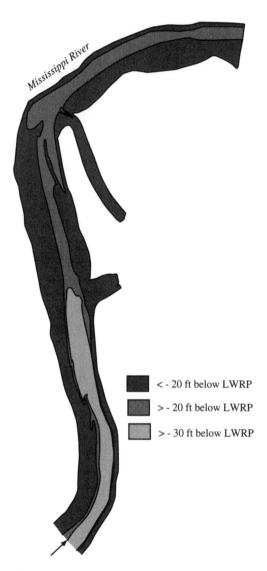

< - 20 ft below LWRP

> - 20 ft below LWRP

> - 30 ft below LWRP

Figure CS.11.1.1. Mississippi riverbed nead the Old River Control Complex.

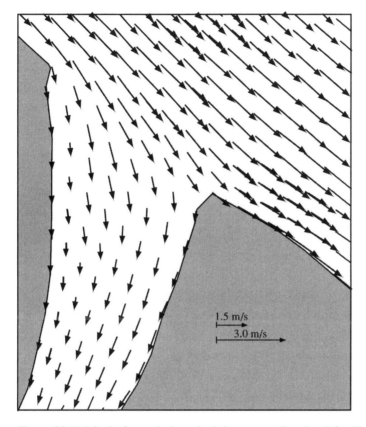

Figure CS.11.1.2. Surface-velocity calculation near a diversion (after U.S. Army Corps of Engineers, 1999).

Selected results are shown here to illustrate the type of information that can be gained from physical and mathematical models. Figure CS.11.1.1 shows two results from the physical model. The position of the thalweg is extremely important for navigation purposes. Comparing the results illustrates the sediment accumulation and the nonuniform thalweg depths downstream of sharp bends as opposed to the more uniform thalweg depths obtained for gently curved channels. The surface flow velocities calculated with the model CH3D-SED are shown in Fig. CS.11.1.2 and, as expected, maximum flow velocities can be found near the concave bank of the diversion channel. The sediment concentration of fine sand in suspension is shown in Fig. CS.11.1.3. The higher sediment concentrations of fine sand on the point bars are the result of secondary flows, as expected.

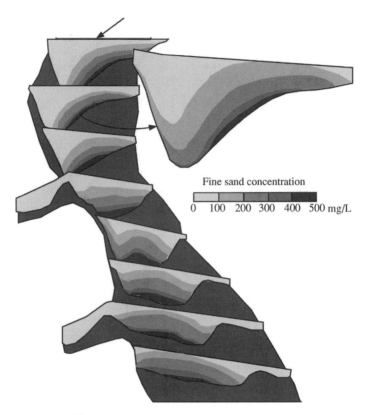

Figure CS.11.1.3. Calculated sediment concentration profiles of the Mississippi River (after U.S. Army Corps of Engineers, 1999).

♦Exercise 11.1

Plot the modules of the amplification factor $|A|$ of Eq. (11.14) on Fig. 11.2(b) as a function of $2 < N < 20$ for values of $C_r = 0.25, 0.5, 0.75, 1.0, 1.25, 1.5,$ and 2.0. Show that $|A|$ approaches unity as $N \to \infty$ for all values of C_r.

♦♦Exercise 11.2

With reference to Example 11.1, repeat the calculations for $\Delta x = 500$ m and $\Delta x = 250$ m for values of $C_r = 0.5$ and 2.0, respectively. Compare the results with those of Tables E.11.1.1 and E.11.1.2. Plot the results at $t = 2,000$ s for $C_r = 0.5$ and $\Delta x = 2,000, 1,000, 500,$ and 250 m and show that the results converge. Also, plot the corresponding results for $C_r = 2.0$ and show that the results diverge as Δx decreases even if the scheme is consistent.

♦Exercise 11.3

Reconcile the results obtained in Exercise 11.2 with the fact that $|A|$ approaches unity as N is large, as obtained in Exercise 11.1.

♦Exercise 11.4

With reference to Subsection 11.2.1, demonstrate that $\Delta x = 2K_d/V(1 - C_u)$ and $\Delta t = 2K_d C_u/V^2(1 - C_u)$ from $K = K_{\text{num}}$ in Eq. (11.27) and $C_u = V\Delta t/\Delta x$. Also, calculate Δx and Δt for $V = 1$ m/s, $K_d = 50$ m^2/s, and $C_u = 0.9$. Compare the results with those of Subsection 11.2.1.

Exercise 11.5

Derive Eq. (11.32b) from a stable Leonard scheme with $C_u = 0.8$ without oscillations and with a grid Peclet number $P_\Delta = 2$.

Exercise 11.6

Consider the following time approximation:

$$\frac{\partial h}{\partial t} \simeq \frac{h_j^{k+1} - \tilde{h}_j^k}{\Delta t} \text{ with } \tilde{h}_j^k = \frac{1}{2}\alpha\left(h_{j+1}^k + h_{j-1}^k\right) + (1 - \alpha)h_j^k,$$

where α is a weighting coefficient between a central difference and the nodal value at h_j^k. Define the finite-difference approximation, Eq. (11.6), with a central difference in time. Show that when $\alpha = C_r$, the scheme reduces to a simple FTBS scheme, or $h_j^{k+1} = h_j^k - c(\Delta t/\Delta x)(h_j^k - h_{j-1}^k)$.

♦♦Computer Problem 11.1

Calculate the floodwave propagation of Example 5.4 at a grid size $\Delta x = 32.5$ km and $\Delta t = 2$ h and use the diffusive-wave approximation.

Answer: in Table E.5.4.1

♦Computer Problem 11.2

With reference to Example 11.2, what would be the maximum concentration 10 km downstream of the release if the first pulse only lasted 9 min at a concentration of 100,000 mg/l^2. Also, compare the results with and without the second pulse.

♦♦Computer Problem 11.3

With reference to Computer Problem 4.1, calculate sediment transport for $d_s = 0.3$ mm over the 25-km reach and use the aggradation–degradation algorithm to calculate changes in bed elevation through time. Repeat the calculations and correct for the water- surface elevation changes through time. Provide graphical output of sediment-transport capacity, bed elevation, and hydraulic grade line at three different times. (*Hint*: The results are somewhat similar to those of Fig. 11.8.)

12

Waves and tides in river estuaries

This chapter relates to river features observed in wide, open areas and in river estuaries. Section 12.1 presents the theory of surface waves with applications to wind waves. Section 12.2 specifically deals with tides in river estuaries. Section 12.3 presents a brief discussion of saline wedges in river estuaries.

12.1 Surface waves

As sketched in Fig. 12.1, let us consider short gravity waves propagating in a smooth canal of depth h and unit width. The fluid is incompressible and the motion is irrotational. The velocity potential Φ is defined such that the velocity component in the downstream x direction is $v_x = (-\partial\Phi/\partial x)$ and $v_z = (-\partial\Phi/\partial z)$ in the vertical z direction. The equation of continuity can be rewritten as

$$\frac{\partial^2\Phi}{\partial x^2} + \frac{\partial^2\Phi}{\partial z^2} = 0. \tag{12.1}$$

To obtain a solution to the equation of continuity, the method of separation of variables is considered, and the wavelength λ is assumed to propagate in time t at a celerity c:

$$\Phi = f(z)\cos\frac{2\pi}{\lambda}(x - ct). \tag{12.2}$$

Substituting Eq. (12.2) into Eq. (12.1) yields an equation of the form

$$\frac{d^2f(z)}{dz^2} - f(z) = 0. \tag{12.3}$$

The solution to this equation requires the use of hyperbolic functions. The hyperbolic sine, cosine, and tangent are defined by the three relations

$$\sinh z = \frac{e^z - e^{-z}}{2}, \cosh z = \frac{e^z + e^{-z}}{2}, \tanh z = \frac{\sinh z}{\cosh z} = \frac{e^z - e^{-z}}{e^z + e^{-z}}, \tag{12.4}$$

Table 12.1. *Properties of hyperbolic functions*

$f(z)$	Value at $z \to 0$	Value at $z \to \infty$	Value at $z \to -\infty$	$\dfrac{\mathrm{d}f(z)}{\mathrm{d}z}$	$\int f(z)\mathrm{d}z^z$
$\sinh z$	z	$^1\!/_2\,e^z$	$-^1\!/_2\,e^{-z}$	$\cosh z$	$\cosh z$
$\cosh z$	1	$^1\!/_2\,e^z$	$^1\!/_2\,e^{-z}$	$\sinh z$	$\sinh z$
$\tanh z$	z	1	-1		

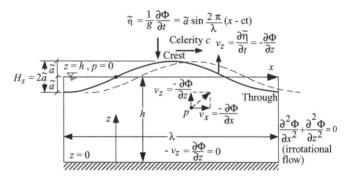

Figure 12.1. Surface-wave diagram.

respectively, which should be compared with the relations

$$\sin z = \frac{e^{iz} - e^{-iz}}{2i}, \cos z = \frac{e^{iz} + e^{-iz}}{2}, \tan z = \frac{\sin z}{\cos z}, \quad (12.5)$$

where $i = \sqrt{-1}$ for the circular functions. It follows at once from their definitions that $\cosh^2 z - \sinh^2 z = 1$. We note that, like their circular cousins, $\sinh z$ is an odd function and $\cosh z$ is even. In addition we have, by definition, $\operatorname{cosech} z = 1/\sinh z$, $\operatorname{sech} z = 1/\cosh z$, and $\coth z = 1/\tanh z$. It is clear from the table that $\sinh z$ and $\cosh z$ are solutions of Eq. (12.3). Other elementary properties of the hyperbolic functions which follow at once are summarized in Table 12.1 and sketched in Fig. 12.2.

Therefore the complete solution for the potential function can be written as

$$\Phi = \left(\tilde{A} \cosh \frac{2\pi z}{\lambda} + \tilde{B} \sinh \frac{2\pi z}{\lambda}\right) \cos \frac{2\pi}{\lambda}(x - ct), \quad (12.6)$$

where coefficients $\tilde{A}$ and $\tilde{B}$ and celerity c can be determined from the boundary conditions.

The first boundary condition is that the vertical velocity is zero, or

$$v_z = \frac{\partial \Phi}{\partial z} = 0 \text{ at } z = 0. \quad (12.7)$$

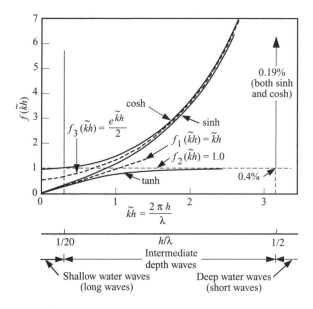

Figure 12.2. Properties of hyperbolic functons.

Substituting Eq. (12.6) into Eq. (12.7) imposes the condition that $\tilde{B} = 0$. There are two remaining boundary conditions to define the celerity c in Subsection 12.1.1 and the coefficient $\tilde{A}$ as a function of wave amplitude in Subsection 12.1.2. The concept of wave energy is presented in Subsection 12.1.3, followed by group velocity in Subsection 12.1.4 and wave power in Subsection 12.1.5. Finally, applications to wind waves are presented in Subsection 12.1.6.

12.1.1 Wave celerity

An expression for the wave celerity can be obtained from the boundary condition at the free surface. The expression emerges from the equation of motion without friction. For 2D flow ($v_y = 0$), the equation of motion in the vertical z direction is

$$\frac{\partial v_z}{\partial t} + v_x \frac{\partial v_z}{\partial x} + v_z \frac{\partial v_z}{\partial z} = -g - \frac{1}{\rho} \frac{\partial p}{\partial z}. \tag{12.8}$$

In the case of irrotational flow, $(\partial v_z / \partial x) = (\partial v_x / \partial z)$, the following is obtained from the flow potential definition $v_z = (-\partial \Phi / \partial z)$:

$$\frac{\partial}{\partial z} \left[\frac{-\partial \Phi}{\partial t} + \frac{1}{2} \left(v_x^2 + v_z^2 \right) + gz + \frac{p}{\rho} \right] = 0. \tag{12.9}$$

Considering that the velocity of the water is small, we can neglect the term in v^2. The potential that is due to gravity is $g(z + \tilde{\eta})$ and the equation of pressure becomes

$$\frac{p}{\rho} = \frac{\partial \Phi}{\partial t} - g\,(z + \tilde{\eta}) \tag{12.10a}$$

or

$$\tilde{\eta} = \frac{1}{g}\frac{\partial \Phi}{\partial t} - z - \frac{p}{\rho g}. \tag{12.10b}$$

Differentiating the reduced equation of motion with respect to time gives

$$v_z = \frac{\partial \tilde{\eta}}{\partial t} = \frac{1}{g}\frac{\partial^2 \Phi}{\partial t^2} - \frac{1}{\rho g}\frac{\partial p}{\partial t}. \tag{12.11a}$$

At the free surface $z = h$, the atmospheric pressure p is constant and the vertical velocity v_z can be obtained as

$$v_z = \underbrace{-\frac{\partial \Phi}{\partial z}}_{\text{velocity potential}} = \underbrace{+\frac{1}{g}\frac{\partial^2 \Phi}{\partial t^2}}_{\text{equation of motion}}. \tag{12.11b}$$

After substituting the potential function from Eq. 12.6 with $\tilde{B} = 0$, into Eq. (12.11b), we obtain at $z = h$

$$-\frac{2\pi}{\lambda}\sinh\frac{2\pi h}{\lambda} = -\frac{1}{g}\left[\frac{2\pi c}{\lambda}\right]^2 \cosh\frac{2\pi h}{\lambda}$$

or $\qquad\qquad\qquad\qquad\qquad\qquad\qquad\qquad\qquad$ (12.12a) ♦♦

$$c^2 = \frac{g\lambda}{2\pi}\tanh\frac{2\pi h}{\lambda}.$$

This is the relationship for wave celerity c, also known as the phase velocity, derived by Airy. There are two particular cases of interest: (1) long waves, also called shallow-water waves, where $\lambda \gg h$; and (2) short waves, also called deep-water waves, where $\lambda \ll h$.

For shallow-water waves, the wavelength λ is very long compared with the flow depth h ($\lambda \gg h$):

$$c^2 \simeq \frac{g\lambda}{2\pi}\frac{2\pi h}{\lambda} = gh. \tag{12.12b}$$

This is the case of tidal waves, in which the velocity of propagation is independent of the wavelength.

For deep-water waves, the wavelength λ is very short compared with the flow depth h ($\lambda \ll h$):

$$c^2 \simeq \frac{g\lambda}{2\pi}. \tag{12.12c}$$

This is the case of deep sea waves that are due to the wind. As the celerity is proportional to the square root of the wavelength, a high wind gives rise to longer waves than does a gentle breeze.

The wave period T relates to the wavelength λ and the celerity c as $\lambda = cT$, and the following identities are obtained directly from Eqs. (12.12):

$$T = \sqrt{\frac{2\pi\lambda}{g} \coth \frac{2\pi h}{\lambda}}, \tag{12.13a} \blacklozenge$$

$$\lambda = \frac{gT^2}{2\pi} \tanh \frac{2\pi h}{\lambda}. \tag{12.13b}$$

The relationship among celerity, period, and flow depth is illustrated in Fig. 12.3. It is sometimes convenient to define the wave number $\tilde{k} = (2\pi/\lambda)$ and the angular frequency $\tilde{\sigma} = \tilde{k}c = (2\pi/T)$ to define the surface waves of amplitude $\tilde{a}$ as $\tilde{\eta} = \tilde{a}\sin(\tilde{k}x - \tilde{\sigma}t)$.

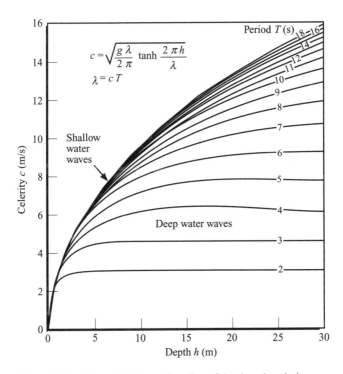

Figure 12.3. Wave celerity as a function of depth and period.

12.1.2   Displacement and velocity

The velocity components are obtained directly from the velocity potential equation (12.6) with $\tilde{B} = 0$:

$$v_x = \frac{\partial \tilde{\xi}}{\partial t} = -\frac{\partial \Phi}{\partial x} = \frac{2\pi \tilde{A}}{\lambda} \cosh \frac{2\pi z}{\lambda} \sin \frac{2\pi}{\lambda}(x - ct), \qquad (12.14a)$$

$$v_z = \frac{\partial \tilde{\eta}}{\partial t} = -\frac{\partial \Phi}{\partial z} = -\frac{2\pi \tilde{A}}{\lambda} \sinh \frac{2\pi z}{\lambda} \cos \frac{2\pi}{\lambda}(x - ct), \qquad (12.14b)$$

and integrating over time gives the horizontal $\tilde{\xi}$ and vertical $\tilde{\eta}$ displacements;

$$\tilde{\xi} = \frac{\tilde{A}}{c} \cosh \frac{2\pi z}{\lambda} \cos \frac{2\pi}{\lambda}(x - ct), \qquad (12.15a)$$

$$\tilde{\eta} = \underbrace{\frac{\tilde{A}}{c} \sinh \frac{2\pi z}{\lambda}}_{\text{| wave amplitude } \tilde{\mathbf{a}} \text{ |}} \sin \frac{2\pi}{\lambda}(x - ct). \qquad (12.15b)$$

This relationship defines the constant $\tilde{A}$ as a function of three wave characteristics: (a) wave amplitude $\tilde{\mathbf{a}}$, (b) wavelength λ, and (c) wave celerity c. From $\cos^2 + \sin^2 = 1$, the path of a fluid element is seen to be an ellipse:

$$\frac{\tilde{\xi}^2}{\left[\dfrac{\tilde{A}}{c} \cosh \dfrac{2\pi z}{\lambda}\right]^2} + \frac{\tilde{\eta}^2}{\left[\dfrac{\tilde{A}}{c} \sinh \dfrac{2\pi z}{\lambda}\right]^2} = 1, \qquad (12.15c) \blacklozenge$$

with its major axis horizontal because $\cosh > \sinh$. As the hyperbolic sine and cosine approach the same limiting value for large values of the argument, this ellipse becomes a circle near the surface of deep-water waves, as shown in Fig. 12.4.

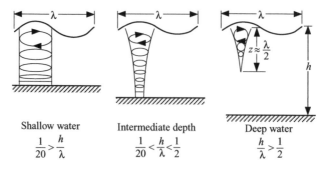

Figure 12.4. Particle motion in water waves.

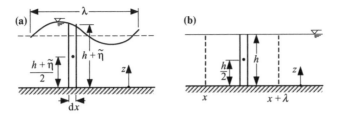

Figure 12.5. PE definition sketch.

12.1.3 Wave energy

The total energy of a wave is divided into potential-energy (PE) and kinetic-energy (KE) components generally expressed in terms of average energy over a complete wavelength per unit surface area.

The PE that is due to the progressive waveform on the free surface is obtained from subtracting the PE without the wave from the PE with the wave, as sketched in Fig. 12.5(a); the incremental PE per unit width, dPEa, in a small column of water is the height of the center of gravity times the mass increment dm. In a water column $h + \tilde{\eta}$ high and dx long for water of specific weight γ, we obtain dPE$_a$ = (height to center of gravity) g dm = $[(h + \tilde{\eta})^2/2]\gamma$d$x$. The average PE per unit surface area is obtained from integration of the incremental PE over one wavelength λ and one wave period T:

$$\overline{PE_a} = \frac{\gamma}{2\lambda T} \int_t^{t+T} \int_x^{x+\lambda} (h + \tilde{\eta})^2 dx dt. \tag{12.16}$$

Using $\tilde{\eta} = \tilde{a}\sin(\tilde{k}x - \tilde{\sigma}t)$ from Eq. (12.15b), we find that Eq. (12.16) becomes

$$\overline{PE_a} = \frac{\gamma}{2\lambda T} \int_t^{t+T} \int_x^{x+\lambda} [h^2 + 2\tilde{a}\sin(\tilde{k}x - \tilde{\sigma}t) + \tilde{a}^2 \sin^2(\tilde{k}x - \tilde{\sigma}t)] dx dt$$

or simply

$$\overline{PE_a} = \frac{\gamma h^2}{2} + \frac{\gamma \tilde{a}^2}{4},$$

which is the average PE per unit surface area of all the water above $z = 0$.

The PE in the absence of a wave, as sketched in Fig. 12.5(b), is

$$\overline{PE_b} = \frac{\gamma}{2\lambda T} \int_t^{t+T} \int_x^{x+\lambda} h^2 dx dt = \frac{\gamma h^2}{2}.$$

The average PE per unit area that is attributable to the wave is

$$\overline{PE_a} - \overline{PE_b} = \frac{\gamma \tilde{a}^2}{4}. \tag{12.17}$$

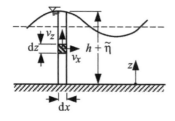

Figure 12.6. KE definition sketch.

From Fig. 12.6, the KE per unit width of a small element dx long and dz high with velocity components v_x and v_z is given by

$$d(KE) = \frac{1}{2}\left(v_x^2 + v_z^2\right)dm$$

$$= \frac{1}{2}\rho\left(v_x^2 + v_z^2\right)dzdx.$$

The average KE per unit surface area is then given after integration over the flow depth, one wavelength, and one wave period

$$\overline{KE} = \frac{\rho}{2\lambda T}\int_t^{t+T}\int_x^{x+\lambda}\int_0^{h+\tilde\eta}\left(v_x^2 + v_z^2\right)dzdxdt. \tag{12.18}$$

Using the velocity components compatible with the progressive wave $\tilde\eta = \tilde{a}\sin(\tilde{k}x - \tilde\sigma t)$ and velocity components given by Eqs. (12.14), we find that Eq. (12.18), then becomes

$$\overline{KE} = \frac{\rho}{2\lambda T}\int_t^{t+T}\int_x^{x+\lambda}\int_0^{h+\tilde\eta}\tilde{k}^2\tilde{A}^2\left[\begin{array}{l}\cosh^2 \tilde{k}z \sin^2(\tilde{k}x - \tilde\sigma t) \\ + \sinh^2 \tilde{k}z \cos^2(\tilde{k}x - \tilde\sigma t)\end{array}\right]dzdxdt.$$

By using the following identities: (1) $\cosh^2 kz = \tilde{k}z = \frac{1}{2}[1 + \cosh 2\tilde{k}z]$; (2) $\sinh^2 \tilde{k}z = -\frac{1}{2}(1 - \cosh 2\tilde{k}z)$; (3) $\cos^2(\tilde{k}x - \tilde\sigma t) - \sin^2(\tilde{k}x - \tilde\sigma t) = \cos 2(\tilde{k}x - \tilde\sigma t)$; and (4) $\cos^2(\tilde{k}x - \tilde\sigma t) + \sin^2(\tilde{k}x - \tilde\sigma t) = 1$, the average KE density becomes

$$\overline{KE} = \frac{\rho\tilde{k}^2\tilde{A}^2}{4\lambda T}\int_t^{t+T}\int_x^{x+\lambda}\int_0^{h+\tilde\eta}[\cosh 2\tilde{k}z - \cos 2(\tilde{k}x - \tilde\sigma t)]dzdxdt$$

$$= \frac{\rho\tilde{k}\tilde{A}^2}{8}\sinh 2\tilde{k}h.$$

It reduces further from: (1) $\sinh 2\tilde{k}h = 2\sinh \tilde{k}h \cosh \tilde{k}h$; (2) $\tilde{A} = \tilde{a}c/\sinh \tilde{k}h$ from Eq. (12.15b); and (3) $c^2 = (g/\tilde{k})\tanh \tilde{k}h$ from relations (12.12):

$$\overline{KE} = \frac{\gamma\tilde{a}^2}{4}. \tag{12.19}\;\blacklozenge$$

The KE of small-amplitude waves [Eq. (12.19)] thus equals the PE [Eq. (12.17)].

The average total energy $\tilde{E}$ per unit surface area is the sum of the average potential of Eq. (12.17) and average KE of Eq. (12.19):

$$\tilde{E} = \overline{PE} + \overline{KE} = \frac{\gamma\tilde{a}^2}{2}. \tag{12.20}\;\blacklozenge\blacklozenge$$

12.1.4 Group velocity

Several waves can be superposed simply by the addition of the flow-potential functions, which results in adding the wave amplitudes as

$$\tilde{\eta} = \tilde{a}_1 \sin \tilde{\sigma}_1 t + \tilde{a}_2 \sin \tilde{\sigma}_2 t, \tag{12.21a}$$

$$\tilde{\eta} = \tilde{a}_1 (\sin \tilde{\sigma}_1 t + \sin \tilde{\sigma}_2 t) + (\tilde{a}_2 - \tilde{a}_1) \sin \tilde{\sigma}_2 t. \tag{12.21b}$$

In deep water, celerity is a function of wavelength, and two sets of waves of slightly different wavelengths travel at different velocities. This causes local reinforcement and interference as one set gains on the other. The velocity with which the regions of reinforcement or interference advance is known as the group velocity. To find this velocity, consider two trains of simple harmonic waves of the same amplitude but of slightly different wavelength and frequency, such as

$$\tilde{\eta}_1 = \tilde{a} \sin \frac{2\pi}{\lambda_1} (x - c_1 t), \tag{12.22a}$$

$$\tilde{\eta}_2 = \tilde{a} \sin \frac{2\pi}{\lambda_2} (x - c_2 t). \tag{12.22b}$$

The following identity is used: $\sin x + \sin y = 2 \sin \frac{1}{2} (x + y) \cos \frac{1}{2} (x - y)$. The resultant water elevation η is obtained from Eqs. (12.21) and (12.22):

$$\tilde{\eta} = \tilde{\eta}_1 + \tilde{\eta}_2 = 2 \tilde{a} \left\{ \cos \pi \left[\left(\frac{1}{\lambda_1} - \frac{1}{\lambda_2} \right) x - \left(\frac{c_1}{\lambda_1} - \frac{c_2}{\lambda_2} \right) t \right] \right\}$$
$$\left\{ \sin \pi \left[\left(\frac{1}{\lambda_1} + \frac{1}{\lambda_2} \right) x - \left(\frac{c_1}{\lambda_1} + \frac{c_2}{\lambda_2} \right) t \right] \right\}. \tag{12.23}$$

As the wavelengths and the velocities are only sightly different, we may replace λ_1 and λ_2 with the mean wavelength λ and c_1 and c_2 with the mean celerity c in the second trigonometrical function in braces:

$$\tilde{\eta} \simeq 2\tilde{a} \underbrace{\left\{ \cos \pi \left[d \left(\frac{1}{\lambda} \right) x - d \left(\frac{c}{\lambda} \right) t \right] \right\}}_{\text{group envelope}} \sin \frac{2\pi}{\lambda} (x - ct). \tag{12.24}$$

This expression represents a wave traveling with phase velocity c with an amplitude of $2\tilde{a}$ and a group envelope in braces. At a given instant the amplitude plotted against x is represented by a cosine curve, the distance from one region of maximum amplitude to the next being the reciprocal d $(1/\lambda)$. Moreover, this region of maximum amplitude is progressing in the positive x direction with

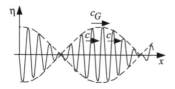

Figure 12.7. Group-velocity
sketch.

the group velocity c_G:

$$c_G = \frac{d\left(\dfrac{c}{\lambda}\right)}{d\left(\dfrac{1}{\lambda}\right)} = \frac{\dfrac{\lambda dc - cd\lambda}{\lambda^2}}{-\dfrac{d\lambda}{\lambda^2}}$$

$$= c - \lambda\frac{dc}{d\lambda}. \tag{12.25}$$

Therefore we have a wave within a wave, as represented in Fig. 12.7. The full-line wave curve advances with the celerity c whereas the dashed curve advances with the group velocity c_G. If c increases with λ, the phase wave travels more rapidly than the group wave. In this case the full-line waves pass through the group from rear to front, each crest disappearing at the front of the group while a new crest appears at the rear.

The group velocity obtained from substituting the wave celerity from relations (12.12) into Eq. (12.25) is

$$c_G = \frac{c}{2}\left(1 + \frac{2\tilde{k}h}{\sinh 2\tilde{k}h}\right), \tag{12.26} \blacklozenge\blacklozenge$$

which asymptotically reduces to $c/2$ when $\tilde{k}h$ is large (deep-water waves) and reduces to c when $\tilde{k}h$ is very small (shallow-water waves). The group velocity for shallow-water waves is identical to the wave celerity because the wave celerity is independent of wavelength.

12.1.5   Wave power

The increment of power in a frictionless fluid is obtained by the product of the elementary force $a_x dm$ and velocity in the same direction $v_x = -\partial\Phi/\partial x$. Considering that the channel banks and bed do not move and that the pressure is atmospheric at the free surface, the increment of power can then be integrated over the flow depth during one period T to give the total power $\tilde{P}$ of the wave per unit surface area:

$$\tilde{P} = \frac{1}{T}\rho\int_t^{t+T}\int_0^{h+\tilde{\eta}}\frac{\partial\Phi}{\partial t}\left(\frac{-\partial\Phi}{\partial x}\right)dzdt. \tag{12.27a}$$

Substituting the flow-potential function Φ from Eq. (12.6) into Eq. (12.27a) gives

$$\tilde{P} = \frac{\gamma\tilde{a}^2}{2}c_G. \tag{12.27b} \blacklozenge$$

It is concluded that the energy of a wave previously calculated in Eq. (12.20) propagates at a velocity corresponding to the group velocity.

12.1.6 Wind waves

The minimum wind speed to generate gravity waves is approximately 6 m/s. The characteristics of wind waves include a full spectrum of wavelengths and amplitudes. The significative wave height H_s represents the average height of the large waves in the upper third of the spectrum. The wave height H_s is given as twice the wave amplitude $\tilde{a}$. The average wavelength of these waves determines the significative wavelength λ_s. Under a constant wind speed U_w, the wave height increases with the downwind distance, called the fetch length $\tilde{F}$, and with time until equilibrium is obtained. In deep water, either the wind duration or the fetch length can limit the growth of waves. In shallow water, the flow depth h can also limit the growth of wind waves.

In rivers, the wind waves are usually limited by the fetch length $\tilde{F}$. The flow depth may also limit the wave height during large storms when the fetch length exceeds $\sim$3–5 km. The significative wave height H_s can be estimated as a function of wind speed U_w, flow depth h, and fetch length $\tilde{F}$ from Fig. 12.8.

The development of wind waves under a constant wind speed U_w over a fetch length $\tilde{F}$ is sketched in Fig. 12.8. To determine the wave height $\tilde{H}_s$, it can be assumed that the work done by the shear stress $\tau \sim \rho U_w^2$ over the fetch length $\tilde{F}$ corresponds to the energy of the wave $\tilde{E} \sim \rho g H_s^2$ from Eq. (12.20). After the work done and the energy are equated, the wave height $\tilde{H}_s$ is related to the fetch length $\tilde{F}$ as

$$\tilde{H}_s \simeq 0.003\, U_w \sqrt{\tilde{F}/g}. \tag{12.28}$$

We can estimate the windstorm duration $\tilde{t}_s$ required for reaching equilibrium by considering the ratio of the fetch length $\tilde{F}$ to the group velocity c_G. In shallow water, the storm duration can be approximated by $\tilde{t}_s = \tilde{F}/\sqrt{gh}$.

Waves are fetch limited as long as $\tilde{H} < 0.3\,h$, or $U_w < 100\sqrt{gh^2/\tilde{F}}$. In this case, the significative wave height is approximated by Relation 12.28 and the wavelength is $\lambda \simeq 15\,\tilde{H}_s$. The period is obtained from Eq. (12.13a). The flow depth limits the growth of surface waves when $h < 0.01\, U_w\sqrt{\tilde{F}/g}$. In the case of depth-limited waves, the wave height is limited to approximately 30% to 40% of the flow depth, or $\tilde{H}_s \simeq 0.35\,h$. Fetch-limited waves are likely to cause erosion of river banks whereas depth-limited waves will likely also cause resuspension of the bed material.

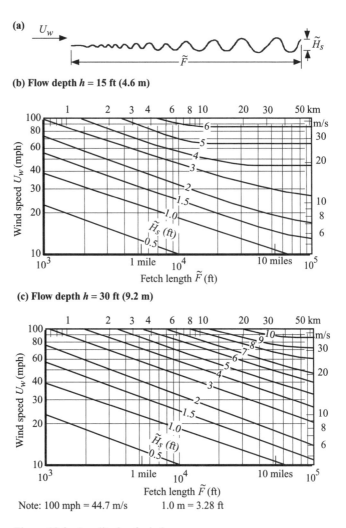

Figure 12.8. Amplitude of wind waves.

12.2   Tides in river estuaries

Tides are complex oscillatory waves in large water masses. The most obvious tidal period corresponds to approximately a half daily cycle, more specifically 12 h 25 min. The tidal range and time of high water not only vary from place to place but also throughout the month and the year. Tidal oscillations result from massive gravitational and centrifugal force balances among Earth, Moon, and Sun.

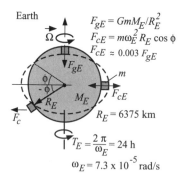

Earth

$$F_{gE} = GmM_E/R_E^2$$
$$F_{cE} = m\omega_E^2 R_E \cos\phi$$
$$F_{cE} \approx 0.003\, F_{gE}$$

$R_E = 6375$ km

$$T_E = \frac{2\pi}{\omega_E} = 24\ h$$

$$\omega_E = 7.3 \times 10^{-5}\ rad/s$$

Figure 12.9. Earth rotation sketch.

Considering first the Earth alone, we find that the gravitational attraction is proportional to the universal gravitation constant $G_u = 6.673 \times 10^{-11}$ Nm^2/kg^2. At the surface of the Earth of mass $m_E = 5.976 \times 10^{24}$ kg and radius $R_E = 6,371$ km, the gravitational acceleration is $g = G_u m_E/R_E^2 = 9.8$ m/s^2. The centrifugal acceleration is proportional to the square of the angular velocity multiplied by the radius of rotation. As sketched in Fig. 12.9, the centrifugal acceleration that is due to the Earth's rotation is $a_{cE} = \omega_E^2 R_E \cos\phi = 0.034$ m/s$^2 \cos\phi$, where $\omega_E = 7.272 \times 10^{-5}$ rad/s is the angular velocity of the Earth and ϕ is the latitude of the point considered.

Suppose the Earth is entirely covered with water. A unit mass of water on the surface is simultaneously subjected to gravitational and centrifugal accelerations. At the poles, there is no local centrifugal action, whereas at the equator the centrifugal acceleration is of the order of 0.034 m/s^2, or 0.3% of the gravitational acceleration. This combined action varies with latitude and draws the water toward the equator. Although there is in fact a preferential accumulation of water near the equator, the daily rotation of the Earth does not cause tides but contributes to tides through Coriolis acceleration. The presentation of the Coriolis acceleration (Subsection 12.2.1) is followed by tidal acceleration (Subsection 12.2.2) and tide amplitude and propagation (Subsection 12.2.3).

12.2.1   Coriolis acceleration

Moving water at the surface of the Earth is subjected to acceleration, known as the Coriolis acceleration, that is due to the rotation of the Earth. The Coriolis acceleration is a function of the angular velocity vector of the Earth's rotation Ω and the flow-velocity vector V of moving water in a frame of reference at the center of mass of the Earth and rotating with the Earth (x to the East, y North, and z vertical up).

The Coriolis acceleration that is due to the Earth's rotation $\Omega = \hat{y}\omega_E \cos\phi + \hat{z}\omega_E \sin\phi$ that is applied at the center of mass of the Earth and is due to the velocity vector V with v_x to the East, v_y to the North, and v_z vertical

up, is then defined as

$$a_{cor} = -2\Omega \times V$$
$$= 2\,\omega_E[\hat{x}(v_y \sin\phi - v_z \cos\phi) - \hat{y}v_x \sin\phi + \hat{z}v_x \cos\phi]. \qquad (12.29)$$

For horizontal flow at the surface of the Earth ($v_z = 0$), the Coriolis acceleration a_{cor} depends on the latitude angle ϕ. Fluid motion is deflected to the right in the northern hemisphere and to the left in the southern hemisphere. Given the angular velocity of the Earth, $\omega_E = 7.3 \times 10^{-5}$ rad/s and the latitude angle ϕ, the magnitude of the Coriolis acceleration is

$$|a_{cor}| \simeq 2\,\omega_E \sin\phi V. \qquad (12.30)$$

The ratio of the Coriolis acceleration to the downstream gravitational acceleration is defined as $2\,\sin\phi\omega_E V/(gS)$. Given $\omega_E = 7.3 \times 10^{-5}$ rad/s, this ratio is small (ratio <0.01) in most rivers and the Coriolis acceleration can be neglected in all but a few large rivers where S is very small, e.g., $S = 5 \times 10^{-5}$, and where the flow velocities are relatively large, e.g., $V > 2$ m/s. Example 12.1 illustrates one case in which the Coriolis acceleration is not negligible. The Coriolis acceleration can also be compared with the centrifugal acceleration in a river bend of radius R_1 as $2\sin\phi\,\omega_E R_1/V$. This ratio is usually negligible in all but fairly straight large rivers. Although negligible in rivers unless the surface slope is very small, the Coriolis force plays a dominant role in large water masses and in the atmosphere, in which it generates geostrophic winds.

Example 12.1.   Calculation of the Coriolis acceleration. Consider the lower Mississippi River at a latitude $\phi = 30°$ in a river bend at a radius of curvature $R_1 = 4$ miles (6.4 km). The water-surface slope of the river is $S_w \approx 6 \times 10^{-5}$ is applied on and the water moves horizontally to the south at 10 ft/s (3 m/s). The westward Coriolis acceleration due to the rotation of the Earth is $|a_{cor}| = 2\,\omega_E \sin\phi V = 2 \times 7.3 \times 10^{-5} \times \sin 30° \times 10 = 7.3 \times 10^{-4}$ ft/s² (2.22×10^{-4} m/s²). The centripetal acceleration in the river bend is calculated from $a_{cent} = V^2/R_1 = 100\,\text{ft}^2/\text{s}^2/4 \times 5{,}280\ \text{ft} = 4.7 \times 10^{-3}$ ft/s² (1.43×10^{-3} m /s²). The Coriolis acceleration is $\sim 15\%$ of the centripetal acceleration in this river bend. In comparison, the downstream gravitational acceleration component is $gS = 32.2\ \text{ft/s}^2 \times 6 \times 10^{-5} = 1.9 \times 10^{-3}$ ft/s² (5.8×10^{-4} m/s²). The Coriolis acceleration is a significant fraction of the downstream gravitational acceleration. Indeed, the Coriolis acceleration is $\sim 38\%$ of the downstream gravitational acceleration and should not be neglected in this case.

Field velocity measurements of the Mississippi River that use the acoustic Doppler current profilers at the Old River Control Complex show that the random turbulent-velocity fluctuations mask the effects of the Coriolis acceleration. Winkley (1989) nevertheless concludes that the Coriolis force exerts an influence on the geometry of the Lower Mississippi River. The Coriolis effects become even more significant for oscillations of longer periods, such as tides.

12.2.2 Tidal accelerations

The Earth and the Moon may be considered as a single system, consisting of two bodies having a common center both of mass and rotation C, as sketched in Fig. 12.10. The common center of mass is just inside the surface of the Earth – as the Earth diameter and mass are respectively approximately 3.7 and 81 times those of the Moon and their distance apart is $\sim$30 Earth diameters. In a frame of reference rotating in space about C with a period of $\sim$27.3 days, the water is subject to the gravitational accelerations of the Earth and the Moon and the centripetal acceleration.

The centrifugal and gravitational accelerations exerted by the Moon are equal and opposite at the center of mass of the Earth. Locally at the Earth surface, the gravitational acceleration attracts water toward the Moon at point A while the centrifugal acceleration about point C forces the water away from the Moon toward point B. Their resultant force causes the water surface to form a spheroid that is oblate along the Earth–Moon axis. The above argument can be applied

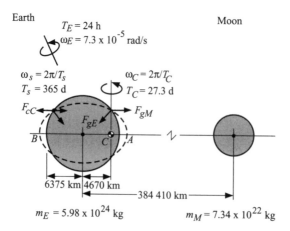

Figure 12.10. Earth–Moon influence on tides.

to the effect of the Sun's giving differential gravitation effects that are 2.17 times smaller than those for the Moon. Consequently, maximal tides should be obtained during new and full Moons.

Also, the gravitational and the centripetal accelerations at the poles do not change during the day. The largest variations are obtained at the equator with, for instance, the gravitational force of the Sun near the east at sunrise and near the west at sunset. In simplified form, the differential in latitudinal accelerations on a daily basis generates flow velocities that are proportional to the cosine of the latitude. The maximum differential is obtained when the Sun is in the equatorial plane, which should increase the magnitude of fall and spring tides at the times of the equinoxes. The magnitude of the Coriolis acceleration from relation (12.30) is therefore proportional to $\sin \phi_e \cos \phi_e$. Accordingly, tides should be small near the poles and the equator and should be maximum at a 45° latitude.

The exact analysis of tides is further complicated by the facts that: (1) the axis of rotation of the Earth is at 66°33 minutes from the plane of Earth's rotation around the Sun; and (2) the plane of the Moon is also inclined at 5°9′ minute from the plane of the Earth's rotation around the Sun. High-water levels do not usually coincide with the maximum lunar attraction at a given locality. Displacements in harmonic systems are out of phase with accelerations. Tidal currents are not necessarily in phase with the displacements because of the finite wave-propagation speed and its dependence on depth. For reasons of boundary shape and friction effects, local systems rotate around amphidromic points where no tide elevation exists. An example is shown in Fig. 12.11.

12.2.3   Tide amplitude and propagation

The propagation of tides in estuaries with variable widths is calculated on the basis that the effects of friction and wave reflections remain negligibly small. Thus the law of conservation of energy provides a convenient way of expressing the effect of changes in width and depth on tide amplitude. The total energy for the entire surface of the wave from Eg. (12.20) is

$$W_0\lambda_0\tilde{E} = W_0\lambda_0\frac{1}{2}\gamma\tilde{a}_0^2 = W_x\lambda_x\frac{1}{2}\gamma\tilde{a}_x^2, \tag{12.31}$$

where W is the width of the estuary, λ is the tidal wave length and $\tilde{a}$ is the tide amplitude. The subscript 0 refers to the initial or reference position and the subscript x to any location. Because the celerity c of shallow-water waves $c = \sqrt{gh_x}$ and because the period $T = \lambda/c$ is independent of any deformation

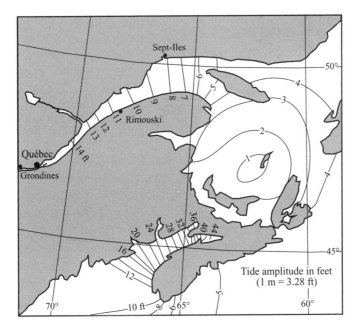

Figure 12.11. Tide amplitude of the St. Lawrence estuary.

of the wave, the tide amplitude $\tilde{a}_x$ can be determined after $\lambda_x = T\sqrt{gh_x}$ is substituted into Eq. (12.31) to obtain

$$\frac{\tilde{a}_x}{\tilde{a}_0} = \left(\frac{W_0}{W_x}\right)^{1/2}\left(\frac{h_0}{h_x}\right)^{1/4}. \qquad (12.32) \blacklozenge$$

This is known as Green's law, which assumes conservation of energy for shallow-water waves. In the case of intermediate-or deep-water waves other than tidal waves, the more general relationship from conservation of power [Eq. 12.27(b)] is recommended. The case of tide propagation with friction is best analyzed with numerical models. From Fig. 12.11 we can examine the tidal amplitude changes as it propagates up the St. Lawrence estuary. It is clear that the increase in amplitude is caused by the reduction in channel width and flow depth.

12.3   Saline wedges in river estuaries

As rivers enter seas and oceans, the density difference between clear freshwater and saltwater often causes density stratification. A saline wedge refers to the saline layer underlying freshwater in a river communicating with a tideless sea.

The still form is called an arrested saline wedge, and the length and the shape of arrested saline wedges are described below.

Let ρ_{sea} be the density of saltwater (usually $\rho_{sea} \simeq 1{,}035$ kg/m^3), and let ρ_m be the mass density of the river freshwater, including suspended sediment. Then a densimetric velocity V_Δ may be defined as

$$V_\Delta = \sqrt{\frac{\Delta\rho}{\rho_m} gh}, \tag{12.33}$$

where $\Delta\rho = \rho_{sea} - \rho_m$, g is the gravitational acceleration, and h is the river depth. The physical significance of V_Δ is linked to the propagation velocity of internal waves of large wavelength. The initial velocity of saline fronts moving in still water is also proportional to V_Δ. A saline wedge extends laterally as long as $V_\Delta > V_1$.

As sketched in Fig. 12.12, we can calculate the height of the wedge at the river mouth h_{s1} by defining the position where the local velocity V_1 equals the local densimetric velocity $V_{\Delta 1}$. The densimetric velocity is applied at the river mouth:

$$V_{\Delta 1}^2 = V_1^2 = \frac{\Delta\rho}{\rho_m} g\,(h - h_{s1}). \tag{12.34}$$

If mixing at the interface is small, the continuity condition for freshwater in a river of constant width gives

$$V_1\,(h - h_{s1}) = Vh. \tag{12.35}$$

Eliminating V_1 between expressions (12.34) and (12.35) and solving for h_{s1}/h gives

$$\frac{h_{s1}}{h} = 1 - \left(\frac{V}{V_\Delta}\right)^{2/3}. \tag{12.36}$$

The length L_Δ of an arrested saline wedge has been studied experimentally and the following empirical relationship is suggested:

$$\frac{L_\Delta}{h} \simeq 6.0 \left(\frac{V_\Delta h}{\nu}\right)^{1/4} \left(\frac{V_\Delta}{V}\right)^{2.5}. \tag{12.37}$$

The length of an arrested saline wedge depends primarily on the river depth h and the density difference $\Delta\rho$.

The shape of the arrested saline wedge sketched in Fig. 12.12 is defined from the height of the saline wedge h_s, the height of the wedge at the river mouth h_{s1}, and the distance L in terms of the wedge length L_Δ. The shape of the arrested saline wedge is practically independent of seawater salinity, river

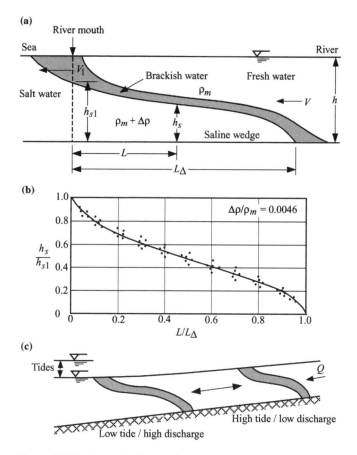

Figure 12.12. Arrested saline wedges.

velocity, water depth, channel width, and viscosity, as shown in Fig. 12.12(b). Case Study 12.1 provides an example of interaction between the saline wedge and tides in river estuaries.

Case Study 12.1   Salmon capture in the Matamek River estuary, Canada.
The Matamek River is a small tributary of the St. Lawrence River estuary near Sept-Iles. When leaving rivers to enter the sea, juvenile Atlantic salmon (*salmo salar*), called smolts, spend 20–30 days in the saline wedge of river estuaries to gradually adapt to saltwater. To study the migration pattern in the Matamek River, the capture–recapture method was used to estimate the population of smolts in the estuary during the migration period in June and early July. Captures were made during high tides at a fixed location (point A

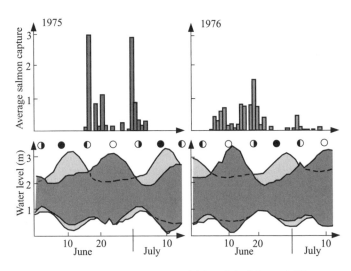

Figure CS.12.1.1. Salmon capture and tides of the Matamek River.

in Fig. 4.2) in the estuary for a period of 2 yrs. The daily average number of smolts captured per seine haul is shown in Fig. CS.12.1.1. The double peak in fish migration intrigued scientists. At the time, arguments were supporting two different fish population's migrating at slightly different times. Scientific studies were pursued to determine whether there could be genetic differences in the fish population. When the moon cycle in Fig. CS.12.1.1 is plotted against the salmon capture chart, the peaks correspond to half-moon periods. Could the Moon also exert influence on fish capture?

When the water levels in the estuary were plotted as a function of time, Fig. CS.12.1.1 shows that the captures were low when tides were very high. The saline wedge is located at the capture site during half-moons when the tide is moderately high. The saline wedge was located upstream during higher tides and downstream for lower tides, as sketched in Fig. 12.12(c). It became clear that the number of fish captured depended primarily on whether or not the seine was hauled in the saline wedge or not. A simple regression equation based on the high- and the low-water levels was set up to determine ahead of time whether or not fishing would be good for any given day. So much for the influence of the Moon on fish behavior and for the two salmon populations in the Matamek River.

Case Study 12.2   Tidal data for the St. Lawrence River, Canada. This case study presents tide-propagation data in the St. Lawrence River. The data presented in Fig. CS.12.2.1 and Table CS.12.2.1 illustrate tidal wave

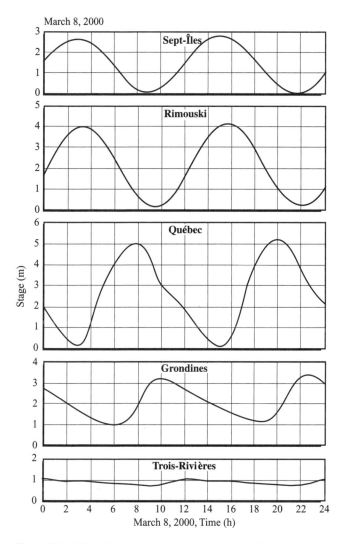

Figure CS.12.2.1. Tide propagation in the estuary of the St. Lawrence River.

propagation in river estuaries. The database serves for specific calculations in Problem 12.5.

♦**Exercise 12.1**

Satisfy Eq. (12.6) with the boundary condition that the vertical velocity is zero at the bottom of the canal [Eq. (12.7)] to show that $\tilde{B} = 0$.

Table CS.12.2.2.   *Tides of the St. Lawrence River*

Date	Sept-Iles		Rimouski		Quebec		Grondines	
	m	Time	*m*	Time	*m*	Time	*m*	Time
March	2.1	0h44	3.2	1h17	4.1	5h23	2	8h06
2000	0.5	6h31	1	7h09	0.5	12h15	0.5	15h03
4	2.7	12h57	3.8	13h26	4.8	17h29	2.6	19h58
	0.4	19h23	0.7	19h53	0.3	0h52	0.4	4h16
5	2.2	1h19	3.4	1h50	4.4	5h59	2.2	8h32
	0.5	7h11	0.8	7h45	0.4	12h58	0.5	16h03
	2.8	13h32	4	13h59	5	18h07	2.8	20h34
	0.3	19h53	0.6	20h21	0.2	1h30	0.5	4h57
6	2.4	1h52	3.6	2h22	4.7	6h34	2.4	8h59
New	0.2	7h49	0.6	8h19	0.3	13h39	0.6	16h57
Moon	2.8	14h06	4.1	14h32	4.8	18h45	2.5	21h14
	0.2	20h19	0.5	20h43	0.2	2h07	0.6	5h36
7	2.5	2h25	3.8	2h54	4.9	7h09	2.7	9h29
	0.1	8h26	0.5	8h54	0.2	14h19	0.7	17h47
	2.8	14h39	4.2	15h05	5.3	19h24	3	21h45
	0.1	20h53	0.4	21h20	0.2	2h43	0.7	6h14
8	2.6	2h59	3.9	3h28	5.1	7h45	2.9	10h04
	0.1	9h05	0.5	9h30	0.2	14h59	0.8	18h33
	2.7	15h14	4.1	15h41	5.3	20h03	3	22h23
	0.1	21h25	0.4	21h53	0.2	3h19	0.8	6h51
9	2.7	3h35	4	4h04	5.2	8h21	3.1	10h42
	0.1	9h45	0.5	10h09	0.1	15h39	0.8	19h18
	2.6	15h51	4	16h20	5.1	20h44	3	23h05
	0.1	21h59	0.4	22h28	0.2	3h56	0.8	7h27
10	2.7	4h14	4	4h43	5.2	9h00	3.1	11h25
	0.2	10h29	0.6	10h53	0.2	16h21	0.9	20h03
	2.4	16h31	3.8	17h03	4.9	21h28	2.8	23h51
	0.2	22h37	0.6	23h07	0.3	4h33	0.9	8h04
11	2.7	4h57	3.9	5h27	5.1	9h42	3.1	12h12
	0.4	11h20	0.7	11h42	0.3	17h05	0.9	20h50

♦♦**Exercise 12.2**

Substitute Eq. (12.6) into Eq. (12.11b) to obtain wave celerity relationships (12.12).

Exercise 12.3

From the definition of velocity in Eqs. (12.14), demonstrate that the convective-acceleration terms in Eq. (12.9) are small compared with the local acceleration for waves of small amplitude ($\tilde{a} \ll \lambda$). [*Hint:* Examine the ratio $(v_x^2 + v_z^2)/\partial\Phi/\partial t$.]

Exercise 12.4

Demonstrate the relationship for the group velocity in Eq. (12.26) from substituting wave celerity relationships (12.12) into Eq. (12.25). Also show that $c_G = c/2$ for deep-water waves.

♦Exercise 12.5

Demonstrate that the energy of gravity waves propagates at the group velocity by substituting Eq. (12.6) into Eq. (12.27a). [*Hint:* Use the identities shown in Eq. (12.19).]

♦♦Exercise 12.6

Calculate the length of the saline wedge for the clearwater of the Matamek River. The flow velocity is 1 m/s and the flow depth is $\sim$4 m.

> *Answer:* $L_\Delta = 1{,}650$ m, $V_\Delta = 1.17$ m/s, and $h_{S1} = 0.4$ m, with $\rho_{\text{sea}} = 1{,}035$ kg/m^3.

♦♦Problem 12.1

Estimate the wave height, wavelength, and waveperiod in a river that is 10 m deep with at fetch length of 10 km. Consider a hurricane with a wind speed of 75–90 mph ($\sim$40 m/s).

> *Answer:* $\tilde{H}_s \simeq 3.5$ m, $\lambda_s \simeq 50$ m, and the period $T \simeq 6$ s. The celerity $c = \lambda_s/T = 8.3$ m/s, and the waves are depth limited.

♦Problem 12.2

Estimate the wave height, wavelength, and waveperiod in a river that is 4 m deep, with a fetch length of 1 km. Consider a storm with a wind speed of 65–75 mph ($\sim$30 m/s).

♦♦Problem 12.3

Compare the magnitude of the Coriolis acceleration in the Rhine River with the downstream gravitational acceleration and also with the centrifugal acceleration in a 2-km-radius river bend near Nijmegen in The Netherlands. (*Hint:* Consider the field measurements $h = 10$ m, $V \simeq 2$ m/s, and $S \simeq 1 \times 10^{-4}$).

Answer: $a_{cor} = 2 \times 10^{-4}$ m/s^2, $a_{cor}/gS \simeq 22\%$, and $a_{cor}/a_{cent} \simeq 10\%$.

♦♦Problem 12.4

Consider a branch of the Mississippi River delta with a discharge of 270,000 ft^3/s (7,640 m^3/s), a branch width of 1,500 ft (460 m), and a depth of 45 ft (13.7 m). Estimate the densimetric velocity V_Δ and the length L_Δ of the salinity wedge.

Answer: Assume that $\Delta\rho/\rho_m = 0.035$, $V = 4.0$ ft/s $= 1.2$ m/s, and $V_\Delta = 7.11$ ft/s $= 2.17$ m/s. The length $L_\Delta \simeq 16.5$ miles (26 km).

♦♦Problem 12.5

With reference to the tide propagation data in Case Study 12.2, examine the following characteristics:

(a) Estimate the floodwave celerity from the shallow-water equation and compare with field observations. (*Hint*: Assume a flow depth of ~10 m near Grondines, ~50 m near Québec, up to 100 m near Rimouski, and up to 200 m near Sept-Iles.)

(b) Compare the shape of tidal waves between Sept-Iles and Trois-Rivières and discuss the effects of bed roughness on tides.

(c) From Table CS.12.2.1, examine the days when the highest tides occur at various locations. Discuss the observations.

♦♦Computer Problem 12.1

From the data in Case Study 12.2, calculate at 30-min intervals the flow velocity at Québec on March 8, 2000. Assume a simple prismatic approximation for the river channel 120 km long with a width of 2 km, depth 15 m, constant inflow discharge of 10,000 m^3/s, and constant stage near Trois-Rivières. Why is there flow reversal at Québec?

Bibliography

AASHTO Task Force on Hydrology and Hydraulics, 1982, Highway Subcommittee on Design, *Hydraulic Analyses for the Location and Design of Bridges*, Vol. VII, Highway Drainage Guidelines, American Association of State Highway and Transportation Officials, 444 N. Capitol St., NW, Suite 225, Washington, D.C., 20001.

Abbott, M. B. and D. R. Basco, 1989, *Computational Fluid Dynamics*, Addison-Wesley Longman, London.

Abed, L. M., 1991, "Local scour around bridge piers in pressure flow," Ph.D. dissertation, Colorado State University, Fort Collins, CO.

Ables, J. H., Jr., 1978, "Filling and emptying system for Bay Springs Lock, Tennessee-Tombigbee Waterway, Mississippi," Tech. Rep. H-78-19, U.S. Army Corps of Engineer, Waterways Experiment Station.

Abt, S. R., M. S. Khattak, J. D. Nelson, J. F. Ruff, A Shaikh, R. J. Wittler, D. W. Lee, and N. E. Hinkle, 1987, "Development of riprap design criteria by riprap testing in flumes: phase I," NUREG/CR-4651 ORNL/TM-10100, Division of Waste Management, U.S. Nuclear Regulatory Commission, Washington, D.C.

Ackers, P. and F. G. Charlton, 1970, "Dimensional analysis of alluvial channels with special reference to meander length," J. Hydraul. Res., IAHR, **8**, 287–314.

Akashi, N. and T. Saitou, 1986, "Influence of water surface on scour from vertical jets," J. Hydrosci. Hydraul. Eng., Jpn. Soc. Civil Eng. 55–69.

Albertson, M. L., Y. B. Dai, R. A. Jensen, and H. Rouse, 1950, "Diffusion of submerged jets," Trans. ASCE, **115**, 639–697.

Altinbilek, H. D. and Y. Basmaci, 1973, "Localized scour at the downstream of outlet structures," in *Proceedings of the Eleventh Congress on Large Dams*, Intl. Commission on Large Dams, pp. 105–121.

Anderson, A. G., 1967, "On the development of stream meanders," in *Proceedings of the Twelfth Congress*, IAHR, Delft, The Netherlands, Vol. 1, pp. 370–378.

Anthony, D. J., 1992, "Bedload transport and sorting in meander bends, Fall River, Rocky Mountain National Park, Colorado," Ph.D. dissertation, Dept. of Earth Resources, Colorado State University, Fort Collins, CO.

Anthony, D. J. and M. D. Harvey, 1991, "Stage-dependent cross-section adjustments in a meandering reach of Fall River, Colorado," Geomorphology **4**, 187–203.

Arcement, G. K. and V. R. Schneider, 1984, "Guide for selecting Manning's roughness coefficients for natural channels and flood plains," USGS Water Supply Paper 2339, U.S. Geological Survey, Washington, D.C.

409

Ashmore, P. and G. Parker, 1983, "Confluence scour in coarse braided streams," Water Resources Res., AGU, **19**, 392–402.

Bagnold, R. A., 1960, "Some aspects of the shape of river meanders, physiographic and hydraulic studies of rivers," USGS Prof. Paper 282E, pp. 135–181, U.S. Geological Survey, Washington, D.C.

Barkau, R. L., 1993, "UNET – one dimensional unsteady flow through a full network of open channels," Rep. CPD-66, U.S. Army Corps of Engineering, Hydrologic Engineering Center, Davis, CA.

Bauer, T., C. Leon, G. Richard, and P. Y. Julien, 2000, Middle Rio Grande, Bernalillo Bridge to San Acacia, hydraulic geometry, discharge and sediment data base and report, 3 vols. prepared for the U.S. Bureau of Reclamation, Albuquerque, NM.

Bernard, R. S. and M. L. Schneider, 1992, "Depth-averaged numerical modeling for curved channels," Tech. Report HL-92-9, U.S. Army Engineer Waterways Experiment Station, Vicksburg, MS.

Biedenharn, D. S., P. G. Combs, G. J. Hill, C. F. Pinkard, Jr., and C. B. Pinkiston, 1989, "Relationship between the channel migration and radius of curvature on the Red River," in *Proceedings of the International Symposium on Sediment Transport Modeling*, ASCE, New York, pp. 536–541.

Blaisdell, F. W. and C. L. Anderson, 1989, "Scour at cantilevered outlets: plunge pool energy dissipator design criteria," ARS-76, U.S. Dept. of Agriculture, Agricultural Research Service, Washington, D.C.

Blench, T., 1969, *Mobile-Bed Fluviology*, Univ. of Alberta Press, Edmonton, Canada.

Bogardi, J. L., 1974, *Sediment Transport in Alluvial Streams*, Akademiai Kiado, Budapest.

Bondurant, D. C., 1963, *Missouri River Division*, U.S. Army Corps of Engineers, Omaha, NE.

Bormann, N. E. and P. Y. Julien, 1991, "Scour downstream of grade-control structures," J. Hydraul. Eng., ASCE, **117**, 579–594.

Boyce, R., 1975, "Sediment routing and sediment delivery ratios, in present and prospective technology for predicting sediment yields and sources," USDA-ARS-S-40, pp. 61–65, U.S. Dept. of Agriculture, Washington, D.C.

Bradley, J. B., 1984, "Transition of a meandering river to a braided system due to high sediment concentration flows," in *River Meandering, Proceedings of the Conference River '83*, ASCE, pp. 89–100.

Brahms, A., 1753, Anfangsgründe der Deich und Wasserbaukunst, Zurich.

Bray, D. I., 1982a, "Flow resistance in gravel-bed rivers," in *Gravel-Bed Rivers*, Wiley, New York, Chap. 6, pp. 109–137.

Bray, D. I., 1982b, "Flow resistance in gravel-bed rivers," in *Gravel-Bed Rivers*, Wiley, New York, Chap. 19, pp. 517–552.

Bray, D. I., 1980, "Evaluation of effective boundary roughness for gravel-bed rivers," Can. J. Civ. Eng. **7**, 392–397.

Bray, D. I., 1979, "Estimating average velocity in gravel bed rivers," J. Hydraul. Div., ASCE, **105**, 1103–1122.

Breusers, H. N. C., 1967, "Two-dimensional local scour in loose sediment," in *Closure of Estuarine Channels in Tidal Regions*, Delft Hydraulics Pub. No. 64, Delft, The Netherlands.

Breusers, H. N. C., G. Nicolet, and H. W. Shen, 1977, "Local scour around cylindrical piers," J. Hydraul. Res., IAHR, **15**, 211–252.

Brice, J. C., 1984a, "Assessment of channel stability at bridge sites," Transportation Research Record, Vol. 2, No. 950, TRB, 2101 Constitution Ave., Washington, D.C. 20418.

Brice, J. C., 1984b, "Planform properties of meandering rivers," in *River Meandering: Proceedings of the ASCE, Conference Rivers '83*, edited by C.M. Elliott, ASCE, New York.

Brice, J. C., 1981, "Stability of relocated stream channels," FHWA/RD-80/158, Federal Highway Administration, Washington, D.C.

Brice, J. C. and J. C. Blodgett, 1978, Countermeasures for Hydraulic Problems at Bridges, Vols. 1 and 2, FHWA/RD-78-162 & 163, Federal Highway Administration, U.S. Dept. of Transportation, Washington, D.C.

Brown, S. A., 1985a, "Design of spur-type streambank stabilization structures, final report," FHWA/RD-84-101, Federal Highway Administration, Washington, D.C.

Brown, S. A., 1985b, "Streambank stabilization measures for highway stream crossings," Executive Summary, Federal Highway Administration, Rep. FHWA-RD-80-160, U.S. Dept. of Transportation, Washington, D.C.

Brown, S. A., 1985c, "Streambank stabilization measures for highway engineers," FHWA/RD-84-100, Federal Highway Administration, McLean, VA.

Brown, S. A. and E. S. Clyde, 1989, "Design of riprap revetment," Hydraulic Engineering Circ. No. 11, FHWA/IP-89-016, for the Federal Highway Administration, Washington, D.C.

Brown, S. A., R. S. McQuivey, and T. N. Keefer, 1980, "Stream channel degradation and aggradation: analysis of impacts to highway crossings," FHWA/RD-80-159, Federal Highway Administration, Washington, D.C.

Burrows, R. L., W. W. Emmett, and B. Parks, 1981, "Sediment transport in the Tanana River near Fairbanks, Alaska, 1977–79," USGS Water Resources Investigations 81-20, prepared in cooperation with U.S. Army Engineer District, Anchorage, AK.

Buska, J. S., E. F. Chacho, C. M. Collins, and L. W. Gatto, 1984, "Overview of Tanana River monitoring and research studies near Fairbanks, Alaska," prepared for U.S. Army Engineering District, Anchorage, Alaska, by U.S. Army Cold Regions Research and Engineering Laboratory, Hanover, NH.

Callander, R. A., 1978, "River meandering," Ann. Rev. Fluid Mech. **10**, 129–158.

Callander, R. A., 1969, "Instability and river channels," J. Fluid Mech., CUP, **36**, 465–480.

Carstens, M. R., 1966, "Similarity laws for localized scour," J. Hydraul. Div., ASCE, **92**, 13–36.

Chabert, J., M. Remillieux, and I. Spitz, 1962, "Correction des rivières par panneaux de fond," Bull. Centre Rech. d'Essais de Chatou, France, Sér. A. **1**, 49–63.

Chabert, J. and P. Engeldinger, 1956, "Étude des affouillements autour des piles de ponts," Laboratoire National d'Hydraulique, Chatou, France.

Chang, H. H., 1988, *Fluvial Processes in River Engineering*, Wiley-Interscience, Wiley, NY.

Chang, H. H., 1984, "Analysis of river meanders," J. Hydraul. Div., ASCE, **110**, 37–50.

Chang, H. H., 1980, "Stable alluvial canal design," J. Hydraul. Div., ASCE, **106**, 873–891.

Chang, H. H., 1979a, "Geometry of rivers in regime," J. Hydraul. Div., ASCE, **105**, 691–706.

Chang, H. H., 1979b, "Minimum stream power and river channel patterns," J. Hydrol. **41**, 303.

Charlton, F. G., 1982, "River stabilization and training in gravel-bed rivers," in *Gravel-Bed Rivers*, Wiley, New York, Chap. 23, pp. 635–657.

Chauvin, J. L., 1962, "Similitude des modèles de cours d'eau à fond mobile," Bull. Centre Rech. d'Essais de Chatou, France, Sér. A **1**, 64–91.

Chee, S. P. and E. M. Yuen, 1985, "Erosion of unconsolidated gravel beds," Can. J. Civ. Eng. **12**, 559–566.

Chen, C. L., 1976, "Flow resistance in broad shallow grassed channels," J. Hydraul. Div., ASCE, **102**, 307–322.

Chien, N., 1957, "A concept of the regime theory," Trans. ASCE, **122**, 785–793.

Chien, N., 1954, "Meyer–Peter formula for bedload transport and Einstein bedload function," Rep. 7, Institute of Engineering Research, Univ. of California, Berkeley.

Chiew, Y. M., 1991, "Prediction of maximum scour depth at submarine pipelines," J. Hydraul. Eng., ASCE, **117**, 452–466.

Chiew, Y. M., 1990, "Mechanics of local scour depth at submarine pipelines," J. Hydraul. Eng., ASCE, **116**, 515–529.

Chitale, S. V., 1973, "Theory and relationship of river channel patterns," J. Hydrol. **19**, 285–308.

Chitale, S. V., 1970, "River channel patterns," J. Hydraul. Div., ASCE, **96**, 201–221.

Combs, P. G., 1994, "Prediction of the loop rating curve in alluvial rivers," Ph.D. dissertation, Dept. of Civil Engineering, Colorado State University, Fort Collins, CO.

Copeland, R. R. and W. A. Thomas, 1989, "Corte Madera Creek sedimentation study: numerical model investigation," Tech. Rep. HL-89-6, U.S. Army Engineer Waterways Experiment Station, Vicksburg, MS.

Creager, W. P., J. D. Justin, and J. Hinds, 1945, *General Design*, Vol. 1 of Engineering for Dams Series, Wiley, New York.

Cunge, J. A., 1969, "On the subject of a flood propagation computation method (Muskingum method)," J. Hydraul. Res., IAHR, **7**, 205–230.

Delft Hydraulics, 1996, "Case: gravity based structure," Rep. Q 2225, Delft Hydraulics, Delft, The Netherlands.

DeVriend, H. J., 1977, "A mathematical model of steady flow in curved shallow channels," J. Hydraul. Res., ASCE, **15**, 37–53.

Doehring, F. K. and S. R. Abt, 1994, "Drop height influence on outlet scour," J. Hydraul. Eng., ASCE, **120**, 1470–1476.

Einstein, A., 1926, Die Ursache der Mäanderbildung der Flussläufe und des sogenannten Baerschen Gesetzes, Naturwissenschaften, **11**.

Elliott, Charles M., ed., 1984, *River Meandering, Proceedings of the ASCE Conference Rivers '83*, ASCE, New York.

Engelund, F., 1974, "Flow and bed topography in channel bends," J. Hydraul. Div., ASCE, **100**, 1631–1648.

Engelund, F., 1970, "Instability of erodible beds," J. Fluid Mech., CUP, **42**, 225–244.

Engelund, F. and O. Skovgaard, 1973, "On the origin of meandering and braiding in alluvial streams," J. Fluid Mech., CUP, **57**, 289–302.

Ettema, R., 1980, "Scour at bridge piers," Rep. 216, University of Auckland, Auckland, New Zealand.

Fahlbusch, F. E., 1994, "Scour in rock riverbeds downstream of large dams," Int. J. Hydropower Dams, 1(4), 30–32.

FAP24, 1996, "Floodplain levels and bankfull discharge," River Survey Project, Special Reports No. 6 and No. 7, Government of the People's Republic of Bangladesh, October.

Farhoudi, J., and K. V. H. Smith, 1985, "Local scour profiles downstream of hydraulic jumps," J. Hydraul. Res., IAHR, **23**, 343–358.

Federal Highway Administration, 1989, "Design of riprap revetments," HEC-11, Turner-Fairbank Highway Research Center, VA.

Fennema, R. J. and M. H. Chaudhry, 1990, "Numerical solution of two-dimensional transient free-surface flows," J. Hydraul. Eng., ASCE, **116**, 1013–1034.

Fennema, R. J. and M. H. Chaudhry, 1986, "Explicit numerical schemes for unsteady free-surface flows with shocks," Water Resources Res., AGU, **22**, 1923–1930.

Ferguson, R. I., 1984, "The threshold between meandering and braiding, channels, and channel control structures," in *Proceedings of the First International Conference on Hydraulic Design in Water Resource Engineering*, K. V. H. Smith, ed., Springer-Verlag, New York, pp. 6-15–6-30.

Fortier, S. and F. C. Scobey, 1926, "Permissible canal velocities," Trans. ASCE, **89**, Paper 1588, 940–984.

Franco, J. J. and C. D. McKellar, 1968, "Navigation conditions at Lock and Dam No. 3, Arkansas River, Arkansas and Oklahoma," Tech. Rep. H-68-8, U.S. Army Corps of Engineers Waterways Experiment Station.

Fredsøe, J., 1979, "Unsteady flow in straight alluvial streams: modification of individual dunes," J. Fluid Mech., CUP, **91**, 497–512.

Fredsøe, J., 1978, "Meandering and braiding of rivers," J. Fluid Mech., CUP, **84**, 609–624.

Frenette, M. and B. Harvey, 1971, *Hydraulique Fluviale I*, Notes de Cours, Les Presses de l'Université Laval, Québec, Canada.

Frenette, M. and P. Y. Julien, 1980, "Rapport synthèse sur les caractéristiques hydrophysiques du bassin de la rivière Matamec," Rep. CENTREAU-80-06, Laval University, Québec, Canada.

Fournier, F., 1969, "Transports solides effectués par les cours d'eau," Bull. IASH, **14**, 7–49.

Froehlich, D. C., 1988, "Analysis of on site measurements of scour at piers," in *Proceedings of the ASCE National Hydraulic Engineering Conference*, ASCE, New York.

Fujita, Y. and Y. Muramoto, 1985, "Studies on the process and development of alternate bars," Bull. Disaster Prevention Res. Inst. Kyoto Univ. **35**, 55–86.

Fujita, Y. and Y. Muramoto, 1982, "Experimental study on shear channel processes in alluvial rivers," Bull. Disaster Prevention Res. Inst. Kyoto Univ. **32**, 49–96.

Galay, V. J., 1983, "Causes of river bed degradation," Water Resources Res., AGU, **19**, 1057–1090.

Gatto, L. W., 1984, "Tanana River monitoring and research program: relationships among bank recession, vegetation, soils, sediments, and permafrost on the Tanana

River near Fairbanks, Alaska," Special Rep. 84-21, for U.S. Army Engineer District, Anchorage, Alaska, by the U.S. Army Cold Regions Research and Engineering Laboratory, Hanover, NH.

Gessler, D., B. Hall, M. Spasojevic, F. Holly, H. Pourtaheri and N. Raphelt, 1999, "Application of 3D mobile bed hydrodynamic model," J. Hydraul. Eng., ASCE.

Gill, M. A., 1972, "Erosion of sand beds around spur dikes," J. Hydraul. Div., ASCE, **98**, 1587–1602.

Graf, W. H., 1984, *Hydraulics of Sediment Transport*, McGraw-Hill, New York.

Green, W. H. and G. A. Ampt, 1911, "Studies of soil physics, 1, the flow of air and water through soils," J. Agric. Sci. **4**, 1–24.

Gregory, K. J., 1977, *River Channel Changes*, Univ. of Southampton, Wiley-Interscience, New York.

Griffiths, G. A., 1981, "Stable-channel design in gravel-bed rivers," J. Hydrol. **52**, 291–305.

Grozier, R. V., J. F. McCain, L. F. Lang, and D. R. Merriman, 1976, "The Big Thompson River flood of July 31–August 1, 1976, Larimer County, Colorado," Flood Information Report, U.S. Geological Survey and Colorado Water Conservation Board.

Hagerty, D. J., 1992, "Identification of piping and sapping erosion of streambanks," Contract Rep. HL-92-1, University of Louisville, Louisville, KY, for U.S. Army Engineer Waterways Experiment Station, Vicksburg, MS.

Hagerty, D. J., M. F. Spoor, and J. F. Kennedy, 1986, "Interactive mechanisms of alluvial-stream bank erosion," in *River Sedimentation*, Vol. III of the Proceedings of the Third International Symposium on River Sedimentation, Univ. of Mississippi, Oxford Mississippi, MS, pp. 1160–1168.

Hartley, D. M. and P. Y. Julien, 1992, "Boundary shear stress induced by raindrop impact," J. Hydraul. Res., IAHR, **30**, 341–359.

Hayashi, T., 1970a, "The formation of meanders in rivers," Trans. Jpn. Soc. Civ. Eng., No. 180.

Hayashi, T., 1970b, "On the cause of the initiations of meandering of rivers," Trans. Jpn. Soc. Civ. Eng. **2**, 235–239.

Henderson, F. M., 1963, *Stability of Alluvial Channels*, ASCE, New York, Vol. 128, Part I, No. 3440, pp. 657–686.

Henderson, F. M., 1961, "Stability of alluvial channels," J. Hydraul. Div., ASCE, **87**, 109–138.

Hey, R. D., 1978, "Determinate hydraulic geometry of river channels," J. Hydraul. Div., ASCE, **104**, 869–885.

Hey, R. D. and C. R. Thorne, 1986, "Stable channels with mobile gravel beds," J. Hydraul. Eng., ASCE, **112**, 671–689.

Hey, R. D., J. C. Bathurst, and C. R. Thorne, eds. 1982, *Gravel-Bed Rivers: Fluvial Processes, Engineering and Management*, Wiley, New York.

Hickin, E. J. and G. C. Nanson, 1975, "The character of channel migration on the Beatton River, northeast B. C., Canada," Geol. Soc. Am. Bull. **86**, 487–494.

Hirano, M., 1973, "River-bed variation with bank erosion," Proc. Jpn. Soc. Civ. Eng. No. 210, 13–20.

Hoffmans, G. J. C. M., 1995, "Ontgrondingen rondom brugpijlers en aan de kop van kribben," Rep. W-DWW-94-312, Ministry of Transport, Public Works and Water Management, Road and Hydraulic Engineering Div., Delft, The Netherlands.

Hoffmans, G. J. C. M., 1994b, "Scour due to plunging jets," (in Dutch), Rep. W-DWW-94-302, Ministry of Transport, Public Works and Water Management, Road and Hydraulic Engineering Div., Delft, The Netherlands.

Hoffmans, G. J. C. M., 1994c, "Scour due to submerged jets," (in Dutch), Rep. W-DWW-94-303, Ministry of Transport, Public Works and Water Management, Road and Hydraulic Engineering Div., Delft, The Netherlands.

Hoffmans, G. J. C. M. and K. W. Pilarczyk, 1995, "Local scour downstream of hydraulics structures," J. Hydraul. Eng., ASCE, **121**, 326–340.

Hoffmans, G. J. C. M. and R. Booij, 1993b, "Two-dimensional mathematical modeling of local-scour holes," J. Hydraul. Res., IAHR, **31**, 615–634.

Hoffmans, G. J. C. M. and H. J. Verheij, 1997, *Scour Manual*, Balkema, Rotterdam.

Hooke, J. M., 1979, "An analysis of the processes of river bank erosion," J. Hydrol. **42**, 39–62.

Hooke, R. L., 1975, "Distribution of sediment transport and shear stress in a meander bend," J. Geol. **83**, 543–560.

Hussein, K. and V. H. Smith, 1986, Flow and bed deviation angle in curved open channels, J. Hydraul. Res., IAHR, **24**, 93–108.

Ikeda, S., 1984a, "Flow and bed topography in channels with alternate bars," in *River Meandering, Proceedings of the ASCE Conference Rivers '83*, ASCE, New York, pp. 733–746.

Ikeda, S., 1984b, "Prediction of alternate bar wavelength and height," J. Hydraul. Eng., ASCE, **110**, 371–386.

Ikeda, S., 1982, "Lateral bedload on side slopes," J. Hydraul. Div., ASCE, **108**, 1369–1373.

Ikeda, S., G. Parker, and K. Sawai, 1981, "Bend theory of river meanders, part 1. Linear development," J. Fluid Mech., CUP, **112**, 363–377.

Inglis, C. C., 1949, "The effect of variations in charge and grade on the slopes and shapes of channels," in *Proceedings of the Third Congress*, IAHR, Delft, The Netherlands, Vol. 1, Paper II, pp. II.1.1–II1.10.

Inglis, C. C., 1947, "Meanders and the bearing on river training," Institute of Civil Engineers, Maritime and Waterways, Paper 7.

Isbash, S. V., 1935, "Construction of dams by dumping stones in flowing water," translated by A. Dorijikov, U.S. Army Engineer District, Eastport, ME.

Ivicsics, L., 1975, *Hydraulic Models*, Research Institute for Water Resources Movement, Budapest.

Jain, S. C. and E. E. Fischer, 1980, "Scour around bridge piers at high velocities," J. Hydraul. Div., ASCE, **106**, 1827–1842.

Jansen, P., L. van Bendegom, J. van den Berg, M. de Vries, and A. Zanen, 1979, *Principles of River Engineering: The Non-Tidal Alluvial River*, Pitman, San Francisco.

Jarrett, R. D. and J. E. Costa, 1986, "Hydrology, geomorphology and dam-break modeling of the July 15, 1982 Lawn Lake dam and Cascade Lake dam failures, Larimer County, Colorado," USGS Professional Paper 1369, U.S. Geological Survey, Washington, D.C.

Jarrett, R. D. and J. E. Costa, 1985, "Hydrology, geomorphology and dam-break modeling of the July 15, 1982 Lawn Lake dam and Cascade Lake dam failures, Larimer County, Colorado," USGS Open File Rep. 84-612, U.S. Geological Survey, Washington, D.C.

Johnson, P. A., 1995, "Comparison of pier-scour equations using field data," J. Hydraul. Eng., ASCE, **121**, 626–629.

Johnson, P. A., 1992, "Reliability-based pier scour engineering," J. Hydraul. Eng., ASCE, **118**, 1344–1358.

Jones, J. S., 1989, "Laboratory studies of the effect of footings and pile groups on bridge pier scour," in *Proceedings of the 1989 Bridge Scour Symposium*, Federal Highway Administration, Washington, D.C.

Jones, J. S., 1983, "Comparison of prediction equations for bridge pier and abutment scour," TRB Record 950, Second Bridge Engineering Conference, Vol. 2, Transportation Research Board, Washington, D.C.

Julien, P. Y., 1996, "Transforms for runoff and sediment transport," J. Hydrolog. Eng., ASCE, **1**, 114–122.

Julien, P. Y., 1995, *Erosion and Sedimentation*, Cambridge Univ. Press, New York.

Julien, P. Y., 1989, "Géométrie hydraulique des cours d'eau à lit alluvial," in *Proceedings of the IAHR* Conference, Natural Resources Council, Ottawa, Canada, B9-16.

Julien, P. Y., 1988, "Downstream hydraulic geometry of noncohesive alluvial channels," in *International Conference on River Regime*, Wiley, New York, pp. 9–16.

Julien, P. Y., 1985, Planform geometry of meandering alluvial channels, Report CER84-85PYJ5, Colorado State University, Fort Collins, Colorado, p. 49.

Julien, P. Y., 1982, "Prédiction d'apport solide pluvial et nival dans les cours d'eau nordiques à partir du ruissellement superficiel," Ph.D. dissertation, Department of Civil Engineering, Laval University, Québec.

Julien, P. Y., 1979, "Erosion de bassin et apport solide en suspension dans les cours d'eau nordiques," M.Sc. thesis, Department of Civil Engineering, Laval University, Québec.

Julien, P. Y. and D. Anthony, Bedload motion by size fractions in meander beds, J. Hydraulic Research, IAHR, Delft, The Netherlands (in press).

Julien, P. Y. and M. Frenette, 1987, "Macroscale analysis of upland erosion," Hydrolog. Sci. J. **32**, 347–358.

Julien, P. Y. and M. Frenette, 1986, "LAVSED II – A model for predicting suspended load in northern streams," Can. J. Civ. Eng. **13**, 162–170.

Julien, P. Y. and M. Frenette, 1985, "Modeling of rainfall erosion," J. Hydraul. Eng., ASCE, **111**, 1344–1359.

Julien, P. Y. and M. Gonzalez del Tanago, 1991, "Spatially-varied soil erosion under different climates," Hydrolog. Sci. J. **36**, 511–524.

Julien, P. Y. and D. M. Hartley, 1986, Formation of roll waves in laminar sheet flow, Journal of Hydraulic Research, IAHR, The Netherlands, Vol. 24, No. 1, pp. 5–17.

Julien, P. Y. and D. B. Simons, 1985, "Sediment transport capacity of overland flow," Trans. ASAE, **28**, 755–762.

Julien, P. Y. and D. B. Simons, 1984, "Analysis of hydraulic geometry relationships in alluvial channels," Rep. CER83-84PYJ-DBS45, Dept. of Civil Engineering, Colorado State University.

Julien, P. Y. and J. Wargadalam, 1995, "Alluvial channel geometry: theory and applications," J. Hydraul. Eng., ASCE, **121**.

Julien, P. Y., B. Saghafian, and F. Ogden, 1995, "Raster-based hydrologic modeling of spatially-varied surface runoff," Water Resources Bull., AWRA, **31**, 523–536.

Kalkwijk, J. P. and H. J. DeVriend, 1980, "Computations of the flow in shallow river bends," J. Hydraul. Res., IAHR, **18**, 327–342.

Keefer, T. N., R. S. McQuivey, and D. B. Simons, 1980, "Interim report – Stream channel degradation and aggradation: causes and consequences to highways," FHWA/RD-80/038, Federal Highway Administration, Washington, D.C.

Kellerhals, R., 1967, "Stable channels with gravel-paved beds," J. Waterways Harbors Div., ASCE, 63–84.

Kellerhals, R. and M. Church, 1989, "The morphology of large rivers: characterization and management," in *Proceedings of the International Large River Symposium, 1986*, D. P. Dodge, ed., Dept. of Fisheries and Oceans, Ottawa, Canada, pp. 31–48.

Kennedy, R. G., 1895, "The prevention of silting in irrigation canals," Minutes of Proc. Inst. Civ. Eng. London **119**, 281–290.

Keown, M. P., 1983, "Streambank protection guidelines for landowners and local governments," U.S. Army Corps. Engineers, Waterways Experiment Station, Vicksburg, MS.

Keulegan, G. H., 1938, Laws of turbulent flow in open channels, J. Res. Natl. Bur. Stand. **21**, 707–741; see also U.S. Bureau of Reclamation, Washington, D.C., Research Paper 1151.

Klaassen, G. J., 1992, "Experience from a physical model for a bridge across a braided river with fine sand as bed material," in *Proceedings of the Fifth International Symposium on River Sedimentation*, Karlsruhe, Germany, pp. 509–520.

Klaassen, G. J., 1990, "On the scaling of braided sand-bed rivers," in H. W. Shen, ed., *Mobile Bed Physical Models*, Kluwer Academic, Boston, MA, pp. 56–71.

Klaassen, G. J. R., E. Mosselman, G. Massekik, H. Brühl, M. Huisink, E. Kooman, and A. C. Seymousbergen, 1993, "Plan for changes in large braided sand-bed rivers," Delft Hydraulics Publication No. 480, December.

Klaassen, G. J. R. and B. H. J. van Zarter, 1990, "On cutoff ratios of curved channels," Delft Hydraulics Publication No. 444, December.

Klaassen, G. J. R. and K. Vermeer, 1988, "Confluence scour in large braided rivers with fine bed material," Presented at the *International Conference on Fluvial Hydraulics*, Budapest, Hungary.

Klaassen, G. J. R. and J. J. van der Zwaard, 1974, "Roughness coefficients of vegetated flood plains," J. Hydraul. Res., IAHR, **12**, 43–63.

Knighton, D., 1998, *Fluvial Forms and Processes*, Arnold, Baltimore, MD.

Kobus, H., P. Leister, and B. Westrich, 1979, "Flow field and scouring effects of steady and pulsating jets impinging on a movable bed," J. Hydraul. Res., IAHR, **17**, 175–192.

Koch, F. G. and C. Flokstra, 1980, "Bed level computations for curved alluvial channels," in *Proceeding of the XIX Congress of the International Association for Hydraulic Research*, Delft, The Netherlands, Vol. 2, p. 357.

Kuhnle, R. A., 1993, "Incipient motion of sand gravel sediment mixtures," J. Hydraul. Div., ASCE, **119**, 1400–1415.

Lacey, G., 1929–30, "Stable channels in alluvium," Minutes Proc. Inst. Civ. Eng. London, **229**, 259–292.

Lagasse, P. F., J. D. Schall, F. Johnson, E. V. Richardson, and F. M. Chang, 1995, "Stream stability at highway structures," Hydraulic Engineering Circular No. 20, Report No. FHWA-IP-90-014, Federal Highway Administration, Washington, D.C.

Lai, A. T., 1998, "Bedforms in the Waal River, characterization and hydraulic roughness," M.Sc. thesis, H. E. 024, International Institute for Infrastructural, Hydraulic and Environmental Engineering, IHE Delft, The Netherlands.

Langbein, W. B. and L. B. Leopold, 1966, "River meander—theory of minimum variance," USGS Prof. Paper 422-H, U.S. Geological Survey, Washington, D.C.

Langbein, W. B. and S. A. Schumm, 1958, "Yield of sediment in relation to mean annual precipitation," Trans. AGU, **39**, 1076–1084.

Langhaar, H. L., 1956, *Dimensional Analysis and Theory of Models*, Wiley, New York.

Lane, E. W., 1957, "A study of the shape of channels formed by natural streams flowing in erodible material," MRD Sediment Series No. 9, U.S. Army Corps of Engineers, Missouri River Div., Omaha, Nebraska.

Lane, E. W., 1955a, "Design of stable channels," Trans. ASCE, **120**, 1234–1279.

Lane, E. W., 1955b, "The importance of fluvial geomorphology in hydraulic engineering," Proc. ASCE, **81**, 1–17.

Lane, E. W., 1953, "Progress report on studies on the design of stable channels of the Bureau of Reclamation," Proc. ASCE, **79**.

Larras, J., 1963, "Profondeurs maximales d'érosion des fonds mobiles autour des piles en rivière," Ann. Ponts Chaussées **133**, 410–424.

Laursen, E. M., 1980, "Predicting scour at bridge piers and abutments," General Rep. No. 3, Arizona Dept. of Transportation, Phoenix, AZ.

Laursen, E. M., 1963, "An analysis of relief bridge scour," Proc. ASCE, **89**, 93–118.

Laursen, E. M., 1960, "Scour at bridge crossings," Proc. ASCE, **86**, 39–54.

Laursen, E. M. and M.W. Flick, 1983, "Scour at sill structures," Final Report, Rep. FHWA/AZ83/184, Arizona Dept. of Transportation, Arizona Transportation and Traffic Institute, Tempe, AZ.

Lee, J. K. and D. C. Froehlich, 1989, "Two dimensional finite element modeling of bridge crossings," Rep. FHWA-RD-88-149, Federal Highway Administration, U.S. Dept. of Transportation, Washington, D.C.

Leon, C., G. Richard, T. Bauer, and P. Y. Julien, 1999, "Middle Rio Grande Cochiti to Bernalillo Bridge, hydraulic geometry, discharge and sediment data base and report," Vols. I, II, and III, prepared for the U.S. Bureau of Reclamation, Albuquerque, NM, January.

Leonard, B. P., 1979, "A stable and accurate convective modeling procedure based on quadratic upstream interpolation," Comput. Meth. Appl. Mech. Eng. **19**, 59–98.

Leopold, L. B. and T. Maddock, 1954, *The Flood Control Controversy*, Ronald Press, New York.

Leopold, L. B. and T. Maddock, Jr., 1953, "The hydraulic geometry of stream channels and stone physiographic implications," USGS Prof. Paper 252, U.S. Government Printing Office, Washington, D.C.

Leopold, L. B. and M. G. Wolman, 1960, "River meanders, Geol. Soc. Am. Bull. **71**, 769–794.

Leopold, L. B. and M. G. Wolman, 1957, "River channel patterns: braided, meandering and straight," USGS Prof. Paper 282-B, U.S. Government Printing Office, Washington, D.C.

Leopold, L. B., M. G. Wolman, and J. P. Miller, 1964, *Fluvial Processes in Geomorphology*, Freeman, San Francisco.

Leopold, L. B., R. A. Bagnold, R. G. Wolman, and L. M. Brush, 1960, "Flow resistance in sinuous or irregular channels," USGS Prof. Paper 282-D, U.S. Geological Survey, Washington, D.C., pp. 111–134.

Li, R. M. and H. W. Shen, 1973, "Effect of tall vegetation on flow and sediment," J. Hydraul. Div., ASCE, **99**(5).

Liggett, J. A. and J. A. Cunge, 1975, "Numerical methods of solution of the unsteady flow equations," in K. Mahmood and V. Yevjevich, eds., *Unsteady Flow in Open Channels*, Water Resources Pub., Fort Collins, CO.

Liggett, J. A. and D. A. Woolhiser, 1967, "Difference solutions of shallow-water equations," Proc. ASCE, **93**, 39–71.

Limerinos, J. T., 1970, "Determination of the Manning's coefficient for measured bed roughness in natural channels," USGS Water Supply Paper 1891-B, U.S. Geological Survey, Washington, D.C.

Lindley, E. S., 1919, *Regime Channels*, Punjab Engineering Congress, Punjab, India.

Liu, H. K., F. M. Chang, and M.M. Skinner, 1961, "Effect of bridge construction on scour and backwater," Rep. CER60-HKL22, Dept." of Civil Engineering, Colorado State University.

MacCormack, R. W., 1969, "The effect of viscosity in hypervelocity impact cratering," American Institute of Aeronautics and Astronautics, Paper 69-354.

Maddock, T., 1970, "Indeterminate hydraulics of alluvial channels," J. Hydraul. Div. ASCE, **96**, 2309–2323.

Mason, P. J. and K. Arumugam, 1985, "Free jet scour below dams and flip buckets," J. Hydraul. Eng., ASCE, **111**, 220–235.

May, D. R. and P. Y. Julien, 1998, "Eulerian and Lagrangian correlation structures of convective rainstorms," Water Resources Res., AGU, **34**, 2671–2683.

Maynord, S. T., 1995, Corps riprap design guidance for channel protection in river, coastal and shoreline protection/erosion control using riprap and armourstone, J. Wiley and Sons, pp. 4–53.

Maynord, S. T., 1992, "Riprap stability: studies in near-prototype size laboratory channel," Tech. Rep. HL-92-5, U.S. Army Engineer Waterways Experiment Station, Vicksburg, MS.

Maynord, S. T., 1988, "Stable riprap size for open channel flows," Tech. Rep. HL-88-4, U.S. Army Engineer Waterways Experiment Station, Vicksburg, MS.

Mayerle, R., C. Nalluri, and P. Novak, 1991, "Sediment transport in rigid bed conveyances," J. Hydraul. Res., IAHR, **29**, 475–496.

McCain, J. F., L. R. Hoxit, R. A. Maddox, C. F. Chappell, F. Caracena, R. R. Scnoba, P. W. Schwidt, E. J. Crosby, W. R. Hansen, and J. M. Soule, 1979, "Storm and flood of July 31–August 1, 1976, in the Big Thompson River and Cache la Poudre River basin, Larimer and Weld Counties, Colorado," USGS Professional Paper 1115, U.S. Geological Survey, Washington, D.C.

McCarley, R. W., J. J. Ingram, B. J. Brown, and A. J. Reese, 1990, "Flood-control channel national inventory," Misc. Paper HL-90-10, U.S. Army Engineer Waterways Experiment Station, Vicksburg, MS.

McCuen, R. H., P. A. Johnson, and R. M. Ragan, 1995, "Hydrologic design of highways," Hydraulic Design Series No. 2, U.S. Department of Transportation, Federal Highway Administration, Washington, D.C.

Melville, B. W. and D. M. Dongol, 1992, "Bridge pier scour with debris accumulation," J. Hydraul. Div., ASCE, **118**.

Melville, B. W. and A. J. Sutherland, 1988, "Design method for local scour at bridge piers, J. Hydraul. Div., ASCE, **114**.

Mirtskhoulava, TS. Ye., 1991, "Scouring by flowing water of cohesive and noncohesive beds," J. Hydraul. Res. IAHR, **29**.

Mirtskhoulava, TS. Ye., 1988, "Basic physics and mechanics of channel erosion," Gidrometeoizdat, Leningrad.

Molinas, A., 1990, "Bridge stream tube model for alluvial river simulation (BRI-STARS)," User's Manual, National Cooperative Highway Research Program, Project No. HR15-11, Transportation Research Board, Washington, D.C.

Mollard, J. D. and J. R. Jones, 1984, "Airphoto interpretation and the Canadian landscape," Dept. of Energy, Mines and Resources, Ottawa, Canada, Canadian Government Publishing Centre, Supply and Services, Hull, Québec, Canada.

Molnar, D. K. and P. Y. Julien, 1998, "Estimation of upland erosion using GIS," J. Comput. Geosci. **24**, 183–192.

Murillo-Muñoz, R. E., 1998, "Downstream fining of sediments in the Meuse River," M.Sc. Thesis, H. H. 341, International Institute for Infrastructural Hydraulic and Environmental Engineering, Delft, The Netherlands.

Nanson, G. C. and E. J. Hickin, 1983, "Channel migration and incision on the Beatton River," J. Hydraul. Eng., ASCE, **109**, 327–337.

Nanson, G. C. and E. J. Hickin, 1986, "A statistical examination of bank erosion and channel migration in Western Canada," Bull. Geol. Soc. Am., **97**, 497–504.

Neill, C. R., ed., 1973, *Guide to Bridge Hydraulics*, Roads and Transportation Assn. of Canada, Univ. of Toronto Press, Toronto, Canada.

Neill, C. R. and E. K. Yaremko, 1988, "Regime aspects of flood control channelization," in *Proceeding of the International Conference on River Regime*, W. R. White, ed., Hydraulics Research Ltd., Wallingford, England and Wiley, New York.

Nelson, J. M. and J. D. Smith, 1989, "Flow in meandering channels with natural topography," in *River Meandering*, AGU Water Resources Monograph, AGU, Washington, D.C., pp. 69–102.

Nicollet, G., 1975, "Affouillement au pied des piles de pont eu milieu cohésif," in *Proceedings of the Sixteenth Congress of the IAHR*, International Association for Hydraulic Research, Delft Congress in Brazil, Paper B60, pp. 478–484.

Nezu, I. and H. Nakagawa, 1993, *Turbulence in Open-Channel Flows*, Balkema, Rotterdam.

Nordin, Jr., C. F., 1971, Graphical aids for determining scour depth in long contractions in Richardson et al. (1990).

Nordin, Jr., C. F., 1964, "Aspects of low resistance and sediment transport Rio Grande near Bernalillo, New Mexico," USGS Water Supply Paper 1498-H, U.S. Government Printing Office, Washington, D.C.

Nunnally, N. R. and F. D. Shields, 1985, "Incorporation of environmental features in flood control channel projects," Tech. Rep. E-85-3, U.S. Army Engineer Waterways Experiment Station, Vicksburg, MS.

O'Brien, J. S., P. Y. Julien, and W. T. Fullerton, 1993, "Two-dimensional water flood and mud flow simulation," J. Hydraul. Eng., ASCE, **119**, 244–261.

Odgaard, A. J., 1982, "Bed characteristics in alluvial channel bends," J. Hydraul. Div., ASCE, **108**, 1268–1281.

Odgaard, A. J., 1981, "Transverse slope in alluvial channel bends," J. Hydraul. Eng., ASCE, **107**, 1677–1694.

Ogden, F. L. and P. Y. Julien, 1994, "Two-dimensional runoff sensitivity to radar resolution," J. Hydrol. **128**, 1–18.

Osman, A. M. and C. R. Thorne, 1988, "Riverbank stability analysis I: theory," J. Hydraul. Eng., ASCE, **114**, 134–150.

Ouellet, Y., 1972, *"Compléments d'Hydraulique,"* Les Presses de l'Université Laval, Québec.

Parker, G., 1976, "On the cause and characteristic scales of meandering and braiding in rivers," J. Fluid Mech., CUP, **76**, 457–480.

Parker, G. and E. D. Andrews, 1985, "Sorting of bedload sediments by flow in meander bends," Water Resources Res., AGU, **21**, 1361–1373.

Pemberton, E. L. and J. M. Lara, 1984, "Computing degradation and local scour," *Technical Guidelines for Bureau of Reclamation*, Engineering Research Center, Denver, CO.

Petersen, M. S., 1986, *River Engineering*, Prentice-Hall, Englewood Cliffs, NJ.

PIANC, 1992, "Guidelines for the design and construction of flexible revetments incorporating geotextiles in marine environment," Supplement of PIANC Bulletin 78/79, Brussels.

Pilarczyk, K. W., 1995, Design tools related to revetments including riprap, river, coastal and shoreline protection: erosion control using riprap and armour stone, Wiley, New York.

Pilarczyk, K. W. and R. B. Zeidler, 1996, *Offshore Breakwaters and Shore Evolution Control*, Balkema, Rottermam.

Pitlick, J. C., 1985, "The effect of a major sediment influx on Fall River, Colorado," M.S. thesis, Dept. of Earth Resources, Colorado State University, Fort Collins, CO.

Portland Cement Association, 1984, "Soil-cement slope protection for embankments: planning and design," PCA Publication IS173.02W, Skokie, IL.

Ports, M.A., ed., 1989, *Hydraulic Engineering*, Proceedings of the ASCE National Conference on Hydraulic Engineering, ASCE, New York.

Preissmann, A., 1971, "Modèles pour le calcul de la propagation des crues," La Houille Blanche **26**, 219–224.

Preissmann, A., 1961, "Propagation des intumescences dans les canaux et rivières," *Congress of the French Association for Computation*, France.

Preissmann, A. and J. A. Cunge, 1961, "Calcul du mascaret sur machine électronique," La Houille Blanche **5**, 588–596.

Przedwojski, B., R. Blazejewski, and K. W. Pilarczyk, 1995, *River Training Techniques: Fundamentals, Design, and Applications*, Balkema, Rotterdam.

Rajaratnam, N., 1982, "Erosion by unsubmerged plane water jets," in *Applying Research to Hydraulic Practice*, ASCE, New York, pp. 280–288.

Rajaratnam, N., 1981, "Erosion by plane turbulent jets," J. Hydraul. Res., IAHR, **19**, 339–358.

Rajaratnam, N. and R. K. MacDougall, 1983, "Erosion by plane wall jets with minimum tailwater," J. Hydraul. Eng., ASCE, **109**, 1061–1064.

Rajaratnam, N. and B. A. Nwachukwu, 1983, "Erosion near groyne-like structures," J. Hydraul. Res., IAHR, **21**, 277–287.

Raudkivi, A. J., 1993, *Sedimentation: Exclusion and Removal of Sediment from Diverted Water*, Balkema, Rotterdam.

Raudkivi, A. J., 1986, "Functional trends of scour at bridge piers," J. Hydraul. Div., ASCE, **112**,

Raudkivi, A. J., 1976, *Loose Boundary Hydraulics* (2nd ed.), Pergamon, Oxford, U.K.

Raudkivi, A. J., and R. Ettema, 1985, "Scour at cylindrical bridge piers in armoured beds," J. Hydraul. Eng., ASCE, **111**, 713–731.

Rawls, W. J., D. J. Brakensiek, and N. Miller, 1983, "Green–Ampt infiltration parameters from soils data," J. Hydraul. Eng., ASCE, **109**, 62–70.

Richard, G., C. Leon, and P. Y. Julien, 2000, "Bernardo Reach geomorphic analysis – Middle Rio Grande, New Mexico," Report prepared for the U.S. Bureau of Reclamation.

Richardson, E. V. and P. F. Lagasse, 1996, "Stream stability and scour at highway bridges," Water Intl. **21**, 108–118.

Richardson, E. V. and S. R. Davis, 1995, "Evaluating scour at bridges," Hydraulic Engineering Circular 18, Federal Highway Administration Rep. No. FHWA-HI-96-031, U.S. Department of Transportation, Washington, D.C.

Richardson, E. V. and D. B. Simons, 1984, "Use of spurs and guidebanks for highway crossings," TRB Record 950, Second Bridge Engineering Conference, Vol. 2, Transportation Research Board, Washington, D.C.

Richardson, E. V., D. B. Simons, and P. Julien, 1990, "Highways in the river environment," FHWA-HI-90-016, Federal Highway Administration, U.S. Dept. of Transportation, Washington, D.C.

RIZA, 1999, H. Middelkoop, and C. O. G. van Haseler (eds.), "Twice a river, Rhine and Meuse in the Netherlands," RIZA Rep. No. 99.003, Arnhem, RIZA.

Rozovskii, I. L., 1957, "Flow of water in bends of open channels," Academy of Sciences of the Ukrainian SSR., Institute of Hydrology and Hydraulic Engineering, Kiev., Translation by Y. Prushansky, 1961 Israel Program for Scientific Translations, S. Monson, Jerusalem, PST Catalog No. 363.

Ruff, J. F., S. R. Abt, C. Mendoza, A. Shaikh, and R. Kloberdanz, 1982, "Scour at culvert outlets in mixed bed materials," Rep. FHWA/RD-82/011, Colorado State University, Fort Collins, CO.

Saint-Venant, J. C. B., 1871, "Théorie du mouvement non-permanent des eaux avec application aux crues des rivières et à l'introduction des marées dans leur lit," C. R. Acad. Sci. Paris, No. 73, 148–154, 237–240.

Schumm, S. A., 1991, *To Interpret the Earth, Ten Ways to be Wrong*, CUP, Cambridge, U.K.

Schumm, S. A., 1977, *The Fluvial System*, Wiley, New York.

Schumm, S. A., 1972, "River morphology," Benchmark Papers in Geology, Colorado State University, Fort Collins, CO, Dowden, Hutchinson & Ross, Inc.

Schumm, S., 1969, "River metamorphosis," J. Hydraul. Eng., ASCE, **95**, 255–273.

Schumm, S. A., 1963, "Sinuosity of alluvial rivers on the Great Plains," Bull. Geol. Soc. Am. **74**, 1089–1100.

Schumm, S. A., M. D. Harvey, and C. C. Watson, 1984, *Incised Channels: Morphology, Dynamics and Control*, Water Resources Publications, Littleton, CO.

Schumm, S. A., M. P. Mosley, and W. E. Weaver, 1987, *Experimental Fluvial Geomorphology*, Wiley, New York.

Schwab, G. O., R. K. Frevert, T. W. Edminster, and K. K. Barnes, 1981, *Soil Water Conservation Engineering* (3rd ed.), Wiley, New York.

Shen, H. W., ed., 1971a, *River Mechanics*, Fort Collins, CO, Vols. I and II.

Shen, H. W., 1971b, "Scour near piers," in *River Mechanics*, Colorado State University, Fort Collins, CO, Vol. II, Chap. 23, pp. 23.1–23.25.

Shen, H. W., V. R. Schneider, and S. S. Karaki, 1969, "Local scour around bridge piers," Proc. ASCE, **95**, 1919–1940.

Shields, A., 1936, Anwendung der Aehnlichkeitsmechanik und der Turbulenzforschung auf die Geschiebebewegung, Mitteilungen der preussischen Versuchanstalt für Wasserbau und Schiffbau, Berlin.

Simons, D. B., 1957, "Theory and design of stable channels in alluvial material," Ph.D. dissertation, Colorado State University, Fort Collins, CO.

Simons, D. B. and M. L. Albertson, 1963, "Uniform water conveyance channels in alluvial material," Trans. ASCE, **128**, 65–167.

Simons, D. B. and P. Y. Julien, 1983, "Engineering analysis of river meandering," Keynote paper in *River Meandering, Proceedings of the ASCE Conference on Rivers '83*, American Society of Civil Engineers, New York, pp. 530–544.

Simons, D. B. and F. Sentürk, 1992, "Sediment transport technology, water and sediment dynamics," Water Resources Publications, Littleton, CO.

Stein, O. R., P. Y. Julien, and C. V. Alonso, 1993, "Mechanics of jet scour downstream of a headcut," J. Hydraul. Res., IAHR, **31**, 723–738.

Sternberg, H., 1875, "Untersuchungen über das Lagen-und Querprofil geschiebeführender Flüsse," Z. Bauwesen **25**, 483–506.

Stevens, M. A. and C. F. Nordin, Jr., 1987, "Critique of the regime theory for alluvial channels," J. Hydraul. Eng., ASCE, **113**, 1359–1380.

Stevens, M. A. and D. B. Simons, 1976, "Safety factors for riprap protection," J. Hydraul. Div., ASCE, **102**, 637–655.

Stevens, M. A. and D. B. Simons, 1971, "Stability analysis for course granular material on slopes," in *River Mechanics*, Water Resources Publication, Fort Collins, CO, Chap. 17.

Struiksma, N., K. W. Olesen, C. Flokstra, and H. J. de Vriend, 1985, "Bed deformation in curved alluvial channels," J. Hydraul. Res., IAHR, **21**, 57–79.

Sundborg, A., 1956, "The river Klaralven: a study of fluvial processes," Geografis. Ann. (Stockholm) **XXXVIII**, 127–316.

Talmon, A. M., M. C. L. M. van Mierlo, and N. Struiksma, 1995, "Laboratory measurements of the direction of sediment transport on transverse alluvial-bed slopes," J. Hydraul. Res., IAHR, **22**, 495–517.

Thomas, W. A. and W. H. McAnally, 1985, "Users manual for the generalized computer program system: open channel flow and sedimentation," TABS-2, U.S. Army Corps of Engineers, Waterways Experiment Station, Vicksburg, MS.

Thorne, C. R. and A. M. Osman, 1988, "Riverbank stability analysis II: applications," J. Hydraul. Eng., ASCE, **114**, 151–172.

Thorne, C. R., J. C. Bathurst, and R. D. Hey, eds., 1987, *Sediment Transport in Gravel-Bed Rivers*, Wiley, New York.

U.S. Army Corps of Engineers, 1999, "Lower Mississippi River sediment study," 2 compact discs with report and data files.

U.S. Army Corps of Engineers, 1995, "River analysis system," HEC-RAS User's Manual Version 1.0, Hydrologic Engineering Center, Davis, CA.

U.S. Army Corps of Engineers, 1994a, "Channel stability assessment for flood control projects," Engineer Manual EM 1110-2-1218.

U.S. Army Corps of Engineers, 1994b, "Design of development of bendway weirs for the Dogtooth Bend reach, Mississippi River," Tech. Rep. HL-94-10, U.S. Government Printing Office, Washington, D.C.

U.S. Army Corps of Engineers, 1991a, "Hydraulic design of flood control channels," EM 1110-2-1601, U.S. Government Printing Office, Washington, D.C.

U.S. Army Corps of Engineers, 1991b, "Scour and deposition in rivers and reservoirs," User's Manual, HEC-6, Hydrologic Engineering Center, Davis, CA.

U.S. Army Corps of Engineers, 1991c, "Water surface profiles user's manual," HEC-2, Hydrologic Engineering Center, Davis, CA.

U.S. Army Corps of Engineers, 1990, "Flood hydrograph package user's manual," HEC-1, Hydrologic Engineering Center, Davis, CA.

U.S. Army Corps of Engineers, 1983, "Dredging and Dredged Material Disposal," EM 1110-2-5025.

U.S. Army Corps of Engineers, 1981, "The Streambank Erosion Control Evaluation and Demonstration Act of 1974," Final report to Congress, Executive Summary and Conclusions, Washington, D.C.

U.S. Dept. of Transportation, Federal Highway Administration, 1988a, "Scour at bridges," Tech. Advisory T5140.20, updated by Technical Advisory T514.23, October 28, 1991, "Evaluating scour at bridges," Washington, D.C.

U.S. Dept. of Transportation, Federal Highway Administration, 1988b, "Interim procedures for evaluating scour at bridges, Office of Engineering, Bridge Div., Washington, D.C.

U.S. Hydrologic Engineering Center, 1993, "Scour and deposition in rivers and reservoirs," User's Manual, HEC-6, Davis, CA.

van Rijn, L. C., 1993, *Principles of Sediment Transport in Rivers, Estuaries and Coastal Seas*, Aqua Publications, Amsterdam.

van Rijn, L. C., 1984, "Sediment transport, part II, suspended load transport," J. Hydraul. Div., ASCE, **110**, 1613–1641.

Vanoni, V. A., ed., 1977, *Sedimentation Engineering*, Manuals and Reports on Engineering Practice No. 54, American Society of Civil Engineers, New York.

Vreugdenhil, C. G., 1972, "Mathematical methods for flood waves," DHL Res. Rep. 89-IV, Delft Hyetraulics, Delft, The Netherlands.

Vriend, H. J. de, 1976, "A mathematical model of steady flow in curved open shallow channels," Rep. 76-1, Dept. of Civil Engineering, University of Technology, Delft, The Netherlands.

Vries, M., de, 1973, "Application of physical and mathematical models for river problems," DHL Pub. 112, Delft Hydraulics, The Netherlands.

Wargadalam, J., 1993, "Hydraulic geometry equations of alluvial channels," Ph.D. dissertation, Colorado State University, Fort Collins, CO.

Water Resources Council, Hydrology Committee, 1981, "Guidelines for determining flood frequency," Bulletin 17B, U.S. Water Resources Council, Washington, D.C.

White, W. R., ed., 1988, *International Conference on River Regime, Wallingford, England*, Wiley, New York.

White, W. R., R. Bettess, and E. Paris, 1982, "Analytical approach to river regime," Proc. J. Hydraul. Div., ASCE, **108**, 1179–1193.

Whiting, P. J. and W. E. Dietrich, 1990, "Boundary shear stress and roughness over mobile alluvial beds," J. Hydraul. Div., ASCE, **116**, 1495–1511.

Wiberg P. and J. D. Smith, 1987, "Calculations of the critical shear stress for motion of uniform and heterogeneous sediment," Water Resources Res., AGU, **23**, 1471–1480.

Williams, D. T. and P. Y. Julien, 1989, "Examination of stage-discharge relationships of compound/composite channels," *Presented at the International Conference on Channel Flow and Catchment Runoff*, PN&L, pp. 478–488.

Williams, G. P., 1978, "Bank-full discharge of rivers," Water Resources Res., AGU, **14**, 1141–1154.

Williams, G. P. and M.G. Wolman, 1984, "Downstream effects of dams on alluvial rivers," USGS Prof. Paper 1286, U.S. Government Printing Office, Washington, D.C.

Winkley, 1989, "The Lower Mississippi River and the Coriolis force," in *Proceedings of the National Conference on Hydraulic Engineering*, ASCE, New York, pp. 1114–1119.

Wischmeier, W. H. and D. D. Smith, 1978, "Predicting rainfall erosion losses: a guide to conservation planning," in USDA Agriculture Handbook 53-7, Washington, D.C.

Wohl, E., 2000, "Mountain rivers," Water Resources Monograph 14, AGU, Washington, D.C.

Wolman, M. G. and L. B. Leopold, 1957, "River floodplains: some observations on their formation," USGS Prof. Paper 282-C, U.S. Geological Survey, Washington, D.C.

Woo, H. S., P. Y. Julien, and E. V. Richardson, 1988, "Suspension of large concentrations of sands," J. Hydraul. Eng., ASCE, **114**, 888–898.

Woolhiser, C. A., 1975, "Simulation of unsteady overland flow," in *Unsteady Flow in Open Channels*, Water Resources Publications, Washington, D.C., Chap. 12 pp. 485–507.

Yalin, M. S., 1992, *River Mechanics*, Pergamon, New York.

Yalin, M. S. and E. Karahan, 1979, "Inception of sediment transport," J. Hydraul. Div., ASCE, **105**, 1433–1443.

Yalin, M. S., 1971, *Theory of Hydraulic Models*, Macmillan, London.

Yang, C. T., 1976, "Minimum unit stream power and fluvial hydraulics," J. Hydraul. Div., ASCE, **102**, 919–934.

Yanmaz, A. M. and H. D. Altinbilek, 1991, "Study of time-dependent local scour around bridge piers," J. Hydraul. Eng., ASCE, **117**, 1247–1268.

Yen, C. L., 1970, "Bed topography effect on flow in a meander," J. Hydraul. Div., ASCE, **96**, 57–73.

Yen, C. L. and K. T. Lee, 1995, "Bed topography and sediment sorting in channel bend with unsteady flow," J. Hydraul. Eng., ASCE, **121**, 591–599.

Zeller, J., 1967a, "Flussmorphologische Studie zum Mäanderproblem," Geogr. Helv. **22** (2), 57–95.

Zeller, J., 1967b, "Meandering channels in Switzerland," in *Proceedings of the Symposium on River Morphology*, IAHR.

Index

434   *Index*